Field Manual
No. 3-05.222

*FM 3-05.222 (TC 31-32)
Headquarters
Department of the Army
Washington, DC, 25 April 2003

Special Forces Sniper Training and Employment

Contents

		Page
	PREFACE	iv
Chapter 1	THE SPECIAL FORCES SNIPER	1-1
	Mission	1-1
	Selection of Personnel	1-1
	Qualifications of SOTIC Graduates	1-5
	The Sniper Team	1-5
	Sniper Team Organization	1-6
	Sniper Training	1-6
Chapter 2	EQUIPMENT	2-1
	Sniper Weapon System	2-1
	Telescopic Sights	2-6
	Leupold and Stevens M3A Telescope	2-7
	Ammunition	2-10
	Observation Devices	2-13
	Sniper Team Equipment	2-29
	Care and Cleaning of the Sniper Weapon System	2-33
	Troubleshooting the Sniper Weapon System	2-41

		Page
Chapter 3	**MARKSMANSHIP TRAINING**	3-1
	Firing Positions	3-1
	Team Firing Techniques	3-17
	Sighting and Aiming	3-19
	Breath Control	3-28
	Trigger Control	3-29
	The Integrated Act of Firing One Round	3-31
	Detection and Correction of Errors	3-34
	Application of Fire	3-37
	Ballistics	3-40
	Sniper Data Book	3-48
	Zeroing the Rifle	3-51
	Environmental Effects	3-59
	Slope Firing	3-66
	Hold-Off	3-69
	Engagement of Moving Targets	3-72
	Common Errors With Moving Targets	3-76
	Engagement of Snap Targets	3-76
	Firing Through Obstacles and Barriers	3-77
	Cold Bore First-Shot Hit	3-78
	Limited Visibility Firing	3-78
	Nuclear, Biological, and Chemical Firing	3-79
Chapter 4	**FIELD SKILLS**	4-1
	Camouflage	4-1
	Cover and Concealment	4-10
	Individual and Team Movement	4-12
	Tracking and Countertracking	4-25
	Observation and Target Detection	4-43
	Range Estimation	4-57
	Selection and Preparation of Hides	4-67
	Sniper Range Card, Observation Log, and Military Sketch	4-80
	KIM Games	4-91
Chapter 5	**EMPLOYMENT**	5-1
	Methods	5-1
	Planning	5-2
	Organization	5-5

FM 3-05.222

		Page
	Command and Control	5-7
	Target Analysis	5-10
	Mission Planning	5-14
	Sniper Support in Special Operations Missions and Collateral Activities	5-18
	Countersniper	5-30
	Conventional Offensive Operations	5-32
	Conventional Defensive Operations	5-35
	Civil Disturbance Assistance	5-37
Chapter 6	**SNIPER OPERATIONS IN URBAN TERRAIN**	6-1
	Urban Terrain	6-1
	Sniper Support in Urban Operations	6-9
	Urban Hides	6-14
	Weapons Characteristics in Urban Terrain	6-23
	Engagement Techniques	6-24
Appendix A	**WEIGHTS, MEASURES, AND CONVERSION TABLES**	A-1
Appendix B	**MISSION-ESSENTIAL TASKS LIST**	B-1
Appendix C	**SUSTAINMENT PROGRAM**	C-1
Appendix D	**MISSION PACKING LIST**	D-1
Appendix E	**M82A1 CALIBER .50 SNIPER WEAPON SYSTEM**	E-1
Appendix F	**FOREIGN/NONSTANDARD SNIPER WEAPON SYSTEMS DATA**	F-1
Appendix G	**SNIPER RIFLE TELESCOPES**	G-1
Appendix H	**BALLISTICS CHART**	H-1
Appendix I	**SNIPER TRAINING EXERCISES**	I-1
Appendix J	**RANGE ESTIMATION TABLE**	J-1
Appendix K	**SNIPER'S LOGBOOK**	K-1
Appendix L	**TRICKS OF THE TRADE**	L-1
Appendix M	**SNIPER TEAM DEBRIEFING FORMAT**	M-1
Appendix N	**SNIPER RANGE COMPLEX**	N-1
Appendix O	**AERIAL PLATFORMS**	O-1
	GLOSSARY	Glossary-1
	BIBLIOGRAPHY	Bibliography-1
	INDEX	Index-1

Preface

This field manual (FM) provides doctrinal guidance on the mission, personnel, organization, equipment, training, skills, and employment of the Special Forces (SF) sniper. It describes those segments of sniping that are unique to SF soldiers and those portions of conventional sniping that are necessary to train indigenous forces. It is intended for use by commanders, staffs, instructors, and soldiers at training posts, United States (U.S.) Army schools, and units.

FM 3-05.222 (formerly TC 31-32) addresses three distinct audiences:
- *Commanders.* It provides specific guidance on the nature, role, candidate selection, organization, and employment of sniper personnel.
- *Trainers.* It provides a reference for developing training programs.
- *Snipers.* It contains detailed information on the fundamental knowledge, skills, and employment methods of snipers throughout the entire operational continuum.

The most common measurements that the sniper uses are expressed throughout the text and in many cases are U.S. standard terms rather than metric. Appendix A consists of conversion tables that may be used when mission requirements or environments change.

The proponent of this manual is the United States Army John F. Kennedy Special Warfare Center and School (USAJFKSWCS). Submit comments and recommended changes to Commander, USAJFKSWCS, ATTN: AOJK-DT-SFA, Fort Bragg, NC 28310-5000.

Unless this publication states otherwise, masculine nouns and pronouns do not refer exclusively to men.

- Experience as a hunter or woodsman.
- Experience as a competitive marksman.
- Interest in weapons.
- Ability to make rapid, accurate assessments and mental calculations.
- Ability to maintain an emotionally stable personal life.
- Ability to function effectively under stress.
- Possession of character traits of patience, attention to detail, perseverance, and physical endurance.
- Ability to focus completely.
- Ability to endure solitude.
- Objectivity to the extent that one can stand outside oneself to evaluate a situation.
- Ability to work closely with another individual in confined spaces and under stress.
- Freedom from certain detrimental personal habits such as the use of tobacco products and alcohol. (Use of these is a liability unless the candidate is otherwise highly qualified. These traits, however, should not be the sole disqualifier.)
- First-class APFT scores with a high degree of stamina and, preferably, solid athletic skills and abilities.

Figure 1-3. Personal Qualities

1-6. The first three personal qualities are particularly important when it comes to sustaining sniper skills, because the sniper with these characteristics will have a greater desire to practice these tasks as they are part of his avocation.

1-7. Commanders may implement diagnostic and aptitude testing. Certain testing procedures may be lengthy and tedious and are therefore subject to limitations of time, equipment, and facilities. It is recommended that the psychological evaluation of a candidate be at least partially determined through the use of the MMPI-2. This test, if properly administered, gives the commander a personality profile of the candidate. It helps him decide whether the candidate can function in confined spaces, work independently, and has the potential to be a sniper.

1-8. The tests are more than simple mental analyses. Psychological screening establishes a profile of characteristics that indicate if an individual would be a successful sniper. Testing eliminates candidates who would not perform well in combat. Psychological screening can identify individuals who have problems.

1-9. To select the best candidate, the commander talks to a qualified psychologist and explains what characteristics he is looking for. That way, once a candidate is tested, the psychologist can sit down with the commander and give him the best recommendation based on the candidate's psychological profile.

1-10. After the commander selects the sniper candidate, he must assess the individual's potential as a sniper. He may assess the candidate by conducting a thorough review of the candidate's records, objective tests, and subjective evaluations. The length of time a commander may devote to a candidate's assessment will vary with his resources and the mission. Normally, 2 or 3 days will suffice to complete an accurate assessment.

1-11. Assessment should include both written and practical tests. Practical examinations will actively measure the candidate's physical ability to perform the necessary tasks and subtasks involved in sniping. Written examinations will evaluate the candidate's comprehension of specific details.

1-12. Assessment testing must objectively and subjectively determine an individual's potential as a sniper. Objectivity measures the capacity to learn and perform in a sterile environment. Subjectivity assesses actual individual performance.

1-13. Objective assessment tests are presented as a battery grouped by subject matter and may be presented either as practical or written examinations. Some examples of objective testing are—

- Shooting battery tests that evaluate the theoretical and practical applications of rifle marksmanship.
- Observation and memory battery tests that measure the candidate's potential for observation and recall of specific facts.
- Intelligence battery tests that consist of standard military tests and previously mentioned specialized tests.
- Critical decision battery tests that evaluate the candidate's ability to think quickly and use sound judgment.
- Motor skills battery tests that assess hand-eye coordination.

1-14. Subjective assessment tests allow the assessor to gain insight into the candidate's personality. Although he is constantly observed in the selection and assessment process, specific tests may be designed to identify desirable and undesirable character traits. A trained psychologist (well versed in sniper selection) should conduct or monitor all subjective testing. Examples of possible subjective tests include, but are not limited to, the following:

- The interview—can identify the candidate's motivation for becoming a sniper and examine his expectations concerning the training.
- The suitability inventory—basically compares the candidate to a "predetermined profile" containing the characteristics, skills, motivations, and experience a sniper should possess.

1-15. A committee of assessors conducts the candidate selections at the end of the assessment program. While the commander should monitor all candidate selections, it is important for the committee to make the decision to preserve consistency and to rule out individual bias. The procedure for selection should be accomplished by a quorum during which the candidates are rated on a progressive scale. The committee should choose candidates based on their standing, in conjunction with the needs of the unit. At this time, the best-qualified soldiers should be selected; alternate and future

1-26. The second environment is the Unit Training Special Operations Target Interdiction Course. This course enables the unit commander to fill his needs within his mission parameters. A graduate of this course becomes a Level II sniper and is fully capable of filling an assigned team slot as a sniper. He also meets the requirements established by USASFC 350-1 for two snipers, either Level I or II, to be assigned to each Special Forces operational detachment A (SFODA). Once a sniper (Level I or II) is assigned to an SFODA, he is then a Category (CAT) I sniper for requesting training ammunition, equipment, and ranges.

1-27. Twice a year USAJFKSWCS conducts a 1-week Challenge Course. Level II snipers who successfully complete the Challenge Course will be awarded a SOTIC diploma and the Level I designator.

1-28. The primary differences in Levels I and II snipers are that Level I snipers are required to fire within close proximity of non-combatants and friendly forces in a Close Combat Situation. Level I snipers are required to run the SF Group Sustainment Program, which is usually conducted in conjunction with a Level II train up. Level I are required to train U.S. forces in Level II courses. Level II snipers may not train other U.S. forces to a Level II status. The Level II sniper may conduct sniper training for host nation courses.

1-29. The Level I sniper is tested to the maximum effective range of the M24 Sniper Weapon System 800 meters, while the Level II sniper furthest required range is 600 yards. However, the unit commander may designate his Level II snipers to be trained to a higher degree of efficiency and accuracy. The unit commander may have his Level II course mirror the Level I course or add greater emphasis in areas he feels are necessary to complete his assigned mission parameters. While the Level I course is a program-of-instruction-driven, 6-week course, the unit course may be 2 weeks or longer depending on the requirements of the command.

1-30. The Level I designation is to identify those snipers that have met a specific standard of training. These snipers have been trained by a cadre of instructors that have gone through the Instructor Training Course conducted by the SOTIC designators. The instructor's sole function is to train sniper students; he does not have to participate in the line unit's operations tempo.

1-31. The unit's Level II courses may train to a higher or lower standard depending upon the commander's needs and assessments. The instructors for the unit Level II course need to be identified as soon as possible, usually 6 to 8 weeks out, and permitted to prepare for the coming course. This lead time may not be possible due to the unit's operations tempo and red-cycle requirements. The longer the lead time, the better the course preparation and instruction.

1-32. Once a sniper is trained, whether Level I or II, he must maintain proficiency. His maintenance training must include the school "learning environment" and the unit "training environment." The school must teach a skill and thus remove variables so that the student may learn. Once the sniper has graduated, he must "train" in these skills with the variables added. Only imagination and desire on the sniper's part can limit the training scenarios.

Chapter 2

Equipment

Snipers, by the nature of their mission, must learn to exploit the maximum potential from all their equipment. The organizational level of employment and the mission will determine the type and amount of equipment needed (Appendix D). Snipers will carry only the equipment necessary for successfully completing their mission. Appendix E describes the M82A1 caliber .50 sniper weapon system (SWS). Appendix F describes the types of SWSs in other countries.

SNIPER WEAPON SYSTEM

2-1. The current SWS is the M24 sniper rifle with the Leopold & Stevens (L&S) ultra 10x M3A rifle scope. The M24 is based on the Remington Model 700 long action with an adjustable trigger. The barrel is a heavy, 5 groove, 11.2-inch twist, stainless steel target barrel. The stock is made of fiberglass, graphite, and Kevlar with an adjustable butt plate. The weapon is constructed to be accurate within 1/2 minute of angle (MOA) or 1/2-inch groups at 100 yards. The M24 is currently chambered for the 7.62-millimeter (mm) North Atlantic Treaty Organization (NATO) cartridge. Two M24s are issued per operational detachment. The parts of the M24 rifle include the bolt assembly, trigger assembly, adjustable stock, barreled action (H 700), and telescopic and iron sights. Because this weapon is the sniper's best friend, he must be proficient in inspecting and loading it. However, during the inspection, the extent of the sniper's repairs is limited. Figure 2-1 lists the M24 SWS components.

- Bolt-action rifle
- Fixed 10x telescope, L&S M3A
- System case
- Scope case
- Detachable iron sights (front and rear)
- Deployment case and kit
- Optional bipod
- Cleaning kit
- Soft rifle case
- Operator's manual

Figure 2-1. Components of the M24 SWS

FM 3-05.222

SAFETY

2-2. The safety is located on the right rear side of the receiver and, when properly engaged, provides protection against accidental discharge under normal usage. The sniper should follow the rules below:

- To engage the safety, place it in the "S" position (Figure 2-2).
- Always place the safety in the "S" position before handling, loading, or unloading the weapon.
- When the weapon is ready to be fired, place the safety in the "F" position (Figure 2-2).

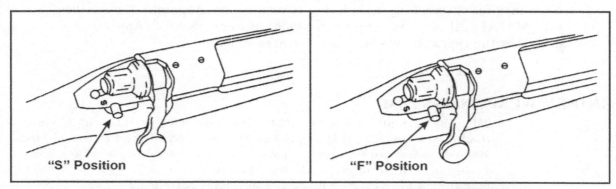

Figure 2-2. The M24 Sniper Weapon System in the SAFE and FIRE Modes

BOLT ASSEMBLY

2-3. The bolt assembly locks the round into the chamber as well as extracts it. The sniper should follow the rules below:

- To remove the bolt from the receiver, place the safety in the "S" position, raise the bolt handle, and pull it back until it stops. Push the bolt stop release up (Figure 2-3, page 2-3) and pull the bolt from the receiver.
- To replace the bolt, ensure the safety is in the "S" position, align the lugs on the bolt assembly with the receiver (Figure 2-4, page 2-3), slide the bolt all the way into the receiver, and then push the bolt handle down.

TRIGGER ASSEMBLY

2-4. Pulling the trigger fires the rifle when the safety is in the "F" position. The sniper may adjust the trigger pull force from a minimum of 2.5 pounds to a maximum of 8 pounds. He can make this adjustment using the 1/16-inch Allen wrench provided in the deployment kit. Turning the trigger adjustment screw (Figure 2-5, page 2-3) clockwise will increase the force needed to pull the trigger. Turning it counterclockwise will decrease the force needed. This change is the only trigger adjustment the sniper will make. The trigger cannot be adjusted less than 2.5 pounds. The screw compresses an independent spring that increases the required pressure to make the sear disengage.

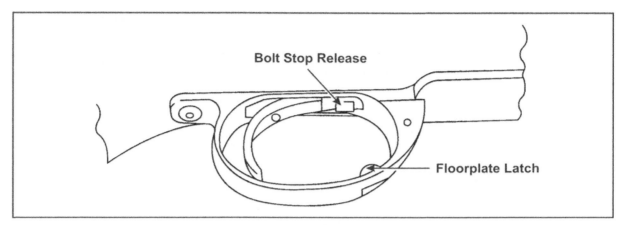

Figure 2-3. Bolt Stop Release

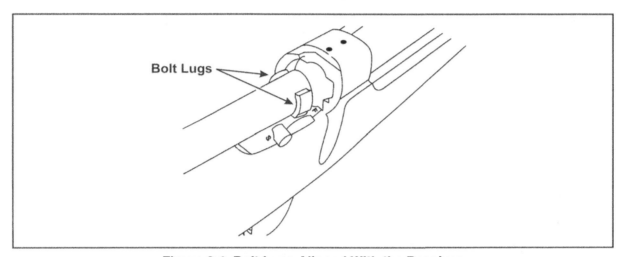

Figure 2-4. Bolt Lugs Aligned With the Receiver

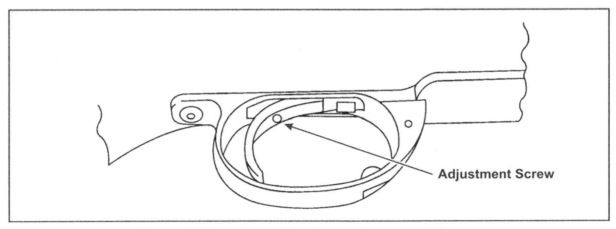

Figure 2-5. Location of the Trigger Adjustment Screw

STOCK ADJUSTMENT

2-5. The M24 has a mechanism for making minor adjustments in the stock's length of pull. The thick wheel provides this adjustment. The thin wheel is for locking this adjustment (Figure 2-6). The sniper should turn the thick wheel clockwise to lengthen the stock or counterclockwise to shorten the stock. To lock the position of the shoulder stock, he should turn the thin wheel clockwise against the thick wheel. To unlock the position of the shoulder stock, he should turn the thin wheel counterclockwise away from the thick wheel. The sniper can adjust the length of pull so that the stock may be extended, but no more than three finger widths. Beyond this, the butt plate becomes unstable.

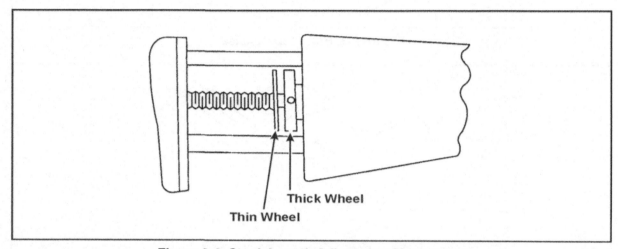

Figure 2-6. Stock Length Adjustment Mechanism

IRON SIGHTS

2-6. The M24 has a backup sighting system consisting of detachable front and rear iron sights. To install the iron sights, the sniper must first remove the telescope. The sniper should—

- Align the front sight and the front-sight base dovetail and slide the sight over the base to attach the front sight to the barrel.
- Ensure the fingernail projection of the front sight fits securely into the fingernail groove on the front-sight base.
- Tighten the screw slowly, ensuring that it seats into the recess in the sight base (Figure 2-7, page 2-5.).
- To attach the rear sight to the receiver, remove one of the three setscrews, and align the rear sight with the rear-sight base located on the left rear of the receiver (Figure 2-8, page 2-5). Tighten the screw to secure the sight to the base. There are three screw holes and two positions for the sight screw to facilitate adjusting shooter eye relief.

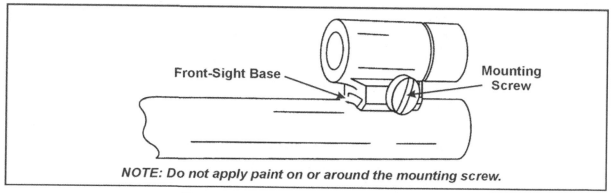

Figure 2-7. Front-Sight Mounting Screw

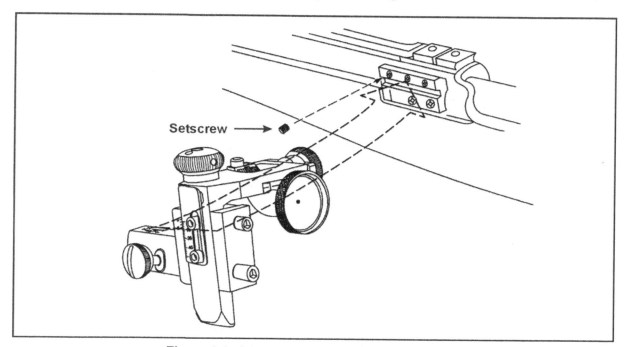

Figure 2-8. Attaching the Rear Sight Assembly

2-7. The sniper should also make sure the other setscrews are below the level of the face of the rear sight base. If not, he should remove and store them.

INSPECTION

2-8. The M24's design enables the sniper to make some repairs. Deficiencies that the sniper is unable to repair will require manufacturer warranty work. When inspecting the M24, the sniper should check the—

- Appearance and completeness of all parts.
- Bolt to ensure it has the same serial number as the receiver and that it locks, unlocks, and moves smoothly.

- Safety to ensure it can be positively placed into the "S" and "F" positions without being too difficult or too easy to move.
- Trigger to ensure the weapon will not fire when the safety is in the "S" position and that it has a smooth, crisp trigger pull.
- Action screws (front of the internal magazine and rear of the trigger guard) for proper torque (65 inch-pounds).
- Telescope mounting ring nuts for proper torque (65 inch-pounds).
- Stock for any cracks, splits, or any contact it may have with the barrel.
- Telescope for obstructions such as dirt, dust, moisture, and loose or damaged lenses.

LOADING

2-9. The M24 has an internal, five-round capacity magazine. To load the rifle, the sniper should—

- Point the weapon in a safe direction.
- Ensure the safety is in the "S" position.
- Raise the bolt handle and pull it back until it stops.
- Push five rounds of 7.62-mm ammunition one at a time through the ejection port into the magazine. Ensure that the bullet end of the rounds is aligned toward the chamber.
- Push the rounds fully rearward in the magazine.
- Once the five rounds are in the magazine, push the rounds downward while slowly pushing the bolt forward over the top of the first round.
- Push the bolt handle down. The magazine is now loaded.
- To chamber a round, raise the bolt and pull it back to fully seat the round. Stopping the bolt early will cause an override situation.
- Push the bolt forward. The bolt strips a round from the magazine and pushes it into the chamber.
- Push the bolt handle down until it is fully seated. Failure will cause a light strike on the primer and a misfire.

2-10. To fire the rifle, place the safety in the "F" position and squeeze the trigger.

NOTE: See TM 9-1005-306-10, *Operator's Manual for 7.62-mm M24 Sniper Weapon System (SWS)*, for shipping uncorrectable maintenance items.

TELESCOPIC SIGHTS

2-11. A telescopic sight mounted on the rifle allows the sniper to detect and engage targets more effectively than he could by using the iron sights. Unlike sighting with iron sights, the target's image in the telescope is on the same focal plane as the aiming point (reticle). This evenness allows for a clearer picture of the target and reticle because the eye can focus on both simultaneously. However, concentration on the reticle is required when engaging a target.

2-12. Another advantage of the telescope is its ability to magnify the target, which increases the resolution of the target's image, making it clearer and more defined. The average unaided human eye can distinguish detail of about 1 inch at 100 yards or 3 centimeters at 100 meters (1 MOA). Magnification combined with well-designed optics permits resolution of this 1 inch divided by the magnification. Thus, a 1/4 MOA of detail can be seen with a 4x scope at 100 meters, or 3 centimeters of detail can be seen at 600 meters with a 6x scope.

2-13. In addition, telescopic sights magnify the ambient light, making shots possible earlier and later during the day. Although a telescope helps the sniper to see better, it does not help him to shoot well. Appendix G provides further information on sniper rifle telescopes.

LEUPOLD AND STEVENS M3A TELESCOPE

2-14. The M3A is a fixed 10x telescope with a ballistic drop compensator dial for bullet trajectory from 100 to 1,000 meters. The elevation knob is marked in 100-meter increments to 600 meters, in 50-meter increments 600 to 1,000 meters, and has 1 MOA elevation adjustment. The windage 1/2-MOA increments and a third knob provides for focus and parallax adjustment. The reticle is a duplex crosshair with 3/4-MOA mil dots (Figure 2-9, page 2-8). The mil dots are 1 mil apart, center to center, with a possible 10 mils vertical and 10 mils horizontal. The sniper uses mil dots for range estimation, holdover, windage holds, mover leads, and reference point holds.

2-15. The M3A consists of the telescope, a fixed mount, a detachable sunshade for the objective lens, and dust covers for the objective and ocular (eyepiece) lens. The telescope has a fixed 10x magnification that gives the sniper better resolution than with the adjustable ranging telescope (ART) series. There are three knobs located midway on the tube—the focus/parallax, elevation, and windage knobs (Figure 2-10, page 2-8).

ADJUSTMENTS

2-16. The sniper should always focus the reticle to his eye **first**. He should turn the ocular eyepiece to adjust the reticle until it is sharp, but should not force-focus his eye. He can adjust the eyepiece by turning it in or out of the tube until the reticle appears crisp and clear. The sniper should focus the eyepiece after mounting the telescope on the rifle. He should grasp the eyepiece and back it away from the lock ring. He should not attempt to loosen the lock ring first; it will automatically loosen when the eyepiece is backed away (no tools are needed). The sniper should rotate the eyepiece several turns to move it at least 1/8 inch. He will need this much change to achieve any measurable effect on the reticle clarity. The sniper then looks through the scope at the sky or a blank wall and checks to see if the reticle appears sharp and crisp. He must do this before adjusting the focus and parallax.

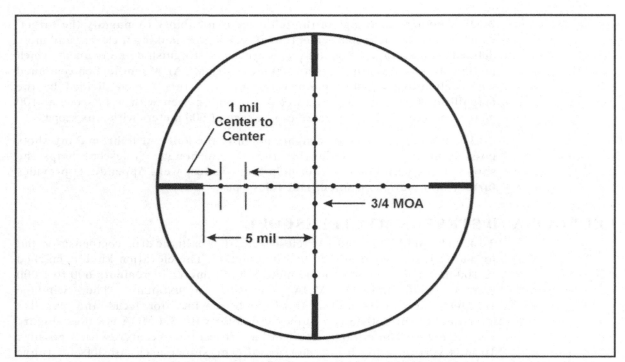

Figure 2-9. The M24 Optical Day Sight Reticle

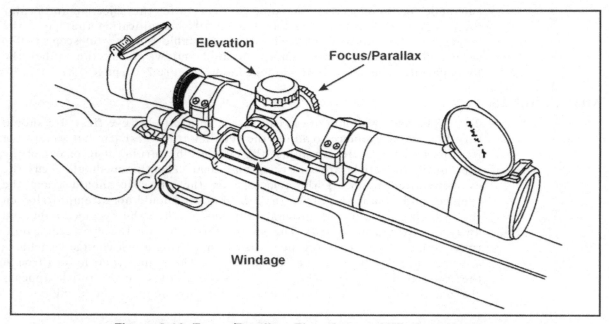

Figure 2-10. Focus/Parallax, Elevation, and Windage Knobs

2-17. The focus/parallax knob sits on the left side of the tube. The sniper uses it to focus the target's image onto the same focal plane as the reticle, thereby reducing parallax to a minimum. Parallax is the apparent movement of the sight picture on the reticle when the eye is moved from side to side or up and down. The parallax adjustment knob has two extreme positions indicated by the infinity mark and the largest of four dots. Adjustments between these positions focus images from less than 50 meters to infinity. These markings are for reference only, after the sniper has initially adjusted his scope for parallax. He then slips the scale to match his requirements (for example, big ball references 100 or 200 meters). The sniper then writes each item and its distance in his log for reference whenever he engages targets at that range. Any change in reticle focus requires the sniper to readjust the focus/parallax setting.

2-18. The elevation knob sits on top of the tube. This knob has calibrated index markings from 1 to 10. These markings represent the elevation setting adjustments needed at varying distances; for example, 1 = 100 meters, 10 = 1,000 meters. There are small hash marks between the 100-meter increments after 600 meters; these represent 50-meter increments. Each click of the elevation knob equals 1 MOA.

2-19. The windage knob sits on the right side of the tube. The sniper uses this knob for lateral adjustments. Turning the knob in the direction indicated moves the point of impact (POI) in that direction. Each click on the windage knob equals 1/2 MOA.

LEUPOLD VARI-X III, M3A-LR

2-20. Incorporating the best features of the Mark 4 M3 and Vari-X III scopes, the Leupold Vari-X III 3.5-10 x 40-mm Long-Range M3 features M3-style adjustment dials that are specially calibrated and interchangeable for bullet drop compensation. Adjustment increments of 1-MOA elevation and 1/2-MOA windage allow for easy adjustment. A parallax adjustment dial allows parallax elimination from a shooting position. This scope has a 30-mm tube diameter, a mil-dot reticle, and multicoated lens.

SCOPE MOUNT

2-21. The scope mount consists of a baseplate with four screws and a pair of scope rings (each with an upper and lower ring half) with eight ring screws (Figure 2-11, page 2-10). The sniper mounts the baseplate to the rifle by screwing the four baseplate screws through the plate and into the top of the receiver. He should have two short and two long baseplate screws. The long screws go to the rear mounting points, the short screws go to the front. The screws must not protrude into the receiver so they do not interrupt the functioning of the bolt. Medium-strength "Loctite" may be used on these four baseplate screws for a more permanent attachment. After mounting the baseplate, he then mounts the scope rings.

2-22. When the sniper mounts the scope rings, he should select one of the slots on the mounting base and engage the ringbolt spline with the selected slot. He should push the ring forward to get spline-to-base contact as the mount ring nut is tightened. He checks the eye relief. If the telescope needs to be adjusted, the sniper loosens the ring nuts and aligns the ringbolts with the other set of slots on the base; he then repeats the process. He makes sure that the crosshairs are perfectly aligned (vertically and horizontally) with the rifle. Any cant will cause misses at longer ranges. To ensure that the reticle is not canted in the rings, the sniper will need a level and plumb line. He uses the level to ensure the weapon is indeed level left to right. Once leveled, he hangs a plumb line on a wall and matches the reticle to the plumb line. When satisfied with the eye relief obtained (approximately 3 to 3 1/2 inches), the sniper then tightens the ring nuts to 65 inch-pounds using the T-handle torque wrench (found in the deployment case).

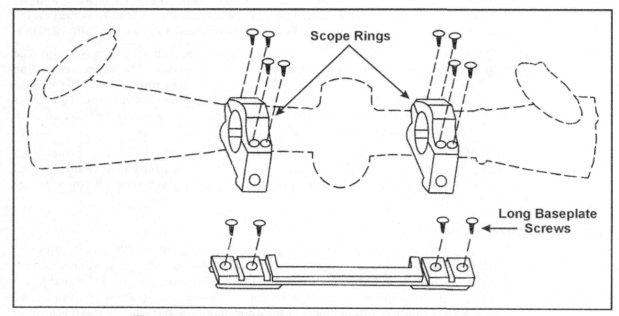

Figure 2-11. The M3A Leupold and Stevens Scope Mount

OPERATION

2-23. When using the telescope, the sniper simply places the reticle on the target, determines the distance to the target by using the mil dots on the reticle, sets parallax, and then adjusts the elevation knob for the estimated range. He then places the crosshair on the desired POI or quarters the target. The sniper then gives the observer a "READY" and awaits the wind call.

AMMUNITION

2-24. Snipers should always attempt to use match-grade ammunition when available because of its greater accuracy and lower sensitivity to environmental effects. However, if match-grade ammunition is not available, or if the situation requires, he may use a different grade of ammunition. Standard-grade ammunition may not provide the same level of accuracy or POI as match-grade ammunition. In the absence of match-grade ammunition, the sniper should conduct firing tests to determine the most accurate lot of ammunition available. Once he identifies a lot of ammunition as meeting the requirements, he should use this lot as long as it is available.

TYPES AND CHARACTERISTICS

2-25. The sniper should use 7.62- x 51-mm (.308 Winchester) NATO M118 Special Ball (SB), M852 National Match, or M118 Long Range (LR) ammunition with the SWS. He must rezero the SWS every time the type or lot of ammunition changes. The ammunition lot number appears on the cardboard box, metal can, and wooden crate that it is packaged in. The sniper should maintain this information in the weapon's data book.

M118 Special Ball

2-26. The M118SB bullet consists of a metal jacket and a lead antimony slug. It is a boat-tailed bullet (the rear of the bullet is tapered to reduce drag) and has a nominal weight of 173 grains. The tip of the bullet is not colored. The base of the cartridge is stamped with the NATO standardization mark (circle and crosshairs), manufacturer's code, and year of manufacture. Its primary use is against personnel. Its accuracy standard requires a 10-shot group to have an extreme spread of not more than 12 inches at 600 yards or 33 centimeters at 550 meters (2 MOAs) when fired from an accuracy barrel in a test cradle. The stated velocity of 2,550 feet per second (fps) is measured at 78 feet from the muzzle. The actual muzzle velocity of this ammunition is 2,600 fps. M118SB is the primary choice for the M24 SWS because the telescopic sights are ballistically matched to this ammunition out to 1,000 meters. This ammunition is being replaced by M118LR.

M852 National Match (Open Tip)

2-27. As of October 1990, the Department of State, Army General Counsel, and the Office of the Judge Advocate General concluded that the use of open-tip ammunition does not violate the law-of-war obligation of the United States. The U.S. Army, Navy, and Marine Corps may use this ammunition in peacetime or wartime missions.

2-28. The M852 bullet (Sierra Match King) is boat-tailed, 168 grains in weight, and has an open tip. The open tip is a small aperture (about the diameter of the wire in a standard-sized straight pin or paper clip) in the nose of the bullet. Describing this bullet as a hollow point is misleading in law-of-war terms. A hollow-point bullet is typically thought of in terms of its ability to expand upon impact with soft tissue. Physical examination of the M852 open-tip bullet reveals that its opening is small in comparison to the aperture of hollow-point hunting bullets. Its purpose is to improve the ballistic coefficient of the projectile. The swaging of the bullet from the base by the copper gilding leaves a small opening in the nose; it does not aid in expansion. The lead core of the M852 bullet is entirely covered by the copper bullet jacket.

2-29. Accuracy standard for the M852 ammunition is 9.5 inches average extreme spread (or slightly over 1.5 MOAs) at 600 yards. Other than its superior long-range accuracy capabilities, the M852 was examined with regard to its performance upon impact with the human body or in artificial material that approximates soft human tissue. In some cases, the bullet would break up or fragment after entry into soft tissue. Fragmentation depends on many factors, including the range to the target, velocity at the time of impact, degree of yaw of the bullet at the POI, or the distance traveled point-first within the body before yaw is induced. The M852 was not designed to yaw intentionally or break up upon impact. There was little discernible difference in bullet fragmentation between the M852 and other military small-arms bullets. Some military ball ammunition of foreign manufacture tends to fragment sooner in human tissue or to a greater degree, resulting in wounds that would be more severe than those caused by the M852 bullet.

NOTE: M852 is the best substitute for M118 taking the following limitations into consideration:

- The M852's trajectory is not identical to the M118's; therefore, it is not matched ballistically with the M3A telescope. The difference to 600 meters is minimal, predictable 700 becomes 725, and 800 requires 850. These are start-point ranges only.
- The M852 is not suited for target engagement beyond 700 meters because the 168-grain bullet is not ballistically suitable. This bullet will drop below the sound barrier just beyond this distance. The turbulence that it encounters as it becomes subsonic affects its accuracy at distances beyond 700 meters.

M118 Long-Range (Open Tip)

2-30. The M118LR bullet (Sierra Match King) is boat-tailed, 175 grains in weight, and has an open tip. The open tip is the same as the M852.

2-31. Accuracy standard for the M118LR ammunition is an average extreme horizontal spread of 10.3 inches and an average extreme vertical spread of 14.0 inches at 1,000 yards or slightly over 1 MOA horizontal and 1.4 MOAs vertical extreme spread. This data is stated in the Detail Specifications dated 3 March 1998. The trajectory of the M118LR will closely match the M118SB. Complete information is not available at this time. This is new ammunition being developed through the Navy and Marine Corps. It is scheduled to replace all lots of the M118SB and M852.

M82 Blank

2-32. Snipers use the M82 blank ammunition during field training. It provides the muzzle blast and flash that trainers can detect during the exercises that evaluate the sniper's ability to conceal himself while firing his weapon and activates the multiple integrated laser engagement system (MILES) training devices. MILES devices are an excellent tool for training the commander on the use of a sniper. However, these devices can cause problems in the sniper's training, because he does not have to lead targets or compensate for wind or range.

ALTERNATIVES

2-33. If match-grade ammunition is not available, snipers can use the standard 7.62- x 51-mm NATO ball ammunition. However, the M3A bullet drop compensator (BDC) is designed for M118SB, so there would be a significant change in zero. Snipers should always test-fire standard ammunition and record the ballistic data in the data book. They should use standard ball ammunition in an emergency situation only. Snipers should test-fire all ammunition for accuracy. Even match-grade ammunition can have a bad lot.

M80/M80E1 Ball

2-34. The M80 and M80E1 ball cartridge bullet consists of a metal jacket with a lead antimony slug. It is boat-tailed and weighs 147 grains. The tip of the bullet is not colored. This bullet is primarily used against personnel. Its accuracy standard requires a 10-shot group to have an extreme spread of not more than 4 MOAs or 24 inches at 600 yards (66 centimeters at 550 meters) when fired from an accuracy barrel in a test cradle. The muzzle velocity of this ammunition is 2,800 fps. The base of the cartridge is stamped with the NATO standardization mark, manufacturer's initials, and the date of manufacture. The sniper should test-fire several lots before using them due to the reduced accuracy and fluctuation in lots. The most accurate lot that is available in the largest quantity (to minimize test repetition) should be selected for use.

M62 Tracer

2-35. The M62 tracer bullet consists of a metal-clad steel jacket, a lead antimony slug, a tracer subigniter, and igniter composition. It has a closure cap and weighs 141 grains. The bullet tip is painted orange (NATO identification for tracer ammunition). It is used for observation of fire, incendiary, and signaling purposes. Tracer ammunition is manufactured to have an accuracy standard that requires 10-shot groups to have an extreme spread of not more than 6 MOAs or 36 inches at 600 yards (99 centimeters at 550 meters). The base of the cartridge is stamped with the NATO standardization mark, manufacturer's initials, and date of manufacture. The amount of tracer ammunition fired through the SWS should be minimized because of its harmful effect on the precision-made barrel.

ROUND COUNT BOOK

2-36. The sniper maintains a running count of the number and type of rounds fired through the SWS. It is imperative to accurately maintain the round count book. The SWS has shown to have a barrel life of about 8,000 to 10,000 rounds. The sniper should inspect the barrel at this time, or sooner if a loss of accuracy has been noted. He inspects the barrel for throat erosion and wear, and if excessive, schedules the SWS to be rebarreled. This inspection should be accomplished IAW the deployment schedule of the unit and the required break in time needed for the new barrel.

OBSERVATION DEVICES

2-37. Aside from the rifle and telescopic sight, the sniper's most important tools are optical devices. The categories of optical equipment that snipers normally use are binoculars, telescopes, night vision devices (NVDs), and range finders. The following paragraphs discuss selected optical equipment for special purposes.

FM 3-05.222

BINOCULARS

2-38. Every sniper should be issued binoculars; they are the sniper's primary tool for observation. Binoculars provide an optical advantage not found with telescopes or other monocular optical devices. The binoculars' typically larger objective lens, lower magnification, and optical characteristics add depth and field of view to an observed area. Many types of binoculars are available. Snipers/observers should take the following into account when selecting binoculars:

- *Durability.* The binoculars must be able to withstand rough use under field conditions. They must be weatherproofed and sealed against moisture that would render them useless due to internal fogging. Binoculars with individually focused eyepieces can more easily be made waterproof than centrally focused binoculars. Most waterproof binoculars offered have individually focused eyepieces.
- *Size.* A sniper's binoculars should be relatively compact for ease of handling and concealment.
- *Moderate magnification.* Binoculars of 6 to 8 power are best suited for sniper work. Higher magnifications tend to limit the field of view for any given size of objective lens. Also, higher magnifications tend to intensify hand movements during observation and compress depth perception.
- *Lens diameter.* Binoculars with an objective lens diameter of 35 to 50 mm should be considered the best choice. Larger lenses permit more light to enter; therefore, the 50-mm lens would be more effective in low-light conditions.
- *Mil scale.* The binoculars should have a mil scale incorporated into the field of view for range estimation.

2-39. The M22 binoculars are the newest in the inventory and are general issue. These binoculars have the same features as the M19, plus fold-down eyepiece cups for personnel who wear glasses, to reduce the distance between the eyes and the eyepieces. They also has protective covers for the objective and eyepiece lenses. The binoculars have laser-protective filters on the inside of the objective lenses. **Direct sunlight reflects off these lenses!** The reticle pattern (Figure 2-12, page 2-15) is different from the M19's reticle. Laser filter also lowers the M22's light transmittance, which lowers its ability to gather light at dusk and dawn. Characteristics of the M19 and M22 are as follows:

- M19 Optical Characteristics:
    - Objective lens: 50 mm.
    - Magnification: 7x.
    - Field of view: 130 mils—130 meters at 1,000 meters.
- M19 Physical Characteristics:
    - Width (open position): 190.5 mm/7.5 inches.
    - Length: 152.4 mm/6 inches.
    - Weight: 966 kg/2.125 pounds.
    - Thickness: 63.5 mm/2.5 inches.

- M22 Optical Characteristics:
  - Objective lens: 50 mm.
  - Magnification: 7x.
  - Field of view: 130 mils—130 meters at 1,000 meters.
  - Depth of field: 12.5 meters to infinity.
- M22 Physical Characteristics:
  - Width (open position): 205 mm/8.1 inches.
  - Length: 180 mm/7.1 inches.
  - Weight: 1.2 kg/2.7 pounds.

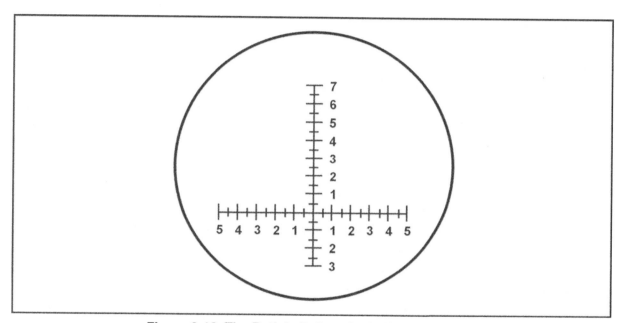

Figure 2-12. The Reticle Pattern in the M22 Binoculars

Method of Holding Binoculars

2-40. Binoculars should be held lightly, resting on and supported by the heels of the hands. The thumbs are positioned to block out light that would enter between the eyes and the eyepieces. The eyepieces are held lightly to the eyes to avoid transmitting body movement. Whenever possible, a stationary rest should support the elbows. An alternate method for holding the binoculars is to move the hands forward, cupping them around the sides of the objective lenses. This method keeps light from reflecting off the lenses, which would reveal the sniper's position. The sniper should **always** be aware of reflecting light. He should operate from within shadows or cover the lens with an extension or thin veil, such as a nylon stocking.

Adjustments

2-41. Interpupillary distance is the space between the eyes. Interpupillary adjustment is moving the monocles to fit this distance. The monocles are hinged together for ease of adjustment. The hinge is adjusted until the field of vision ceases to be two overlapping circles and appears as a single, sharply defined circle. The setting on the hinge scale should be recorded for future use.

2-42. Each eye of every individual requires a different focus setting. The sniper should adjust the focus for each eye as follows:

- With **both** eyes open, look at a distant object, then through the binoculars at this same object.
- Place one hand over the objective lens of the right monocle and turn the focusing ring of the left monocle until the object is sharply defined.
- Uncover the right monocle and cover the left one. Rotate the focusing ring of the right monocle until the object is sharply defined.
- Uncover the left monocle. The object should be clear to both eyes.

The sniper should glance frequently at the distant object during this procedure to ensure that his eyes are not compensating for an out-of-focus condition. He then reads the diopter scale on each focusing ring and records the reading for future reference. Correctly focused binoculars will prevent eyestrain when observing for extended periods.

Eye Fatigue

2-43. Prolonged use of the binoculars or telescope will cause eye fatigue, reducing the effectiveness of observation. Periods of observation with optical devices should be limited to 30 minutes followed by a minimum of 15 minutes rest. A sniper can minimize eyestrain during observation by glancing away at green grass or any other subdued color.

M48/M49 OBSERVATION TELESCOPES AND TRIPOD

2-44. The M48/M49 observation telescopes are prismatic optical instruments of 20x magnification. Both scopes are essentially identical and this manual will refer mainly to the M49 from this point on. The lenses are coated with magnesium fluoride for improved light-transmitting capability. The sniper team carries the M49 when needed for the mission. The designated observer uses the telescope to assist in observation and selection of targets while the sniper is in the fire position. Properly used, the M49 telescope can significantly enhance the success of the team's mission by allowing it to conduct a superior target analysis, read the current environmental conditions, and make spot corrections by observing bullet trace and impact. The high magnification of the telescope makes observation, target detection, and target identification possible where conditions such as range would otherwise prevent identification. Camouflaged targets and those in deep shadows are more readily detected. Characteristics of the M48/M49 are as follows:

- M48 Observation Telescope:
    - Tripod: M14.
    - Carrying case (scope): M26.

FM 3-05.222

- Carrying case (tripod): M31.
- Magnification: 19.6x.
- Field of view: 37.2 mils.
- Exit pupil: 0.100 inches.
- Effective focal length (EFL) of objectives: 13.004 inches.
- EFL of eyepiece: 0.662 inches.
- Length: 13.5 inches.
- M49 Observation Telescope:
  - Tripod: M15.
  - Carrying case (scope): M27.
  - Carrying case (tripod): M42.
  - Magnification: 20x.
  - Field of view: 38.37 mils.
  - Exit pupil: 0.108 inches.
  - EFL of objectives: 14.211 inches.
  - EFL of eyepiece: 0.716 inches.
  - Length: 14.5 inches.

Operating the M49

2-45. An eyepiece cover cap and objective lens cover protect the optics when the telescope is not in use. Snipers must take care to prevent cross-threading of the fine threads. They should turn the eyepiece focusing-sleeve clockwise or counterclockwise until the image is clearly seen.

Operating the M15 Tripod

2-46. The sniper uses the height adjusting collar to maintain a desired height for the telescope. The sniper keeps the collar in position by tightening the clamping screw. He uses the shaft rotation locking thumbscrew to clamp the tripod shaft at any desired azimuth. The elevating thumbscrew enables the sniper to adjust the cradle of the tripod and to increase or decrease the angle of elevation of the telescope. He can then tighten the screw nut at the upper end of each leg to hold the tripod legs in an adjusted position.

Setting Up the M49 and Tripod

2-47. The sniper spreads the tripod legs and places it in a level position on the ground so the cradle is level with the target area. He places the telescope through the strip loop of the tripod and tightens the strap to keep the telescope steady and in place. If the tripod is not carried, he uses an expedient rest for the scope. The sniper should always make sure the scope is in a steady position to maximize its capabilities and minimize eyestrain.

M144 OBSERVATION TELESCOPE

2-48. The M144 observation telescope is the new U.S. Army observation telescope and has a variable power eyepiece. The sniper/observer can adjust the eyepiece from 15x to 45x. This range permits the observer to adjust for a wider field of view or for magnification for clearer target identification. The observer should ensure that while reading winds or spotting trace, the scope should not be placed at a higher magnification than 20x. He can use the M144 in the same manner as the M49 scope that it replaces. Characteristics of the M144 are as follows:

- Objective lens: 60 mm.
- Magnification: 15x to 45x.
- Field of view: 125 ft at 15x and 62 ft at 45x at 100 meters.
- Focus range: 30 ft to infinity.
- Exit pupil: 4 mm at 15x and 1.4 mm at 45x.
- Eye relief: 20.5 mm at 15x and 13.5 mm at 45x.

2-49. The Army will soon replace the M144 observation telescope with a newer scope that is waterproof and more durable. The next scope will be a variable and possess better optics.

NIGHT VISION DEVICES

2-50. Snipers use NVDs to accomplish their mission during limited visibility operations. They can use NVDs as observation aids, weapons sights, or both. First- and second-generation NVDs amplify the ambient light to provide an image of the observed area or target. These NVDs require target illumination (with the exception of the NADS 750); they will not function in total darkness because they do not project their own light source. NVDs work best on bright, moonlit nights. When there is no light or the ambient light level is low (as in heavy vegetation), the use of artificial or infrared (IR) light improves the NVD's performance.

2-51. Fog, smoke, dust, hail, or rain limit the range and decrease the resolution of NVDs. NVDs do not allow the sniper to see through objects in the field of view. The sniper will experience the same range restrictions when viewing dense wood lines as he would when using other optical sights.

2-52. Initially, a sniper may experience eye fatigue when viewing for prolonged periods. He should limit initial exposure to 10 minutes, followed by a 15-minute rest period. After several periods of viewing, he can safely extend the observation time limit. To help maintain continuous observation and to reduce eye fatigue, the sniper should often alternate his viewing eyes.

Night Vision Sight, AN/PVS-2

2-53. The AN/PVS-2 is a first-generation NVD (Figure 2-13, page 2-19). It can resolve images in low, ambient light conditions better than second-generation NVDs. However, first-generation NVDs are larger and heavier. Characteristics of an AN/PVS-2 include the following:

- Length: 18.5 inches.
- Width: 3.34 inches.
- Weight: 5 pounds.

- Magnification: 4x.
- Range: varies depending on ambient light conditions.
- Field of view: 171 mils.
- Focus range: 4 meters to infinity.

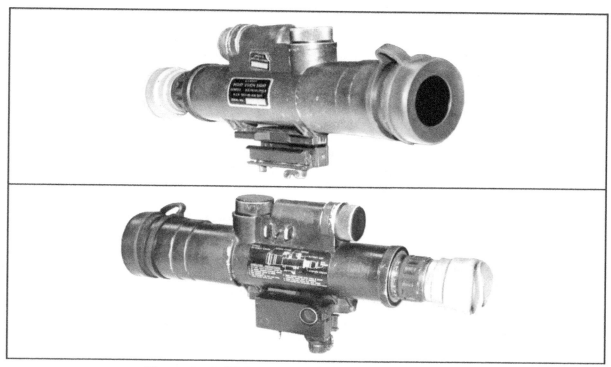

Figure 2-13. Night Vision Sight, AN/PVS-2

Night Vision Sight, AN/PVS-4

2-54. The AN/PVS-4 is a portable, battery-operated, electro-optical instrument that can be used for visual observation or weapon-mounted for precision fire at night (Figure 2-14, page 2-20). The sniper can detect and determine distant targets through the unique capability of the sight to amplify reflected ambient light (moon, stars, or sky glow). The sight is passive; thus, it is free from enemy detection by visual or electronic means. With the correct adapter bracket, the sniper can mount this sight on the M4, M16, M21, or M24. Characteristics of the AN/PVS-4 are as follows:

- Length: 12 inches.
- Width: 3.75 inches.
- Weight: 3.5 pounds.
- Magnification: 3.6x.
- Range: 400 meters/starlight, 600 meters/moonlight, for a man-sized target.
- Field of view: 258 mils.
- Focus range: 20 feet to infinity.

Figure 2-14. Night Vision Sight, AN/PVS-4

2-55. Second-generation NVDs such as the AN/PVS-4 possess the advantage of smaller size and weight over first-generation NVDs. However, they do not possess the extreme low-light capability of the first-generation devices. The AN/PVS-4 also offers advantages of internal adjustments, changeable reticles, and protection from blooming, which is the effect of a single light source, such as a flare or streetlight, overwhelming the entire image.

2-56. When mounted on the M4 or M16 rifle, the AN/PVS-2/4 is effective in achieving a first-round hit out to and beyond 300 meters, depending upon the light and wind conditions. The AN/PVS-2/4 is mounted on the M4 or M16 since the NVD's limited range does not make its use practical for the 7.62-mm SWS. This practice prevents problems that may occur when removing and replacing the NVD. The NVD provides an effective observation capability during limited visibility operations. The NVD does not give the width, depth, or clarity of daytime optics. However, a well-trained sniper can see enough to analyze the tactical situation, detect enemy targets, and engage targets effectively. The sniper team uses the AN/PVS-2/4 to—

- Enhance night observation capability.
- Locate and suppress hostile fire at night.
- Deny enemy movement at night.
- Demoralize the enemy with effective first-round hits at night.

2-57. When given a choice between AN/PVS-2 and AN/PVS-4, snipers should weigh their advantages and disadvantages. The proper training and knowledge with NVDs cannot be overemphasized. The results obtained with NVDs will be directly attributable to the sniper's skill and experience in their use. Generally the PVS-2 is better for very low-light observation while the PVS-4 is better for built-up areas.

KN200(PVS-9)/KN250(PVS-9A) Image Intensifier (SIMRAD)

2-58. The KN200/250 image intensifier (Figure 2-15) increases the use of the existing M3A telescope. It is mounted as an add-on unit and enables the sniper to aim through the eyepiece of the day sight both during day and night—an advantage not achieved with traditional types of NVDs. Sudden illumination of the scene does not affect sighting abilities. Depending on date of manufacture, these image intensifiers can be either second- or third-generation image intensifier tubes. Due to their unique design, the exact position of the image intensifier relative to the day sight is not critical. The mounting procedures take only a few seconds; however, boresighting will be required. The KN200/250 technical specifications include—

- Weight (excluding bracket): 1.4 kg/0.7 kg
- Magnification: 1x, +/– 1 percent.
- Field of view: 177/212 mils.
- Focus range: Fixed and adjustable.
- Objective lens: 100 mm/80 mm.
- Mounting tolerance: +/– 1 degree.
- Battery life: 40 hours at 25 degrees centigrade (C) with two AA alkaline cells.
- Operating temperature: –30 to +50 degrees C.

Figure 2-15. KN200(PVS-9)/KN250(PVS-9A) Image Intensifier (SIMRAD)

FM 3-05.222

NADS 750, 850, 1000 Night Vision Imaging System

2-59. This system (Figures 2-16 and 2-17) is similar to the PVS-9 SIMRAD system as far as mounting and use. Its characteristics are as follows:

- Size (approximately): 4.5 x 7.1; 4.7 x 7.76; 5.8 x 12.1 inches.
- Weight: 2.6; 5.0; 6 lbs.
- Magnification: 1x.
- Field of view: 238, 210, and 120 mils.
- Immersion: 66 feet/2 hours.
- Tube type: Generation (Gen) III.
- Battery life: 24 hours at 73 degrees centigrade (C) with two AA alkaline cells.
- Illuminator: 750 only.

Figure 2-16. NADS 750, 850, and 1000

Figure 2-17. NADS 750 With AN/PEQ 2 IR Pointer/Illuminator Mounted on SR-25

AN/PVS-10 Integrated Sniper Day/Night

2-60. This system (Figure 2-18) requires the sniper to remove his standard day scope and replace it with this system. The system splits the available light and directs part of the light to the daytime scope and part of the light to the night portion of the scope. Then the scope's daytime and nighttime portions do not receive the full available light and thus are not as efficient as stand-alone systems. Characteristics include the following:

- Weight: 4.9 pounds (lbs)/5.5 lbs.
- Magnification: 8.5/12.2x.
- Field of view: 35/26 mils.
- Tube types: Gen II, III, and III+.
- Batteries: 2 AA alkaline cells.

Figure 2-18. AN/PVS-10 Sniper Day/Night Scope

Model 007 "Universal Clip-On" Augmenting Weapon Night Sight

2-61. This clip-on sight (Figures 2-19 and 2-20, page 2-24) is similar to the SIMRAD and NADS. The sight clips onto the front of the day scope through a mounting system attached to the front scope-ring mount. The major difference is the size and weight of the system. Characteristics include the following:

- Magnification: 1x.
- Weight: 1.5 lbs.
- Length (approximately): 6 inches.
- Tube type: Gen III.
- Batteries: 2 AA alkaline cells.
- Boresight deviation upon mounting: < 1 MOA.

FM 3-05.222

Figure 2-19. Universal Clip-on With AN/PEQ 2

Figure 2-20. Universal Clip-on Mounted on SR-25

AN/PVS-17 Mini Night Vision Sight

2-62. This system (Figure 2-21, page 2-25) is designed for the M4 SOPMOD 2 Project and is easily adapted to the M24 for close-in urban work. The sniper must remove the day scope to mount and use this sight. Characteristics include the following:

- Magnification: 2.25x, 4.5x.
- Reticule: internal dot (presently).
- Tube types: Gen III and IV.
- Battery: 1 AA alkaline cell.
- Mount system: single-point, quick-release; two-point.

Figure 2-21. AN/PVS-17 Mini Night Vision Sight

AN/PAS-13 Thermal Weapons Sight

2-63. The sniper can use this passive thermal imager (Figure 2-22) to detect targets in day or night conditions. It is also effective during periods of fog, rain, dust, or other conditions that will hinder the light amplification type of NVDs. The tube type is a Gen II forward-looking infrared (FLIR) and the reticle pattern is the same as the M3A day scope when the sniper uses the PAS-13 Heavy.

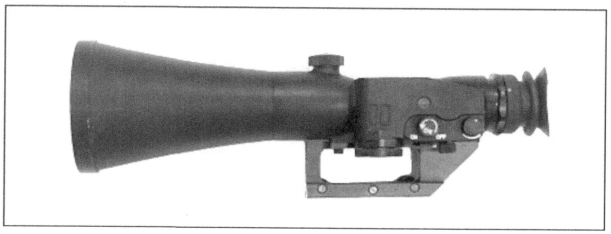

Figure 2-22. AN/PAS-13 Thermal Sight

FM 3-05.222

Night Vision Goggles, AN/PVS-5

2-64. The AN/PVS-5 (Figure 2-23) is a lightweight, passive night vision system that gives the sniper team another means of observing an area during limited visibility. The sniper normally carries the goggles because the observer has the M16 mounted with the NVD. The design of the goggles makes viewing easier. However, the same limitations that apply to the night sight also apply to the goggles.

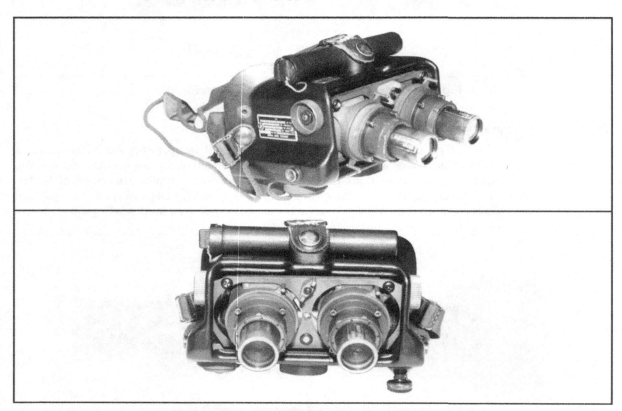

Figure 2-23. Night Vision Goggles, AN/PVS-5

2-65. The sniper can use the AN/PVS-7 (Figure 2-24, page 2-27) instead of the AN/PVS-5 goggles. These goggles provide better resolution and viewing ability than the AN/PVS-5. The AN/PVS-7 series come with a head-mount assembly that allows them to be mounted in front of the face to free both hands. The sniper can also use the goggles without the mount assembly for handheld viewing. TM 11-5855-262-10-1, *Operator's Manual for Night Vision Goggles*, provides additional technical information.

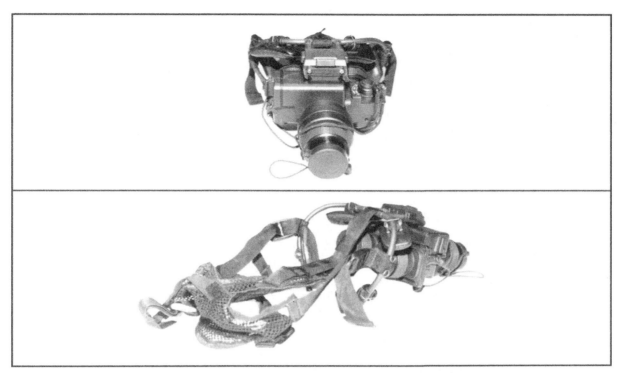

Figure 2-24. Night Vision Goggles, AN/PVS-7 Series

AN/PVS-14 Monocular Night Vision Device

2-66. The AN/PVS-14 (Figure 2-25) is the replacement monocular for the PVS-7. The sniper can use either the 1x as a movement device or the 3x to 5x with an adapter as an observation device. The sniper can wear this NVD with a Kevlar helmet or a head harness for soft headgear. The NVD can also be handheld. It has a Gen III tube with a 40-degree field of view, and uses 2 AA batteries for power.

Figure 2-25. AN/PVS-14 Mounted With Helmet Clip

FM 3-05.222

RANGE FINDERS

2-67. The sniper must use special equipment to reduce the possibility of detection. When necessary, he uses the following equipment to better determine the range to the target and provide greater accuracy upon engagement.

Laser Observation Set, AN/GVS-5

2-68. Depending on the mission, snipers can use the AN/GVS-5 to determine increased distances more accurately. The AN/GVS-5 is an individually operated, handheld, distance-measuring device designed for distances from 200 to 9,990 meters (with an error of +/– 10 meters). A sniper can use it to measure distances by firing an IR beam at a target and measuring the time the reflected beam takes to return to him. The AN/GVS-5 then displays the target distance, in meters, inside the viewer. The reticle pattern in the viewer is graduated in 10-mil increments and has display lights to indicate low battery and multiple target hits. If the beam hits more than one target, the display gives a reading of the closest target hit. The beam that is fired from the set poses a safety hazard; therefore, snipers that plan to use this equipment should be thoroughly trained in its safe operation. The AN/GVS-5 has two filters (red and yellow) that shorten the range of the range finder. The yellow filter is considered safe when viewed through other filtered optics. The red is considered eye safe. The sniper should use the yellow filter when operating near friendly forces.

Mini-Eyesafe Laser Infrared Observation Set, AN/PVS-6

2-69. The AN/PVS-6 (Figure 2-26) contains a mini-eyesafe laser range finder, nonrechargeable BA-6516/U batteries, lithium thionyl chloride, carrying case, shipping case, tripod, lens cleaning compound and tissues, and an operator's manual. The laser range finder is the major component of the AN/PVS-6. It is lightweight, individually operated, and handheld or tripod-mounted. It can accurately determine ranges from 50 to 9,995 meters in 5-meter increments and display the range in the eyepiece. The ranger finder can also be mounted with and boresighted to the AN/TAS-6 or other comparable long-range night observation device.

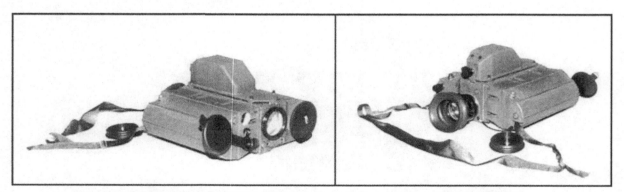

Figure 2-26. Mini-Eyesafe Laser Infrared Observation Set, AN/PVS-6

2-28

into 2 MOAs out to 600 meters (approximately a 33-centimeter group). Various modifications to the weapons themselves or selection of certain types of ammunition may improve the accuracy of the following special weapons:

- Bolt-action target rifles.
- Foreign sniper weapons (procured out of need, compatibility, or to provide a foreign "signature").
- Large-bore, long-range sniper rifles.
- Telescope-mounted handguns (for example, XP100 or the Thompson Center Contender) for easy concealment or used as light multimission SWSs.
- Suppressed weapons.

Suppressors

2-80. The suppressor is a device that snipers can use to deceive observers (forward of the sniper) as to the exact location of the weapon and the sniper. This deception disguises the signature in two ways. First, it reduces the muzzle blast to such an extent that it becomes inaudible a short distance from the weapon. This reduction makes the exact sound location extremely difficult, if not impossible, to locate. Secondly, it suppresses the muzzle flash at night, making visual location equally difficult. Using the suppressor is critical during night operations.

2-81. When the sniper fires a rifle or any high-muzzle velocity weapon, the resulting noise is produced by two separate sources. These sounds are the muzzle blast and the ballistic crack (sonic boom) produced by the bullet:

- The muzzle blast appears when the blast wave (created by the high velocity gases) escapes into the atmosphere behind the bullet. This noise is relatively easy to locate as it emanates from a single, fixed point.
- Ballistic crack results from the supersonic speed of the bullet that compresses the air ahead of it exactly in the same fashion as a supersonic jet creates a sonic boom. The only difference is that the smaller bullet produces a sharp crack rather than a large overpressure wave with its resulting louder shock wave.

Depending on distance and direction from the weapon, the two noises may sound as one or as two different sounds. The further from the weapon the observer is, the more separate the sounds; for example, 600 meters—1 second elapses between the two.

2-82. Unlike the muzzle noise that emanates from a fixed point, the ballistic crack radiates backwards in a conical shape, similar to a bow wave from a boat, from a point slightly ahead of the moving bullet. Thus, the sonic boom created by the supersonic bullet moves at the velocity of the bullet away from the muzzle noise and in the direction of the target. Location and identification of the initial source of the shock wave is extremely difficult because the moving wave strikes the ear at nearly 90 degrees to the point of origin. Attention is thus drawn to the direction from which the wave is coming rather than toward the firing position (Figure 2-27, page 2-32).

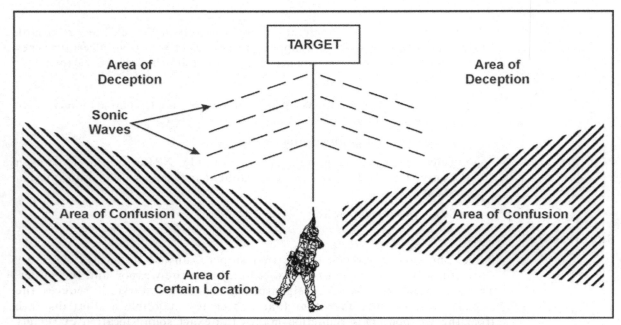

Figure 2-27. Deception Caused by the Sonic Waves of the Bullet
Breaking the Speed of Sound

Surveillance Devices

2-83. In some circumstances, a sniper may use special surveillance devices that will normally involve adding weight and bulk, which can limit his mobility. The sniper should consider using these devices mostly for fixed peacekeeping or Perimeter Force Protection roles. The following paragraphs explain a few of these devices.

2-84. **Single-Lens-Reflector (SLR), Digital, and Video Cameras.** Snipers spend more time observing than shooting. Collecting and reporting intelligence are critical tasks. SLR and digital cameras are important tools that significantly enhance the sniper's ability to meet intelligence collection requirements. Video surveillance kits are being fielded to the SF groups to support operations in urban and rural AOs. These kits are integrated with the sniper's communications package so that sniper teams can provide commanders with "near-real-time" video and still images of EEI. This ability to pass images significantly enhances a sniper team's utility and lethality.

2-85. **100-mm Team Spotting Scope.** This device is a standard team scope for most marksmanship units and should be used for sniper training purposes. The scope's increased field of view will greatly enhance the team's observation capability in static positions. While the Unertl is considered standard, the newer 100-mm Optolyth is clearer and more compact, as well as durable.

2-86. **Crew-Served NVDs.** Snipers commonly use these devices in conjunction with crew-served weapons (typified by the AN/TVS-5) or night observation (typified by the AN/TVS-4). These NVDs offer a significant advantage over their smaller counterparts in surveillance, target acquisition, and night observation (STANO). However, their weight and bulk normally limit their use to static operations.

2-87. **Thermal Imagery.** This relatively new tool is now available to the sniper team. Equipment such as the AN/PAS-7 offers a thermal imagery device in a portable package. Thermal imagery can enhance STANO operations when used with more conventional equipment, or it can provide continuous surveillance when ambient light conditions (such as starlight and moonlight) do not exist for light-intensification devices. Thermal devices offer an option when there is an abundance of light that would cause white out conditions with NVDs.

2-88. **Radars and Sensors.** Just as the sniper's surveillance operations should be integrated into the overall surveillance plan, the sniper should strive to make maximum use of any surveillance radars and sensors in the area of operation. Snipers will normally not use these items themselves, but through coordination with using or supporting units. The snipers may be able to use the target data that the radars and sensors can acquire. However, they must keep in mind that these devices are subject to human error, interpretation, and enemy countermeasures. Total reliance on the intelligence data obtained by using these devices could prove detrimental or misleading.

CARE AND CLEANING OF THE SNIPER WEAPON SYSTEM

2-89. Maintenance is any measure taken to keep the system in top operating condition. It includes inspection, repair, cleaning, and lubrication. Inspection reveals the need for the other measures. The sniper couples his cleaning with a program of detailed inspections for damage or defects. He uses the following maintenance items:

- One-piece plastic-coated caliber .30 cleaning rod with jags (36 inches).
- Field cleaning kit such as Kit and caboodle cleaning cable-with Muzzle Guard-Field.
- Bronze-bristled bore brushes (calibers .30 and .45).
- Muzzle guide.
- Cleaning patches (small and large sizes).
- Shooter's Choice Bore Solvent (SCBS) carbon cleaner.
- Sweets 7.62 Copper Remover (copper cleaner). (Shooter's Choice Copper Remover is the second choice.)
- Shooter's Choice Rust Prevent.
- Cleaner, lubricant, preservative (CLP). (Note: Do not use lubricating oil, weapons semifluid, Breakfree, or WD40 in the bore.)
- Rifle grease.
- Bore guide (long action).
- Q-tips or swabs.
- Pipe cleaners.
- Medicine dropper.
- Shaving brush.
- Toothbrush.

- Pistol cleaning rod.
- Rags.
- Camel-hair brush.
- Lens paper.
- Lens cleaning fluid or denatured alcohol.

NOTE: Never place cleaning fluid directly on lens surface. Use lens paper or cleaning pencil and place cleaning fluids on the tissue or pen.

WHEN TO CLEAN

2-90. Snipers must regularly inspect any weapon sheltered in garrison and infrequently used to detect dirt, moisture, and signs of corrosion and must clean it accordingly. However, a weapon in use and subject to the elements requires no inspection for cleanliness. The fact that it's used and exposed is sufficient evidence that it requires repeated cleaning and lubrication.

Before Firing

2-91. The sniper must always clean the rifle before firing. Firing a weapon with a dirty bore or chamber will multiply and speed up any corrosive action. Oil in the bore and chamber of even a clean rifle will cause pressures to vary and first-round accuracy will suffer. Hydrostatic pressure will also cause cases to blow or jam in the chamber. The sniper should clean and dry the bore and chamber before departing on a mission and use extreme care to keep the rifle clean and dry en route to the objective area. Before the sniper fires the weapon, he should ensure that the bore and chamber are still clean, dry, and no strings are left from the cleaning patches. Firing a rifle with oil or moisture in the bore will cause a puff of smoke that can disclose the firing position. It can also cause damage to the weapon system.

After Firing

2-92. The sniper must clean the rifle after it has been fired, because firing produces deposits of primer fouling, powder ashes, carbon, and metal fouling. Although modern ammunition has a noncorrosive primer that makes cleaning easier, the primer residue can still cause rust if not removed. Firing leaves two major types of fouling that requires different solvents to remove: **carbon** fouling and **copper jacket** fouling. The sniper must clean the rifle within a reasonable interval—a matter of hours—after a cessation of firing. Common sense should preclude the question as to the need for cleaning between rounds. Repeated firing will not damage the weapon if it is properly cleaned before the first round.

2-93. The M24 SWS will be disassembled only when absolutely necessary, not for daily cleaning. An example would be to remove an obstruction that is stuck between the stock and the barrel. When disassembly is required, the recommended procedure is to—

- Place the weapon so that it is pointing in a safe direction.
- Ensure the safety is in the "S" position.
- Remove the bolt assembly.

- Loosen the two mounting ring nuts (Figure 2-28) on the telescope and remove the telescope. (Not necessary when only cleaning the weapon.)
- Remove the two trigger action screws (Figure 2-29).
- Lift the stock from the barrel assembly.

NOTE: Always reassemble the weapon in the same sequence as the last time it was reassembled. This will keep the weapon zeroed to within .5 MOA. For further disassembly, refer to TM 9-1005-306-10.

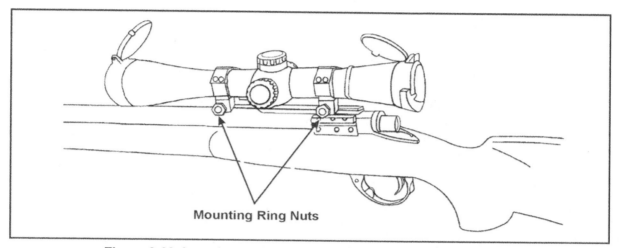

Figure 2-28. Location of the Mounting Ring Nuts on the M24 SWS

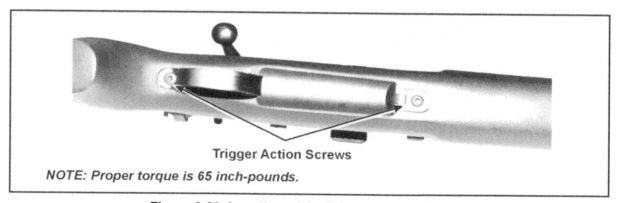

Figure 2-29. Location of the Trigger Action Screws

HOW TO CLEAN

2-94. The sniper cleans the rifle by laying it on a cleaning table or other flat surface with the muzzle away from the body and the sling down. He makes sure not to strike the muzzle or telescopic sight on the table. The cleaning cradle is ideal for holding the rifle, or the sniper can use the bipod to support the weapon.

2-95. The sniper should always clean the bore from the chamber toward the muzzle, attempting to keep the muzzle lower than the chamber to prevent bore cleaner from running into the receiver or firing mechanism. When in garrison, he should always use the chamber guide to move the one-piece steel rod from chamber to muzzle. When in the field, he should use the muzzle guide and insert the one-piece cable down the bore to the chamber and pull the patches through to the muzzle. The sniper must be careful not to get solvents between the receiver and the stock. Solvents soften the bedding compound. When the rifle is fired, the action shifts in the soft bedding, which decreases accuracy and increases wear and tear on the bedding material. Solvents contribute to the accumulation of debris between the action and the stock interfering with barrel harmonics.

NOTE: The sniper should always use a bore guide to keep the cleaning rod centered in the bore during the cleaning process.

2-96. The sniper first pushes several patches saturated with SCBS through the barrel to loosen the powder fouling and begin the solvent action on the copper jacket fouling. He then saturates the bronze-bristled brush (**Never use stainless steel bore brushes—they will scratch the barrel!**) with SCBS (shake bottle regularly to keep the ingredients mixed) using the medicine dropper to prevent contamination of the SCBS. He runs the bore brush through the barrel approximately 20 times. He makes sure that the bore brush passes completely through the barrel before reversing its direction; otherwise the bristles can break off.

NOTE: The sniper should never stick the bore brush into the bottle of SCBS. This will contaminate the fluid.

2-97. Using a pistol cleaning rod and a caliber .45 bore brush, the sniper cleans the chamber by rotating the patch-wrapped brush 8 to 10 times. He should **NOT** scrub the brush in and out of the chamber. He then pushes several patches saturated with SCBS through the bore to push out the loosened powder fouling.

2-98. The sniper continues using the bore brush and patches with SCBS until the patches come out without traces of the black/gray powder fouling and become increasingly green/blue. This process indicates that the powder fouling has been removed and that only the copper fouling remains. He then removes the SCBS from the barrel with several clean patches. This is important because the different solvents should never be mixed in the barrel.

2-99. The sniper pushes several patches saturated with Sweets through the bore, using a scrubbing motion to work the solvent into the copper. He lets the solvent work for 10 to 15 minutes. (**Never leave Sweets in the barrel for more than 30 minutes!**)

2-100. While waiting, the sniper scrubs the bolt with the toothbrush moistened with SCBS and wipes down the remainder of the weapon with a cloth. He pushes several patches saturated with Sweets through the barrel. The patches will appear dark blue at first, indicating the amount of copper fouling removed. He continues this process until the saturated patches come out without a trace of blue/green. If the patches continue to come out dark blue after several treatments with Sweets, he should run patches with SCBS through the bore deactivating the sweets and start the cleaning process over again.

2-101. When the barrel is completely clean, the sniper then dries it with several tight fitting patches. He should also dry out the chamber using the caliber .45 bore brush with a patch wrapped around it. The sniper then runs a patch saturated with Shooter's Choice Rust Prevent (not CLP) down the barrel and chamber if the weapon is to be stored for any length of time. He should also be sure to remove the preservative by running dry patches through the bore and chamber before firing.

NOTE: Stainless steel barrels are not immune from corrosion.

2-102. The sniper places a small amount of rifle grease on the rear surfaces of the bolt lugs. This grease will prevent galling of the metal surfaces. He should also place grease on all wear points (the shiny areas) of the bolt. The sniper then wipes down the complete weapon exterior (if it is not covered with camouflage paint) with a CLP-saturated cloth to protect it during storage.

Barrel Break-in Procedure

2-103. To maximize barrel life and accuracy and to minimize the cleaning requirement, the sniper must use the following barrel break-in procedure. This procedure is best done when the SWS is new or newly rebarreled. The break-in period "laps-in" the barrel by polishing the barrel surface under heat and pressure. The sniper must first completely clean the barrel of all fouling, both powder and copper. He dries the barrel and fires one round. He then completely cleans the barrel using Shooter's Choice Solvent, followed by Sweets 7.62 copper remover. Again, the barrel must be completely cleaned and another round fired. This procedure of firing one shot, then cleaning, must be done for a total of 10 rounds. After the 10th round, the sniper tests the SWS for groups by firing three-round shot groups, with a complete barrel cleaning between shot groups for a total of five shot groups (15 rounds total). The barrel comes from the factory with 60 test-fire rounds already through it. The barrel is now broken-in and will provide superior accuracy and a longer usable barrel life. It also will be easier to clean because the surface is smoother. Although the full accuracy potential may not be noticed until after 100 rounds or more have been fired, again, the sniper should clean the barrel at least every 100 rounds to maximize barrel life.

Storage

2-104. The M24 SWS should be properly stored to ensure it is protected and maintained at a specific level. The sniper should—

- Clear the SWS, close the bolt, and squeeze the trigger.
- Place all other items in the system case (M24).
- Transport the weapon in the system case during nontactical situations.
- Protect the weapon at all times during tactical movement.

OPTICAL EQUIPMENT MAINTENANCE

2-105. Dirt, rough handling, or abuse of optical equipment will result in inaccuracy and malfunction. When not in use during field conditions, the sniper should case the rifle and scope and cap the lenses.

Cleaning the Lenses

2-106. The sniper should coat the lenses with a special magnesium fluoride reflection-reducing material. The coat should be very thin and the sniper must take great care to prevent damaging the lenses. To remove dust, lint, or other foreign matter from the lens, he brushes it lightly with a clean camel-hair brush.

2-107. The sniper must also remove oil or grease from all the optical surfaces. He applies a drop of lens cleaning fluid or denatured alcohol to a lens tissue and carefully wipes off the lens surface in circular motions, from the center to the outside edge. He dries off the lens with a clean lens tissue. In the field, if the proper supplies are not available, the sniper can breathe heavily on the glass and wipe with a soft, clean cloth.

Handling Telescopes

2-108. Telescopes are delicate instruments and the sniper must handle them with great care. The following precautions will prevent damage. The sniper should—

- Check the torque on all mounting screws periodically and always before any operation. He should also be careful not to change coarse windage adjustment.
- Keep lenses free from oil and grease and never touch them with the fingers. Body grease and perspiration can also injure them. Keep lenses capped.
- Not force elevation and windage screws or knobs.
- Not allow the telescope to remain in direct sunlight and avoid letting the sun's rays shine through the lens. Lenses magnify and concentrate sunlight into a pinpoint of intense heat, which is focused on the mil-scale reticle. This exposure may damage the telescope internally. Keep the lenses covered and the entire telescope covered when not firing or preparing to fire. Never use the rifle scope for observation purposes only.
- Avoid dropping the telescope or striking it with another object. This blow could damage it severely and permanently, as well as change the zero. When placing the weapon in the carrying case, he should place the scope away from the hinges. This will help protect the scope from vibration and dropping.
- Not allow just anyone to handle the equipment. The sniper or armorer should really be the only personnel that handle the telescope or any other sniper equipment.
- Once the scope is zeroed, note the reticle position on a bore scope grid for future reference.

WEAPON MAINTENANCE AND CARE

2-109. Maintenance is any measure that the sniper takes to keep the SWS in top operating condition. A sniper may have to operate in many different environments and every type requires him to care for his weapon in a specific manner. The following paragraphs explain each of these environments.

Cold Climates

2-110. In temperatures below freezing, the sniper must maintain and treat the rifle a specific way. He should—

- Always keep the rifle free of moisture and heavy oil (both will freeze) to prevent working parts from freezing or operating sluggishly.
- Store the rifle in a room with the temperature equal to the outside temperature.
- If the rifle is taken into a warm area, be sure to remove the condensation and thoroughly clean and dry the rifle before taking it into the cold. Otherwise, the condensation will cause icing on exposed metal parts and optics.
- Disassemble the firing pin, clean it thoroughly with a degreasing agent, and then lubricate it with CLP. Rifle grease will harden and cause the firing pin to fall sluggishly.

2-111. In extreme cold, the sniper must take the following care to avoid condensation and the congealing of oil on the weapon. He should—

- If not excessive, remove condensation by placing the instrument in a warm place. Not apply concentrated heat because it will cause expansion and damage.
- Blot moisture from the optics with a lens tissue or a soft, dry cloth.
- In cold temperatures, ensure the oil does not thicken and cause sluggish operation or failure. Remember that focusing parts are particularly sensitive to freezing oils.
- Remember that breathing will form frost, so he must clean the optical surfaces with lens tissue, preferably dampened lightly with lens cleaning fluid or denatured alcohol. Never apply the fluid directly to the glass.

Saltwater Exposure

2-112. Salt water and a saltwater atmosphere have extreme and very rapid corrosive effects on metal. During this type of exposure, the sniper must ensure the rifle is—

- Checked frequently and cleaned as often as possible, even if it means only lubricating the weapon.
- Always well lubricated, including the bore, except when actually firing.
- Thoroughly cleaned by running a dry patch through the bore before firing, if possible. To keep the patches dry, store them in a waterproof container.

Jungle Operations (High Humidity)

2-113. There is no standard jungle. The tropical area may be rain forests, secondary jungles, savannas, or saltwater swamps. When operating in any jungle environment, high temperatures, heavy rainfall, and oppressive humidity become a sniper's concern in maintaining his weapon. He should—

- Use more lubricant.
- Keep the rifle cased when not in use.

- Protect his rifle from rain and moisture whenever possible.
- Keep ammunition clean and dry.
- Clean the rifle, bore, and chamber daily.
- Keep the telescope caps on when not in use. If moisture or fungus develops inside the telescope, he should get a replacement.
- Keep cotton balls between lens caps and lens.
- Clean and dry the stock daily.
- Dry the carrying case and rifle in the sun whenever possible.
- Take an 8- or 9-inch strip of cloth and tie a knot in each end to protect the free-floating barrel of the weapon. Before going on a mission, he should slide the cloth between the barrel and stock all the way to the receiver and leave it there. When in position, he slides the cloth out, taking all restrictive debris and sand with it. **(This procedure should be done in all environments.)**

Desert Operations

2-114. Hot, dry climates are usually dusty and sandy areas. They are hot during daytime hours and cool during the nighttime. Dust and sand will get into the rifle and will cause malfunctions and excessive wear on component working surfaces through abrasive action during the firing operations. When operating in this type of environment, the sniper should—

- Keep the weapon completely dry and free of CLP and grease except on the rear of the bolt lugs.
- Keep the rifle free of sand by using a carrying sleeve or case when not in use.
- Protect the weapon by using a wrap. He should slide the wrap between the stock and barrel then cross over on top of the scope, cross under the weapon (over magazine), and secure. He can still place the weapon into immediate operation but all critical parts are covered. The sealed hard case is preferred in the desert if the situation permits.
- Keep the telescope protected from the direct rays of the sun.
- Keep ammunition clean and protected from the direct rays of the sun.
- Use a toothbrush to remove sand from the bolt and receiver.
- Clean the bore and chamber daily.
- Protect the muzzle and receiver from blowing sand by covering them with a clean cloth.
- Take an 8- or 9-inch strip of cloth and tie a knot in each end to protect the free-floating barrel of the weapon. Before going on a mission, he should slide the cloth between the barrel and stock all the way to the receiver and leave it there. When in position, he can slide the cloth out, taking all restrictive debris and sand with it. **(This procedure should be done in all environments.)**

Hot Climates and Saltwater Exposure

2-115. A hot climate and saltwater atmosphere may cause waves and wind. To keep these environmental hazards from affecting the optical equipment, a sniper must take precautionary measures. He should—

- Protect optics from hot, humid climates and saltwater atmosphere.

- **NOT** expose optical equipment to direct sunlight in a hot climate.
- In humidity and salt air, inspect and clean the optical instruments frequently to avoid rust and corrosion. A light film of oil is beneficial.
- Thoroughly dry and lightly oil optical instruments because perspiration from the hands is a contributing factor to rusting.

TROUBLESHOOTING THE SNIPER WEAPON SYSTEM

2-116. Table 2-1 lists some possible SWS malfunctions, causes, and corrective actions. If a malfunction is not correctable, the complete system must be sent to the proper maintenance/supply channel for return to the contractor. (TM 9-1005-306-10 provides further shipment information.)

Table 2-1. M24 SWS Malfunctions and Corrective Actions

MALFUNCTIONS	CAUSES	CORRECTIONS
Fail to Fire	Safety in "S" position.	Move safety to "F" position.
	Defective ammunition.	Eject round.
	Firing pin damaged.	Change firing pin assembly.
	Firing pin binds.	Change firing pin assembly.
	Firing pin protrudes.	Change firing pin assembly.
	Firing control out of adjustment.	Turn complete system in to the maintenance/supply channel for return to contractor.
	Trigger out of adjustment.	Turn in as above.
	Trigger binds on trigger guard.	Turn in as above.
	Trigger does not retract.	Turn in as above.
	Firing pin does not remain in cocked position with bolt closed.	Turn in as above.
Bolt Binds	Action screw protrudes into bolt track.	Turn in as above.
	Scope base screw protrudes into bolt track.	Turn in as above.
Fail to Feed	Bolt override of cartridge.	Ensure bolt is pulled fully toward the rear.
	Cartridge stems chamber.	Pull bolt fully rearward; remove stemmed cartridge from ejection port area; reposition cartridge fully in the magazine.
	Magazine follower in backward.	Remove magazine spring and reinstall with long-leg follower.
	Weak or broken magazine spring.	Replace spring.
Fail to Eject	Broken ejector.	Turn complete weapon system in to the maintenance/supply channel for return to contractor.
	Fouled ejector plunger.	Inspect and clean bolt face; if malfunction continues, turn in as above.
Fail to Extract	Broken extractor.	Turn in as above.
Bolt Release Fails	Bolt release mechanism fouled.	Disassemble rifle. Remove and clean bolt release mechanism. Lubricate with graphite lube.

Chapter 3
Marksmanship Training

The role of the SF sniper is to engage targets with precision rifle fire. A sniper's skill with a rifle is the most vital skill in the art of sniping. This skill is extremely perishable. Sniper marksmanship differs from basic rifle marksmanship only in the degree of expertise. The sniper, using basic and advanced marksmanship as building blocks, must adapt the conventional methods of firing to meet his unique requirements. The sniper must make first-round hits in a field environment under less than ideal conditions and become an expert in marksmanship. The fundamentals are developed into fixed and correct firing habits that become instinctive. This reaction is known as the "conditioning of the nervous system."

Snipers should maintain their proficiency at the following **minimum** standards:

- 90 percent first-round hits on stationary targets at ranges of 600 meters.
- 50 percent first-round hits on stationary targets at ranges from 600 to 900 meters.
- 70 percent first-round hits on moving targets at ranges to 300 meters.
- 70 percent first-round hits on snap targets at ranges to 400 meters.

FIRING POSITIONS

3-1. A sniper's firing position must be solid, stable, and durable. Solid—not influenced by outside factors; stable—for minimized movement of the weapon; and durable—able to hold the weapon and position for an extended period of time to accomplish the mission. Unlike the target shooter who must fire from different positions of varying stability to satisfy marksmanship rules, the sniper searches for the most stable position possible. He is not trying to see if he can hit the target; he must **know** he can hit the target. A miss could mean a failed mission or his life. A good position enables the sniper to relax and concentrate when preparing to fire.

3-2. Whether prone, kneeling, or standing, the sniper's position should be supported by firing rests or other means. Properly employed, the sling, in all but the standing position, provides a stable, supported position. Firing from a rest helps to minimize human factors such as heartbeat, muscular tension, and fatigue. A rest can support both the front and the rear of the rifle, as in the case of benchrest firing.

3-3. Regardless of the rest selected (tree, dirt, sandbag), the sniper will prevent any objects from contacting the barrel. During the firing process, the barrel vibrates like a tuning fork and any disturbance to this harmonic motion will result in an erratic shot. Also, a hard support will normally cause the rifle to change its POI. The sniper can help eliminate this problem by firing from objects of similar hardness. The sniper's hat, glove, or sock filled with sand or dirt can be placed between the rifle forestock and firing support to add consistency from range to combat. A support or rest greatly helps the sniper and he must use one whenever possible. Accuracy with a rifle is a product of consistency, and a rest aids consistency to firing positions.

3-4. On the battlefield, the sniper must assume a steady firing position with maximum use of cover and concealment. Considering the variables of terrain, vegetation, and the tactical situation, the sniper can use many variations of the basic positions. When assuming a firing position, he should adhere to the following basic rules:

- Use the prone position or its variations whenever possible because it is the most stable.
- Use any solid support available, when the bipod is not available or too short.
- Do not touch the support with the weapon's barrel since it interferes with the barrel harmonics and creates shot displacement.
- Use a cushion between the weapon and the support when not using the bipod.
- Do not allow the side of the weapon to rest against the support. This position will have an effect on the weapon during recoil and may affect the POI.
- Never cant the weapon while firing or aiming. The sniper should tilt his head to the weapon, not the weapon to his head.

ELEMENTS OF A GOOD POSITION

3-5. Three elements of a good position are bone support, muscular relaxation, and a natural point of aim (POA) on the aiming point. The following paragraphs explain each element.

Bone Support

3-6. Proper bone support is a learned process; only through practice (dry fire, live fire) will the sniper gain proficiency in this skill. Positions provide foundations for the rifle, and good foundations for the rifle are important to the sniper. When a sniper establishes a weak foundation (position) for the rifle, the position will not withstand the repeated recoil of the rifle in a string of rapid-fire shots or deliver the support necessary for precise firing. Therefore, the sniper will not be able to apply the marksmanship fundamentals properly.

Muscular Relaxation

3-7. The sniper must learn to relax as much as possible in the various firing positions. Undue muscle strain or tension causes trembling, which is transmitted to the rifle. However, in all positions, a certain amount of controlled muscular tension is needed. For example, in a rapid-fire position there should be pressure on the stock weld. Only through practice and achieving a natural POA will the sniper learn muscular relaxation.

Natural Point of Aim on the Aiming Point

3-8. In aiming, the rifle becomes an extension of the body. Therefore, the sniper must adjust the body position until the rifle points naturally at the target. To avoid using muscles to aim at a target, the sniper must shift his entire firing position to move his natural POA to the desired POI. The sniper reaches this point by—

- Assuming a good steady position.
- Closing both eyes and relaxing as if preparing to fire.
- Opening both eyes to see where the weapon is pointing.
- Leaving the nonfiring elbow in place and shifting the legs, torso, and firing elbow left or right.
- Repeating the process until the weapon points naturally at the desired POI.

If the sniper must push or pull the weapon onto target, he is not on his natural POA regardless of how small a movement is involved. Thus, muscle relaxation is not achieved, either.

3-9. The sniper can change the elevation of a natural POA by leaving the elbows in place and sliding the body forward or rearward. This movement causes the muzzle of the weapon to drop or rise, respectively. Minor adjustments to the natural POA can be made by the right leg (right-handed sniper). The sniper moves the lower leg in the opposite direction that he wants the sight to go. Another consideration is to maintain a natural POA after the weapon has been fired; therefore, proper bolt operation becomes critical. The sniper must practice reloading while in the prone position without removing the butt of the weapon from the firing shoulder.

COMMON FACTORS TO ALL POSITIONS

3-10. Establishing a mental checklist of steady position elements greatly enhances the sniper's ability to achieve a first-round hit. This checklist includes the factors discussed below that are inherent to a good firing position.

Nonfiring Hand

3-11. The sniper should use the nonfiring hand as a support. The nonfiring hand should either support the forestock or the butt of the weapon. The sniper should never grasp the forestock with the nonfiring hand. He should let the weapon rest in the nonfiring hand. If he grasps the weapon, the recoil and muscle tremor will cause erratic shots. If the sniper uses the nonfiring hand to support the butt, he should place the hand next to the chest and rest the tip of the butt on it. He then balls his hand into a fist to raise the butt or loosen the fist to lower the weapon's butt. The sniper can also use a firing sock in place of the fist. He must take care not to squeeze his fist as the trigger is squeezed. The muzzle will drop due to the rising of the stock causing a low shot. The sniper must not rest the nonfiring hand or fingers on the shooting side shoulder. Doing so will increase the transmission of the heartbeat to the weapon and destabilize the position.

Placement of the Rifle Butt

3-12. The sniper should place the rifle butt firmly in the pocket of the shoulder. Proper placement of the butt helps to steady the rifle and lessen recoil. The key to the correct rifle-butt method is consistent rearward pressure by the firing hand and correct placement in the shoulder. A hard hold versus a very light hold may change bullet impact. Again, consistency is important. A firm hold is necessary and using a shooting sock may cause a light hold and erratic groups.

Firing Hand

3-13. The sniper should grasp the small of the stock firmly but not rigidly with the firing hand. He then exerts pressure rearward, mainly with the middle and ring fingers of the firing hand. He should not "choke" the small of the stock. A choking-type grip can cause a twisting action during recoil. The sniper must not steer the rifle with the hand or shoulder. He should make large windage adjustments by altering the natural POA, not by leaning or steering the rifle, which will cause the rifle to steer in that direction during recoil. He can wrap his thumb over the top of the small of the stock and use it to grasp, or he can lay it alongside or on top of the stock in a relaxed manner. He places the index finger on the bottom or the trigger, ensuring that it does not touch the stock of the weapon and does not disturb the lay of the rifle when the trigger is pulled. The sniper must maintain steady rearward pressure on the weapon when firing. This tension will help steady the weapon.

Elbows

3-14. Each sniper must find a comfortable position that provides the greatest support. How a sniper uses his elbows will vary with each individual.

Stock Weld

3-15. The stock weld is the point of firm contact between the sniper's cheek and the stock. The sniper places his cheek on the stock in a position that gives proper eye relief. The stock weld will differ from position to position. However, due to the position of the telescope on the sniper rifle and the necessity to have eye relief, the sniper may not get a normal stock weld. An important factor is to get firm contact so that the head and weapon recoil as one unit, thereby facilitating rapid recovery. The point on the weapon should be a natural point where the sniper can maintain eye relief. The sniper should put his cheek in the same place on the stock with each shot. A change in stock weld tends to cause misalignment with the sights, thus creating misplaced shots. This change is more of a problem when using iron sights than with the telescopic sight that is properly adjusted.

3-16. Once the sniper obtains a spot or stock weld, he should use this same positioning for each shot. He must stay with the weapon, not lift his head from the stock during recoil, and maintain the spot or cheek weld. During the initial period of firing, the cheek may become tender and sore. To prevent this discomfort and to prevent flinching, the sniper should press the face firmly against the stock. Moving the head will only give the weapon a chance to build up speed before it impacts with the sniper's cheek.

TYPES OF POSITIONS

3-17. Due to the importance of delivering precision fire, the sniper makes maximum use of artificial support and eliminates any variable that may prevent adhering to the basic rules. He uses the following types of positions when engaging the target.

Prone Supported Position

3-18. The sniper first selects his firing position. He picks a position that gives the best observation, fields of fire, and concealment. He then assumes a comfortable prone position and prepares a firing platform for his rifle (Figure 3-1). The sniper should use the bipod whenever possible. The rifle platform should be as low to the ground as possible. The rifle should rest on the platform in a balanced position to the rear of the upper sling swivel and forward of the floor plate. The sniper must take care to ensure that the operating parts, the magazine, and the barrel do not touch the support, as contact will cause erratic shots. He then forms a wide, low bipod with his elbows. He grips the small of the stock with his firing hand, thumb over or alongside the small of the stock and the forefinger (just in front of the first joint) on the trigger, and pulls the butt of the rifle into his firing shoulder. He then places the nonfiring hand under the toe of the stock, palm down, and places the lower sling swivel into the web of the thumb and forefinger. The sniper can then adjust his fingers and thumb of the nonfiring hand by curling the fingers and thumb into a fist or relaxing the fingers and thumb and laying them flat. In this manner the sniper can raise or lower the barrel onto the target. He then relaxes into a comfortable supported position, removing his nonfiring hand from the stock when necessary to manipulate the scope. He can reload single rounds into the M24 with the firing hand while supporting the rifle at the toe of the stock with the nonfiring hand. When firing from this position, the sniper must have a clear field of fire because the shot may become erratic if the bullet strikes a leaf, grass, or a twig. For extended periods in the prone position, the sniper should cock the firing side leg up to relieve pressure off of the abdomen and reduce heartbeat pulse.

Figure 3-1. Prone Supported Position

Hawkins Position

3-19. The sniper uses this position when he needs a low silhouette. It is very useful when firing from a small depression, a slight rise in the ground, or from a roof (Figure 3-2). However, the sniper should make sure there are no obstructions above the boreline but below line of sight by removing the bolt and observing the target through the bore. This position is the steadiest of all firing positions. Concealment is also greatly aided by using the Hawkins position because the sniper is lying flat on the ground. The sniper will not use this position on level ground because he cannot raise the muzzle high enough to aim at the target.

3-20. The Hawkins position is similar to the prone supported position, except that the support of the weapon is provided by the nonfiring hand. The sniper grasps the front sling swivel with the nonfiring hand, forming a fist to support the front of the weapon. He makes sure the wrist and elbow are locked straight, and the recoil is taken up entirely by the nonfiring arm. Otherwise, his face will absorb the weapon's recoil. The sniper lies flat on the ground, either directly behind the rifle (Canadian version) or angled off to one side (British version). It will appear as though he is lying on the rifle. He can make minor adjustments in muzzle elevation by tightening or relaxing the fist of the nonfiring hand. If more elevation is required, he can place a support under the nonfiring fist.

3-21. If using the Canadian version, the sniper places the butt of the rifle in the shoulder. If using the British version, he tucks the butt under the armpit. The sniper should always use what is most comfortable.

Figure 3-2. Hawkins Position

Sling-Supported Prone Position

3-22. The sniper faces the target squarely with the sling attached to the nonfiring arm above the bicep and lies down facing the target, legs straight to the rear (Figure 3-3, page 3-7). He extends the nonfiring elbow so it is in line with the body and the target and as far under the rifle as comfortable. With the firing hand, he pushes forward on the butt of the stock and fits it into the pocket of the shoulder. The sniper then places the firing side elbow down wherever it feels natural and grasps the grip of the stock, pulling it firmly into the shoulder. He lets his cheek rest naturally on the stock where he can see through the sights and acquire the target. He draws his firing side knee up to a comfortable position so as to take the weight off of the diaphragm. He can obtain a natural POA by adjusting the elevation. This can be done by sliding his body forward or rearward and adjusting his breathing.

Figure 3-3. Sling-Supported Prone Position

Prone Backward Firing Position (Creedmore Firing Position)

3-23. The terrain or situation dictates when to use this firing position. It provides a higher angle of fire as required when firing uphill and other positions are inadequate. Also, the sniper can use this position when he must engage a target to his rear but cannot turn around because of the enemy situation or hide constrictions. The sniper assumes a comfortable position on his side with both legs bent for support and stability. He places the butt of the SWS into the pocket of his shoulder where it meets the armpit. He attempts to support his head for better stability and comfort. The small exit pupil of the telescope requires the sniper to maintain a solid hold and center the exit pupil in the field of the telescope to minimize the errors in sight alignment. This is an extreme firing position and not recommended under most circumstances.

Sitting Supported Position

3-24. To assume this position, the sniper prepares a firing platform for the rifle or rests the rifle on the raised portion of the position. If a platform is not available, then the sniper can use the observer to improvise this position (Figure 3-4). The sniper must ensure the barrel or operating parts do not touch the support. The sniper assumes a comfortable sitting position to the rear of the rifle, grasps the small of the stock with the firing hand, and places the butt of the rifle into the shoulder pocket. He places the nonfiring hand on the small of the stock to assist in getting a stock weld and the proper eye relief.

Figure 3-4. Sitting Supported Position

3-25. The sniper rests the elbows on the inside of the knees in a manner similar to the standard crossed-leg position. He changes position by varying the position of the elbows on the inside of the knees or by varying the body position. This position may be tiring; therefore, the firing mission should be alternated frequently between the sniper team members.

Sling-Supported Sitting Position

3-26. The sniper faces his body 30 degrees away from the target in the direction of the firing hand. He sits down and crosses his ankles so that the nonfiring side ankle is across the firing side ankle (Figure 3-5, page 3-9). He then adjusts the sling for the sitting position. The sniper uses his firing hand palm to place the butt of the stock into the shoulder while allowing the weapon to rest on the nonfiring hand. He uses his firing hand to pull the stock firmly into his shoulder. He rests his elbows inside the knees and leans his body forward. The sniper must not have direct contact between the points of the elbows and the knees. Avoiding direct contact ensures that the sniper uses bone support. He holds the stock high enough in the shoulder to require only a slight tilt of the head to acquire the sights, without canting the weapon. He lowers and raises the muzzle by moving the nonfiring hand forward and backward on the forestock. The sniper holds his breath when the sights are on the target.

3-27. The sniper assumes the crossed-leg position in the same way as the sitting position, but he faces 45 to 60 degrees away from the target and crosses his legs instead of his ankles.

Figure 3-5. Sling-Supported Sitting Position

Supported Kneeling Position

3-28. The sniper uses the supported kneeling position when it is necessary to quickly assume a position and there is insufficient time to assume the prone position (Figure 3-6). This position can also be used on level ground or on ground that slopes upward where fields of fire or observation preclude using the prone position.

Figure 3-6. Supported Kneeling Position

3-29. The sniper assumes this position in much the same way as the standard kneeling position, except he uses a tree or some other immovable object for support, cover, or concealment. He gains support by contact with the calf and knee of the leading leg, the upper forearm, or the shoulder. He might also rest the rifle on the hand lightly against the support. As with other supported positions, the sniper ensures that the operating parts and the barrel do not touch the support. Since the sniper's area of support is greatly reduced, he must maximize bone support.

3-30. This position differs between right- and left-handed snipers. Right-handed snipers use the following techniques and left-handed snipers do the opposite. The sniper faces 45 degrees to the right of the direction of the target. He kneels down and places the right knee on the ground, keeping the left leg as vertical as possible. He sits back on the right heel, placing it as directly under the spinal column as possible. A variation is to turn the toe inward and sit squarely on the right foot. The sniper grasps the small of the stock with the firing hand, and cradles the fore-end of the weapon in a crook formed with the left arm. He places the butt of the weapon in the pocket of the shoulder, then places the meaty underside of the left elbow on top of the left knee. Reaching under the weapon with the left hand, the sniper lightly grasps the firing arm. He relaxes forward and into the support, using the left shoulder as a contact point. This movement reduces transmission of the pulse beat into the sight picture. The sniper can use a tree, building, or vehicle for support.

Sling-Supported Kneeling Position

3-31. If vegetation height presents a problem, the sniper can raise his kneeling position by using the rifle sling (Figure 3-7, page 3-11). He takes this position by performing the first three steps for a kneeling supported position. With the leather sling mounted to the weapon, the sniper turns the sling one-quarter turn to the left. The lower part of the sling then forms a loop. He places his left arm through the loop, pulls the sling up the arm, and places it on the upper arm above the bicep. He can tighten the sling on the arm by manipulating the upper and lower parts of the sling, if time permits. The sniper then rotates his arm in a clockwise motion around the sling and under the rifle with the sling secured to the upper arm. He places the fore-end of the stock in the "V" formed by the thumb and forefinger of the left hand. He can relax the left arm and let the sling support the weight of the weapon. Then he places the flat part of the rifle behind the point of the left elbow on top of the left knee. To add stability, the sniper can use his left hand to pull back along the fore-end of the rifle toward the trigger guard.

Figure 3-7. Sling-Supported Kneeling Position

Squatting Position

3-32. The sniper uses the squatting position during hasty engagements or when other stable positions would be unacceptable due to inadequate height or concealment. He assumes this position by facing 45 degrees away from his direction of fire, putting his feet shoulder-width apart, and simply squatting. He can either rest his elbows on his knees or wrap them over his body. The sniper prefers this position when making engagements from rotary-winged aircraft as it reduces the amount of body contact with the inherent vibrations of the aircraft. Body configuration will determine the most comfortable and stable technique to use. The sniper can also use solid supports to lean up against or to lean back into.

Supported Standing Position

3-33. The sniper uses this position under the same circumstances as the supported kneeling position, where time, field of fire, or observation preclude the use of more stable positions. It is the least steady of the supported positions; the sniper should use it only as a last resort.

3-34. The sniper assumes this position in much the same manner as the standard standing position, except he uses a tree or some other immovable object for support. He gains support by contact with the leg, body, or arm. He might also rest the rifle lightly against the support. The sniper ensures the support makes no contact with operating parts or the barrel of the rifle.

3-35. This position also allows the sniper to use horizontal support, such as a wall or ledge. The sniper locates a solid object for support. He avoids branches because they tend to sway when the wind is present. He places the fore-end of the weapon on top of the support and the butt of the weapon into the pocket of the shoulder. The sniper forms a "V" with the thumb and forefinger of the nonfiring hand. He places the nonfiring hand, palm facing away, against the support with the fore-end of the weapon resting in the "V" of the hand. This hold steadies the weapon and allows quick recovery from recoil.

3-36. The sniper can also use a vertical support such as a tree, telephone pole, corner of building, or vehicle (Figure 3-8). He locates the stable support, faces 45 degrees to the right of target, and places the palm of the nonfiring hand at arm's length against the support. He then locks the arm straight, lets the lead leg buckle, and places body weight against the nonfiring hand. He should keep the trail leg straight. The sniper places the fore-end of the weapon in the "V" formed by extending the thumb of the nonfiring hand. He should exert more pressure to the rear with the firing hand.

Figure 3-8. Vertically Supported Standing Position

Standing Unsupported or Off-Hand Position

3-37. This position is the least desirable because it is least stable and most exposed of all the positions (Figure 3-9, page 3-13). The situation could dictate that the sniper use this position. The sniper faces perpendicular to the target, facing in the direction of his firing hand, with his legs spread about shoulder-width apart. He grasps the pistol grip of the stock with his firing hand and supports the fore-end with the nonfiring hand. He raises the stock of the weapon so the toe of the stock fits into the pocket of the shoulder and the weapon is lying on its side away from the body. The sniper rotates the weapon until it is vertical and the firing elbow is parallel with the ground. He pulls the nonfiring elbow into the side to support the weapon with the arm and rib cage. He then tilts his head slightly toward the weapon to obtain a natural spot or cheek weld and to align his eye with the sights. If his eye is not aligned with the sights, he adjusts his head position until the front sight and the target can be seen through the rear sight. Once in position, the sniper looks through his sights and moves his entire body to get the sights on target. He does not muscle the weapon onto the target. The sniper rests the rifle on a support to relax his arm muscles after firing the shot and following through.

Figure 3-9. Standing Unsupported or Off-Hand Position

Other Supported Positions

3-38. During fundamental training, positions are taught in a step-by-step process. The sniper follows a series of precise movements until he obtains the correct position. Repetitive training ensures that he knows and correctly applies all the factors that can assist him in holding the rifle steady. As the sniper perfects the standard and supported positions, he can then use his ingenuity to devise other supported positions. Through practice he will gradually become accustomed to the feel of these positions and will know instinctively when his position is correct. This response is particularly important in combat because the sniper must be able to assume positions rapidly and stabilize the position by adapting it to any available artificial support. Figure 3-10 lists some significant nonstandard supported positions. The sniper must remember to adapt the position to his body so the position is solid, stable, and durable.

Foxhole-Supported Position	Used primarily in prepared defense areas where there is time for preparation. In this position, the sling, sandbags, or other material may be used to provide a stable firing platform.
Tree-Supported Position	Used when observation and firing into an area cannot be accomplished from the ground. When using this position, it is important to select a tree that is inconspicuous, is strong enough to support the sniper's weight, and affords concealment. **Remember: Avenues of escape are limited when in a tree.**
Bench Rest Position	Used when firing from a building, a cave, or a deeply shaded area. Sniper can use a built-up platform or table with a sitting aid and a rifle platform for stability. This position is very stable and will not tire the sniper. In this position, the sniper should stay deep in the shadows to prevent detection by the enemy.

Figure 3-10. Nonstandard Supported Positions

FIELD-EXPEDIENT WEAPON SUPPORT

3-39. Support of the weapon is critical to the sniper's success in engaging targets. Unlike a well-equipped firing range with sandbags for weapon support, the sniper will encounter situations where weapon support relies on common sense and imagination. The sniper should practice using the following supports at every opportunity and select the one that best suits his needs. He must train as if in combat to avoid confusion and self-doubt. While he should use the Harris Bipod when possible, the following items are commonly used as field-expedient weapon supports:

Sand Sock

3-40. The sniper may use the sand sock when delivering precision fire at long ranges. He uses a standard-issue, wool sock filled one-half to three-quarters full of sand or rice and knotted off. He places it under the rear-sling swivel when in the prone supported position for added stability. By limiting minor movements and reducing pulse beat, the sniper can concentrate on trigger control and aiming. He uses the nonfiring hand to grip the sand sock, rather than the rear sling swivel. The sniper makes minor changes in muzzle elevation by squeezing or relaxing his grip on the sock. He also uses the sand sock as padding between the weapon and a rigid support. The sniper must remember not to use a loose hold while firing the weapon.

NOTE: When using the sand sock, the sniper must be sure to grip the weapon firmly and hold it against his shoulder.

Rucksack

3-41. If the sniper is in terrain bare of any natural support, he may use his rucksack. He must consider the height and presence of rigid objects within the rucksack. The rucksack must conform to weapon contours to add stability.

Buttpack

3-42. The sniper can use a buttpack if the rucksack would give too high of a profile. He must also remember to consider the contents of the buttpack if he decides to change.

Sandbag

3-43. A sandbag is the simplest field-expedient support. The sniper can fill and empty a sandbag on site.

Tripod

3-44. The sniper can build a field-expedient tripod by tying together three 12-inch-long sticks with 550 cord or the equivalent (Figure 3-11, page 3-15). When tying the sticks, he wraps the cord at the center point and leaves enough slack to fold the legs out into a triangular base. Then he places the fore-end of the weapon between the three uprights. The juncture should be padded with a sand sock. A small camera table tripod padded with a sock full of sand or dirt can also be used.

Cross Sticks

3-45. The sniper can build a field-expedient bipod by tying together two 12-inch-long sticks, thick enough to support the weight of the weapon (Figure 3-11, page 3-16). Using 550 cord or the equivalent, he ties the sticks at the center point, leaving enough slack to fold them out in a scissorlike manner. He then places the weapon between the two uprights. The bipod is not as stable as other field-expedient items, and it should be used only in the absence of other techniques. The sniper should use a sling and grip the crossed stick juncture for stability.

Forked Stake

3-46. The tactical situation determines the use of the forked stake (Figure 3-11, page 3-16). Unless the sniper can drive a forked stake into the ground, this is the least desirable of the techniques; that is, he must use his nonfiring hand to hold the stake in an upright position. Delivering long-range precision fire is a near-impossibility due to the unsteadiness of the position.

SLINGS

3-47. The M1907 National Match leather sling is superior to the standard M16 web sling when used as a firing aid. Snipers who use a sling when firing should be aware of the possibility of a zero change. If the weapon is zeroed using a sling support, the POI may change when or if the sling is removed. This change is most noticeable in rifles with stocks that contact the barrel, such as the M21. The sling must be adjusted for each position. Each position will have a different point in which the sling is at the correct tightness. The sniper counts the number of holes in the sling and writes these down so that he can properly adjust the sling from position to position. An acceptable alternative is the cotton web M14 sling with a metal slide adjuster. The sniper must modify the sling for use.

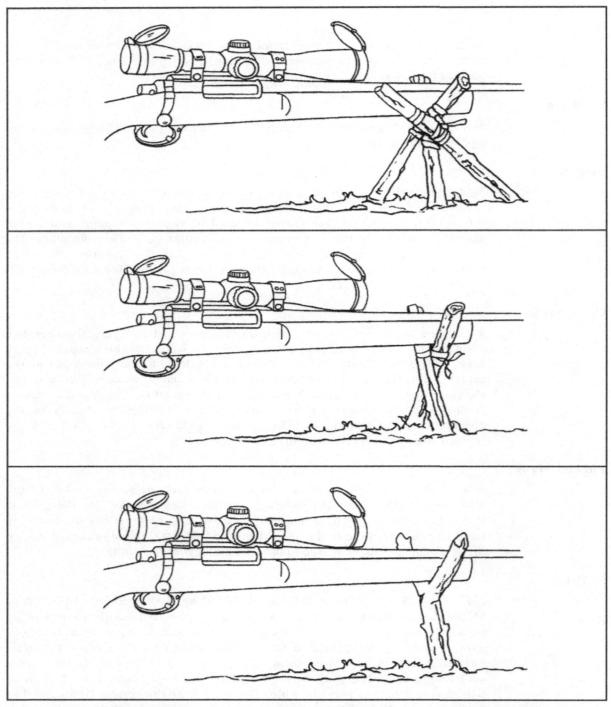

Figure 3-11. Tripod, Cross Sticks, and Forked Field-Expedient Weapon Support

TEAM FIRING TECHNIQUES

3-48. A successful sniper team consists of two intelligent and highly versatile members—the sniper and the observer. Each must be able to move and survive in a combat environment. The sniper's special mission is to deliver precision fire on targets that may not easily be engaged by conventional-fighting forces. The team must also—

- Calculate the range to the target.
- Determine the effects of the environment on ballistics.
- Make necessary sight changes.
- Observe bullet impact.
- Quickly critique performance before any subsequent shots.

3-49. These tasks call for a coordinated, efficient team effort. Mission success occurs only if the sniper and observer thoroughly understand and react in a timely manner to one another.

SNIPER AND OBSERVER RESPONSIBILITIES

3-50. Each member of the sniper team has specific responsibilities when engaged in eliminating a target. Only through repeated practice can the team begin to function properly. Although responsibilities of team members differ, they are equally important.

3-51. The sniper—

- Builds a steady, comfortable position.
- Locates and identifies the target designated by the observer.
- Reads the mil height of the target and gives this to the observer.
- Makes the elevation adjustments given by the observer to engage the target.
- Notifies observer of readiness to fire.
- Takes aim at the designated target as directed by the observer.
- Controls breathing at natural respiratory pause.
- Executes proper trigger control.
- Follows through each action.
- Makes an accurate shot call immediately after the shot.
- Prepares to fire subsequent shots, if necessary.

3-52. The observer—

- Properly positions himself so as not to disturb the sniper's position.
- Selects an appropriate target. The target closest to the team presents the greatest threat. If multiple targets are visible at various ranges, the engagement of closer targets allows the sniper to confirm his zero and ensure his equipment is functioning properly. The observer must consider existing weather conditions before trying a shot at a distant target (effects of weather increase with range).

- Uses the mil reading from the sniper to compute the range to the target and confirms by eye or other means. The observer communicates the elevation adjustment required to the sniper.
- Calculates the effect of existing weather conditions on ballistics. Weather conditions include detecting elements of weather (wind, light, temperature, and humidity) that will affect bullet impact and calculating the mil hold-off to ensure a first-round hit.
- Reports elevation and parallax adjustment to the sniper and when the sniper is ready, gives the windage in a mil hold-off.
- Uses the spotting telescope for shot observation. He aims and adjusts the telescope so that both the downrange indicators and the target are visible.
- Critiques performance. He receives the sniper's shot call and compares sight adjustment data with bullet impact if the target is hit. He gives the sniper an adjustment and selects a new target if changes are needed. If the target is missed, he follows the above procedure after receiving the sniper's shot call so that an immediate mil hold and follow-up shot will ensure a target hit.

SNIPER AND OBSERVER POSITIONING

3-53. The sniper should find a place on the ground that allows him to build a steady, comfortable position with the best cover, concealment, and visibility of the target area. Once established, the observer should position himself out of the sniper's field of view on his firing side.

3-54. The closer the observer gets his spotting telescope to the sniper's gun target line, the easier it is to follow the trace (path) of the bullet and observe impact. A 4 to 5 o'clock position (7 to 8 o'clock for left-handed snipers) off the firing shoulder and close to (but not touching) the sniper is best (Figure 3-12).

Figure 3-12. Positioning of the Observer's Spotting Telescope to the Sniper

SIGHTING AND AIMING

3-55. The sniper's use of iron sights serves mainly as a back-up system to his optical sight. However, iron sights are an excellent means of training for the sniper. The sniper is expected to be proficient in the use of iron sights before he obtains formal sniper training and he must remain proficient. By using iron sights during training, the sniper is forced to maintain his concentration on the fundamentals of firing. For a review of basic rifle marksmanship, see FM 23-9, *M16A1 and M16A2 Rifle Marksmanship*. While this manual is good for a basic review, some modifications in firing techniques must be made.

3-56. The sniper begins the aiming process by assuming a firing position and aligning the rifle with the target. He should point the rifle naturally at the desired POA. If his muscles are used to adjust the weapon onto the POA, they will automatically relax as the rifle fires, and the rifle will begin to move toward its natural POA. Because this movement begins just before the weapon discharges, the rifle is moving as the bullet leaves the muzzle. This movement causes displaced shots with no apparent cause (recoil disguises the movement). By adjusting the weapon and body as a single unit, rechecking, and readjusting as needed, the sniper achieves a true natural POA. Once the position is established, the sniper then aims the weapon at the exact point on the target. Aiming involves three factors: eye relief, sight alignment, and sight picture.

EYE RELIEF

3-57. Eye relief is the distance from the sniper's firing eye to the rear sight or the rear of the telescope tube (Figure 3-13, page 3-20). When using iron sights, the sniper ensures that this distance remains constant from shot to shot to preclude changing what he views through the rear sight. However, relief will vary from firing position to firing position and from sniper to sniper according to—

- The sniper's neck length.
- His angle of head approach to the stock.
- The depth of his shoulder pocket.
- The position of the butt of the stock in the shoulder.
- His firing position.

3-58. This distance is more rigidly controlled with telescopic sights than with iron sights. The sniper must take care to prevent eye injury caused by the rear sight or the telescope tube striking his eyebrow during recoil. Regardless of the sighting system he uses, he must place his head as upright as possible with his firing eye located directly behind the rear portion of the sighting system. This head placement also allows the muscles surrounding his eye to relax. Incorrect head placement causes the sniper to look out of the top or corner of his eye, which can result in blurred vision or eyestrain. The sniper can avoid eyestrain by not staring through the iron or telescopic sights for extended periods. The best aid to consistent eye relief is maintaining the same stock weld from shot to shot; because as the eye relief changes, a change in sight alignment will occur. Maintaining eye relief is a function of the position and stock weld use. Normal eye relief from the rear sight or scope on the M24 is 2 to 3 inches. Once the sniper is ready to fire, it is imperative that he concentrates on the front sight or reticle and not the target.

Figure 3-13. Eye Relief

SIGHT ALIGNMENT

3-59. Sight alignment is the most critical factor in aiming. An error in sight alignment increases proportionately with range and will result in increased misses. The M24 has a hooded front sight that simplifies sight alignment. The front sight hood is centered in the rear sight aperture.

3-60. With iron sights, sight alignment is the relationship between the front and rear sights as seen by the sniper (Figure 3-14, page 3-21). The sniper centers the front sight post horizontally and vertically within the rear aperture. (Centering the two circles is the easiest way for the eye to align the front and rear sights. This method allows the sniper to be consistent in blade location within the rear sight.) With telescopic sights, sight alignment is the relationship between the crosshairs and a full field of view as seen by the sniper. The sniper must place his head so that a full field of view fills the tube, with no dark shadows or crescents to cause misplaced shots. He centers the reticle in a full field of view, ensuring the vertical crosshair is straight up and down so that the rifle is not canted. Again, the center is easiest for the sniper to locate and allows for consistent reticle placement.

Figure 3-14. Proper Sight Alignment With the M24 Sniper Weapon Iron Sight System

SIGHT PICTURE

3-61. With iron sights, the sight picture is the correlation between the front sight blade, the rear aperture, and the target as seen by the sniper (Figure 3-15). The sniper aligns his sights and places the top edge of the blade in the center (center hold) of the largest visible mass of the target (disregard the head and use the center of the torso). With telescopic sights, sight picture is the correlation between the reticle, full field of view, and the target as seen by the sniper (Figure 3-16, page 3-22). The sniper centers the reticle in a full field of view. He then places the reticle center on the largest visible mass of the target (as in iron sights). The center of mass of the target is easiest for the sniper to locate, and it surrounds the intended POI with a maximum amount of target area. When aiming, the sniper concentrates on the front sight, or reticle, not the target. A clear front sight or focusing on the crosshairs is critical to detecting errors in sight alignment and is more important than the sight picture.

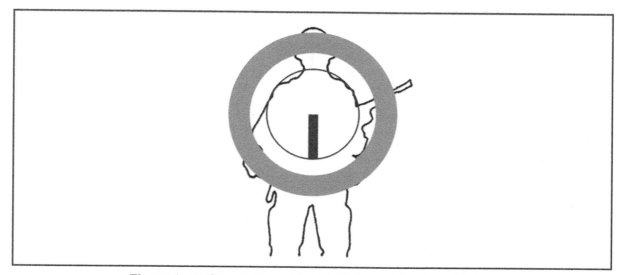

Figure 3-15. Correlation of Sight Picture Using Iron Sights

Figure 3-16. Correlation of Sight Picture Using Telescopic Sights

3-62. When aiming, the sniper has the following choices of where to hold the front sight:

- *Center hold*. This hold places the front sight on the desired POI. The problem with this hold is that the front sight blocks part of the target. **This hold is probably the best sight picture for combat use** because it is the most "natural" for U.S. Army-trained soldiers. Variation is the pimple hold.
- *6 o'clock hold*. This hold places the target on top of the front sight. The main problem is that it is easy for the front sight to "push up" into the target, causing the round to go high. Variation is the flat tire hold.
- *Line-of-white hold*. This hold allows a strip of contrasting color to show between the target and the front sight. The advantage of using this hold is that it permits the sniper to see the entire target and prevents the front sight from going high or low without him noticing it. The disadvantage is when the target and surrounding area blend into each other.
- *Reference point hold*. This hold is used when the sniper cannot see the target but can see a reference point given by the observer. This is the least accurate technique for aiming and should be used with care when using the iron sights. This hold can be used with greater accuracy when using the telescopic sights.

SIGHT ALIGNMENT ERROR

3-63. When sight alignment and sight picture are perfect (regardless of sighting system) and all else is done correctly, the shot will hit center of mass on the target. However, with an error in sight alignment, the bullet is displaced in the opposite direction of the error. Such an error creates an angular displacement between the line of sight (LOS) and the line of bore and is measured in minutes of angle. This displacement increases as range increases;

the amount of bullet displacement depends on the size of alignment error. Close targets show little or no visible error. Distant targets can show great displacement or can be missed altogether due to severe sight misalignment. An inexperienced marksman is prone to this kind of error, since he is unsure of how a correctly aligned sight should look (especially telescopic sights). When a sniper varies his head position (and eye relief) from shot to shot, he is apt to make sight alignment errors while firing (Figure 3-17). When parallax is properly adjusted out of the weapon, then shadowing is not a problem.

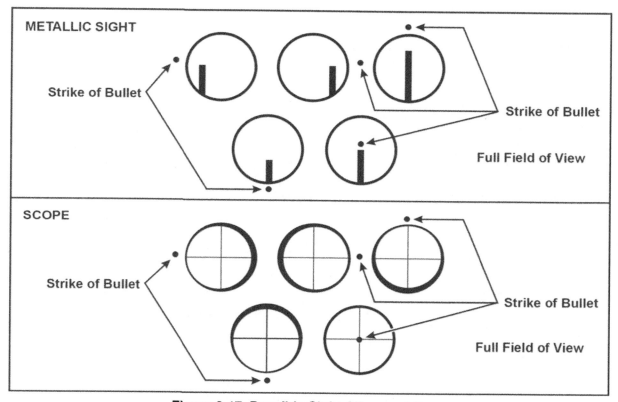

Figure 3-17. Possible Sight Alignment Error

SIGHT PICTURE ERROR

3-64. A sight picture error is an error in the placement of the aiming point. This mistake causes no displacement between the LOS and the line of bore. The weapon is simply pointed at the wrong spot on the target. Because no displacement exists as range increases, close and far targets are hit or missed depending on where the front sight or the reticle is when the rifle fires (Figure 3-18, page 3-24). All snipers face this kind of error every time they fire. Regardless of firing position stability, the weapon will always be moving. A supported rifle moves much less than an unsupported one, but both still move in what is known as a **wobble area**. The sniper must adjust his firing position so that his wobble area is as small as possible and centered on the target. With proper adjustments, the sniper should be able to fire the shot while the front sight blade or reticle is on the target at, or very near, the desired aiming point. How far the blade or reticle is from this point when the weapon fires is the amount of sight picture error all snipers face. Also, the sniper should not attempt to aim for more than 5 or 6 seconds without blinking. Doing so places an additional strain on the eye and "burns" the sight alignment and sight picture into the retina. The result could cause minor changes in sight alignment and sight picture to go unnoticed.

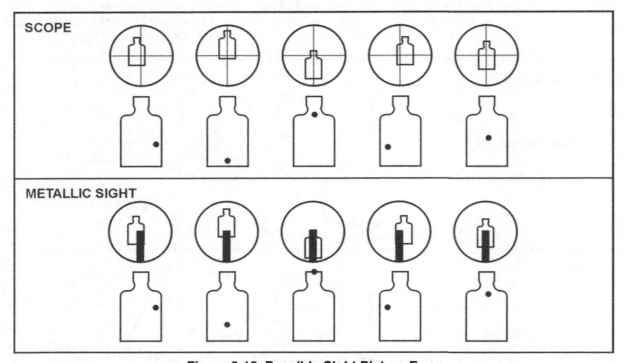

Figure 3-18. Possible Sight Picture Error

DOMINANT EYE

3-65. Some individuals may have difficulty aiming because of interferences from their dominant eye if it is not the eye used in the aiming process. This feature may require the sniper to fire from the other side of the weapon (right-handed sniper will fire left-handed). To determine which eye is dominant, hold an index finger 6 to 8 inches in front of your eyes. Close one eye at a time while looking past the finger at an object; one eye will make the finger appear to move and the other will not. The eye that does not make the finger appear to move is the dominant eye. If the sniper does not have a cross-dominant problem, it is best to aim with both eyes open. Aiming with both eyes allows him to see naturally and helps him relax. Also, with both eyes open the sniper can find targets more quickly in his telescopic sight. Closing one eye puts an unnatural strain on the aiming eye and limits the sniper's protective peripheral vision.

ADVANTAGES OF TELESCOPIC SIGHTS

3-66. Telescopic sights offer many advantages. They provide—
- Extremely accurate aiming, which allows the sniper to fire at distant, barely perceptible, and camouflaged targets not visible to the naked eye.

- Rapid aiming, because the sniper's eye sees the crosshairs and the target with equal clarity in the same focal plane.
- Accurate fire under conditions of unfavorable illumination (such as at dawn and dusk) and during periods of limited visibility (moonlight and fog).

3-67. Despite these advantages, telescopic sights have limitations. The telescopic sight will never make a poor sniper any better. The magnification is also a disadvantage, as it also magnifies aiming and holding errors. Although technically there is no sight alignment with the telescopic sight, shadowing will occur if the eye is not centered on the scope. This error will have the same effect as improper sight alignment when the scope has not been adjusted parallax-free. The bullet will strike at a point opposite the shadow and will increase in error as the distance increases.

3-68. Improper head placement on the stock is the main cause of shadowing. Due to the scope being higher than the iron sights, it is difficult to obtain a good solid stock weld. If this is a problem, temporary cheek rests can be constructed using T-shirts or any material that can be removed and replaced. The rest will assist the sniper in obtaining a good stock weld and will help keep his head held straight for sighting. It is recommended that the sniper learn to establish a solid position without these aids.

AIMING WITH TELESCOPIC SIGHTS

3-69. A telescopic sight allows aiming without using the organic rifle sights. The LOS is the optical axis that runs through the center of the lens and the intersection of the crosshairs. The crosshairs and the image of the target are in the focal plane of the lens (that plane which passes through the lens focus, perpendicular to the optical axis). The sniper's eye sees the crosshairs and the image of the target with identical sharpness and clarity. To aim with a telescope, the sniper must position his head at the exit pupil of the telescope eyepiece so that the LOS of his eye coincides with the optical axis of the telescope. He then centers the crosshairs on the target.

SHADOW EFFECTS

3-70. During aiming, the sniper must ensure that there are no shadows in the field of vision of the telescope. If the sniper's eye does not have proper eye relief, a circular shadow will occur in the field of vision. This straining will reduce the field-of-vision size, hinder observation, and in general, make aiming difficult. If the eye is positioned incorrectly in relation to the main optical axis of the telescope (shifted to the side), crescent-shaped shadows will occur on the edges of the eyepiece. They can occur on either side, depending upon the position of the axis of the eye with respect to the optical axis of the telescope. If these crescent-shaped shadows are present, the bullets will strike to the side away from them when parallax is not adjusted out of the scope. This error is the same as a sight alignment error with iron sights.

HEAD ADJUSTMENTS

3-71. If the sniper notices shadow on the edges of the field of vision during aiming, he must find a head position in which the eye will clearly see the entire field of vision of the telescope. Consequently, to ensure accurate aiming with a telescope, the sniper must direct his attention to keeping his eye on the optical axis of the telescope. He must also have the intersection of the crosshairs coincide exactly with the aiming point. However, his concentration must be on the crosshairs and not the target. It is important not to stare at the crosshairs while aiming.

CANTING

3-72. Canting is the act of tipping the rifle to either side of the vertical crosshair, causing misplaced and erratic shot grouping.

POINT OF AIM

3-73. The POA is mission- and range-dependent and should not be the center of mass unless required by the situation. The best POA between 300 and 600 meters is anywhere within the triangle formed by the base of the neck and the two nipples (Figure 3-19, page 3-27). This point will maximize the probability of hitting major organs and vessels and rendering a clean one-shot kill. The optional POA, at this range if the upper chest hold is not available, is the centerline below the belt. The pelvic girdle is rich in major blood vessels and nerves. A hit here will cause a mechanical collapse or mechanical dysfunction. A strike here is also an advantage if the target is wearing body armor, which usually covers only the upper chest. An alternate POA for closer than 300 meters is the head hold (Figure 3-20, page 3-27). This point is very difficult to achieve because of its size and constant motion. The advantage of the head hold is incapacitation well under 1 second if the correct placement is achieved. This hold is well suited for hostage situations where closer ranges are the norm and instant incapacitation is required. One hold is along the plane formed by the nose and the two ear canals. The target is the brain stem, thus severing the spinal cord from the medulla oblongata. Note that the POA is neither the forehead nor between the eyes, which would result in hits that would be too high. The sniper is best served by imagining a golf ball-sized shape inside the middle of the head. He is to hit that inner ball by aiming through the middle of the head regardless of position horizontally or vertically.

3-74. What the sniper is trying to sever or pulverize is the target's brain stem, the location where the spinal cord connects to the brain. Nerves that control motor function are channeled through here, and the lower third of the stem (the medulla) controls breathing and heartbeat. Hit here, the target will not experience even reflexive motor action. His entire body will instantly experience what is called "flaccid paralysis." The target's muscles will suddenly relax and he will become incapable of any motion of any kind thereafter. The sniper can tell how successful his headshot is by watching how his target falls. If the target goes straight down, limp, there is a high assurance of fatality. If the target falls to the side or is "knocked" down, the target has only been partially incapacitated.

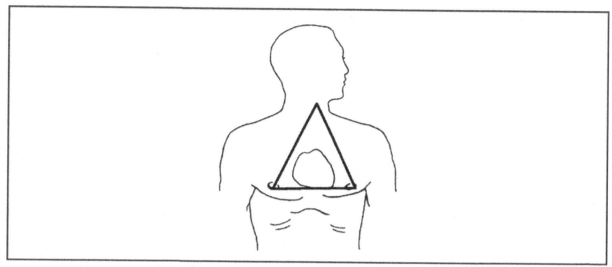

Figure 3-19. Triangular Point of Aim Formed by the Nipples and the Base of the Chin

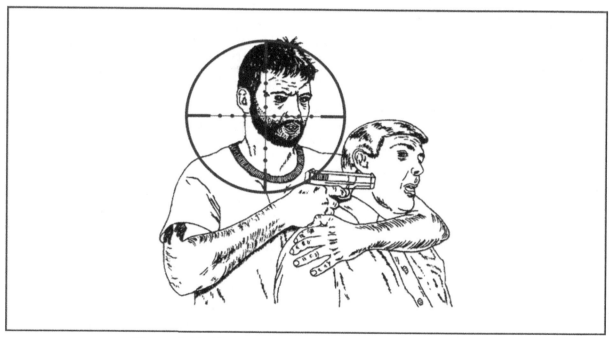

Figure 3-20. View of the Final Point of Aim—Head Hold

3-75. For a chest shot that is ideally placed (mid-sternum), the bullet will strike the largest and hardest of the bones overlying the vital organs. When the bullet strikes and severs the target's spine, his legs will buckle under flaccid paralysis. However, his arms may not be incapacitated instantly. With a chest shot, even though the suspect may technically be "dead" from the devastation of the round, there may be a brief and dangerous delay before he acts dead. His brain may not die for one to two minutes after his heart has ceased to function. During this time, his brain may command his arms to commit a simple, final act. The sniper anticipates these possibilities and delivers an immediate second round if the suspect is not fully down and out or anyone is within his sphere of danger. An alternative and final aim point is any major joint mass. A hit here will cause grave injury, shock, and possible incapacitation. However, if the target is on any type of stimulant, this hit may not have much effect.

BREATH CONTROL

3-76. Breath control is important to the aiming process. If the sniper breathes while trying to aim, the rise and fall of his chest will cause the rifle to move vertically. The sniper breathes while he does sight alignment, but he must be able to hold his breath to complete the process of aiming. To properly hold his breath, the sniper inhales, exhales normally, and stops at the moment of natural respiratory pause. If the sniper does not have the correct sight picture, he must change his position.

3-77. A respiratory cycle lasts 4 to 5 seconds. Inhalation and exhalation require only about 2 seconds. Thus, between each respiratory cycle, there is a pause of 2 to 3 seconds. This pause can be expanded to 12 to 15 seconds without any special effort or unpleasant sensation; however, the maximum **safe** pause is 8 to 10 seconds. The sniper must fire the shot during an extended pause between breaths or start the process over again. During the respiratory pause, the breathing muscles are relaxed and the sniper thus avoids straining the diaphragm (Figure 3-21).

3-78. A sniper should assume his position and breathe naturally until his hold begins to settle. Many snipers then take a slightly deeper breath, exhale and pause, expecting to fire the shot during the pause. If the hold does not settle sufficiently to allow the shot to be fired, the sniper resumes normal breathing and repeats the process.

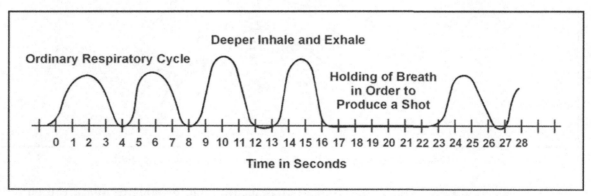

Figure 3-21. A Sniper's Respiratory Pause Before Firing at the Target

3-79. The respiratory pause should never feel unnatural. If the pause is extended for too long, the body suffers from oxygen deficiency and sends out signals to resume breathing. These signals produce slight involuntary movements in the diaphragm and interfere with the sniper's ability to concentrate. The heart rate also increases and there is a decrease of oxygen to the eyes. This lack of oxygen causes the eyes to have difficulty focusing and results in eyestrain. During multiple, rapid-fire engagements, the breathing cycle should be forced through a rapid, shallow cycle between shots instead of trying to hold the breath or breathing. Firing should be accomplished at the forced respiratory pause.

3-80. The natural tendency of the weapon to rise and fall during breathing allows the sniper to fine-tune his aim by holding his breath at the point in which the sights rest on the aiming point.

TRIGGER CONTROL

3-81. Trigger control is an important component of sniper marksmanship fundamentals. It is defined as causing the rifle to fire when the sight picture is at its best, without causing the rifle to move. Trigger squeeze, on the other hand, is defined as the independent action of the forefinger on the trigger, with a uniformly increasing pressure straight to the rear until the rifle fires. Trigger control is the last task to be accomplished before the weapon fires. This task is more difficult to apply when using a telescope or when a firing position becomes less stable. Misses are usually caused by the aim being disturbed as the bullet leaves the barrel or just before it leaves the barrel. This kind of miss results when a sniper jerks the trigger or flinches. The trigger need not be jerked violently to spoil the aim; even a slight, sudden pressure of the trigger finger is enough to cause the barrel to waver and spoil the sight alignment. Flinching is an involuntary movement of the body—tensing of the muscles of the arm, the neck, or the shoulder in anticipation of the shock of recoil or the sound of the rifle firing. A sniper can correct these errors by understanding and applying proper trigger control.

3-82. Proper trigger control occurs when the sniper places his firing finger as low on the trigger as possible and still clears the trigger guard, thereby achieving maximum mechanical advantage. The sniper engages the trigger with that part of his firing finger that allows him to pull the trigger straight to the rear. A firm grip on the rifle stock is essential for trigger control. If the sniper begins his trigger pull from a loose grip, he tends to squeeze the stock as well as the trigger and thus loses trigger control. To avoid transferring movement of the finger to the entire rifle, the sniper should see daylight between the trigger finger and the stock as he squeezes the trigger, straight to the rear. To ensure a well-placed shot, he fires the weapon when the front blade or reticle is on the desired POA.

3-83. The sniper best maintains trigger control by assuming a stable position, adjusting on the target, and beginning a breathing cycle. As the sniper exhales the final breath toward a natural respiratory pause, he secures his finger on the trigger. As the front blade or reticle settles on the desired POA and the natural respiratory pause is entered, the sniper applies initial pressure. He increases the tension on the trigger during the respiratory pause as long as the front blade or reticle remains in the area of the target that ensures a well-placed shot. If the front blade or reticle moves away from the desired POA on the target and the pause is free of strain or tension, the sniper stops increasing the tension on the trigger, waits for the front blade or reticle to return to the desired point, and then continues to squeeze the trigger. The sniper perfects his aim while continuing the steadily increasing pressure until the hammer falls. This is trigger control. If movement is too large for recovery or if the pause has become uncomfortable (extended too long), the sniper should carefully release the pressure on the trigger and begin the respiratory cycle again.

3-84. Most successful snipers agree that the trigger slack should be taken up with a heavy initial pressure. Concentration should be focused on the perfection of the sight picture as trigger control is automatically applied. Concentration, especially on the front sight or reticle, is the greatest aid to prevent flinching and jerking.

3-85. The methods of trigger control involve a mental process, while pulling the trigger is a mechanical process. The sniper uses two methods of trigger control to pull the trigger. They are as follows:

- *Smooth motion/constant pressure trigger pull.* The sniper takes up the slack with a heavy initial pressure and, when the sight picture settles, pulls the trigger with a single, smooth action. This method is used when there is a stationary target and the position is steady. This type of trigger control will help prevent flinching, jerking, and bucking the weapon.
- *Interrupted trigger pull.* The sniper applies pressure to the trigger when the sight picture begins to settle and as long as the sight picture looks good or continues to improve. If the sight picture deteriorates briefly, the sniper maintains the pressure at a constant level and increases it when the picture again begins to improve. He then continues the pressure or repeats this technique until he fires the rifle. The sniper does not jerk the trigger when the sights are aligned and the "perfect" sight picture occurs. This technique is used in the standing position to correct the wavering of the sights around, through, or in the target or aiming point due to the instability of the position.

3-86. Trigger control is not only the most important fundamental of marksmanship but also the most difficult to master. The majority of firing errors stems directly or indirectly from the improper application of trigger control. Failure to hit the target frequently results from the sniper jerking the trigger or applying pressure on both the trigger and the side of the rifle. Either of these actions can produce a miss. Therefore, instructors should always check for indications of improper trigger control, since an error in this technique can start a chain reaction of other errors.

3-87. Trigger control can be developed into a reflex action. The sniper can develop his trigger control to the point that pulling the trigger requires no conscious effort. The sniper will be aware of the pull, but he will not be consciously directing it. Everyone exhibits this type of reflex action in daily living. The individual who walks or drives a car while carrying on a conversation is an example. He is aware of his muscular activity but is not planning it. He is thinking about the conversation.

3-88. Trigger control is taught in conjunction with positions. When positions and trigger control are being taught, an effective training aid for demonstrating the technique of trigger control with reference to the interrupted or controlled pressure is the wobble sight and target simulator. The wobble sight may be used with a fixed target simulator to demonstrate wobble area, adjustment of natural POA, breathing, and trigger control.

3-89. In all positions, dry firing is one of the best methods of developing proper trigger control. In dry firing, not only is the coach able to detect errors, but the individual sniper is able to detect his own errors, since there is no recoil to conceal the rifle's undesirable movements. Where possible, trigger control practice should be integrated into all phases of marksmanship training. The mastery of trigger control takes patience, hard work, concentration, and a great deal of self-discipline.

THE INTEGRATED ACT OF FIRING ONE ROUND

3-90. Once the sniper has been taught the fundamentals of marksmanship, his primary concern is to apply this knowledge in the performance of his mission. An effective method of applying fundamentals is through the use of the integrated act of firing one round. The integrated act is a logical, step-by-step development of the fundamentals whereby the sniper develops habits to fire each shot exactly the same. Thus he achieves the marksmanship goal that a sniper must strive for: one shot—one kill. The integrated act of firing can be divided into the following four phases.

PREPARATION PHASE

3-91. Before departing the preparation area, the sniper ensures that—
- The team is mentally conditioned and knows what mission to accomplish.
- A systematic check is made of equipment for completeness and serviceability including, but not limited to—
  - Properly cleaned and lubricated rifles.
  - Properly mounted and torqued scopes.
  - Zero-sighted systems and recorded data in the sniper data book.
  - The study of weather conditions to determine the effects on the team's performance of the mission.

BEFORE-FIRING PHASE

3-92. On arrival at the mission site, the team exercises care in selecting positions. The sniper ensures that the selected positions complement the mission's goal. During this phase, the sniper—
- Maintains strict adherence to the fundamentals of position. He ensures that the firing position is as relaxed as possible, making the most of available external support. He also makes sure the support is stable, conforms to the position, and allows a correct, natural POA for each designated area or target.

- Once in position, removes the scope covers and checks the field of fire, making any needed corrections to ensure clear, unobstructed firing lanes.
- Checks the boreline for any obstructions.
- Makes dry-firing and natural POA checks.
- Double-checks ammunition for serviceability and completes final magazine loading.
- Notifies the observer he is ready to engage targets. The observer must constantly be aware of weather conditions that may affect the accuracy of the shots. He must also stay ahead of the tactical situation.

FIRING PHASE

3-93. Upon detection, or if directed to a suitable target, the sniper makes appropriate sight changes and aims, and tells the observer he is ready to fire. The observer then gives the needed windage and observes the target. To fire the rifle, the sniper should remember the key word, **BRASS**. Each letter is explained as follows:

- *Breathe*. The sniper inhales and exhales to the natural respiratory pause. He checks for consistent head placement and stock weld. He ensures eye relief is correct (full field of view through the scope, no shadows present). At the same time, he begins aligning the crosshairs or front blade with the target at the desired POA.
- *Relax*. As the sniper exhales, he relaxes as many muscles as possible while maintaining control of the weapon and position.
- *Aim*. If the sniper has a good, natural POA, the rifle points at the desired target during the respiratory pause. If the aim is off, the sniper should make a slight adjustment to acquire the desired POA. He avoids "muscling" the weapon toward the aiming point.
- *Slack*. (Does not apply to the M24 as issued.) The first stage of the two-stage trigger must be taken up with heavy initial pressure. Most experienced snipers actually take up the slack and get initial pressure as they reach the respiratory pause. In this way, the limited duration of the pause is not used up by manipulating the slack in the trigger.
- *Squeeze*. As long as the sight alignment and sight picture is satisfactory, the sniper should squeeze the trigger. The pressure applied to the trigger must be straight to the rear without disturbing the lay of the rifle or the desired POA.

3-94. After the shot, the sniper must remember to follow through with the recoil and recover back on target. He should make sure to call his shot so the observer can record any adjustment made.

AFTER-FIRING PHASE

3-95. The sniper's after-firing actions include observing the target area to certify the hit, observing the enemy reaction, acquiring another target, and avoiding compromise of his position. The sniper must analyze his performance. If the shot impacted at the desired spot (a target hit), it may be assumed that the integrated act of firing one round was correctly followed.

However, if the shot was off call, the sniper and observer must check for the following possible errors:

- Failure to follow the key word BRASS (partial field of view, breath held incorrectly, trigger jerked, rifle muscled into position).
- Target improperly ranged with scope (causing high or low shots).
- Incorrectly compensated-for wind (causing right or left shots).
- Possible weapon or ammunition malfunction (used only as a last resort when no other errors are detected).

3-96. Once the probable reasons for an off-call shot are determined, the sniper must make note of the errors. He should pay close attention to the problem areas to increase the accuracy of future shots.

FOLLOW-THROUGH

3-97. Applying the fundamentals increases the odds of a well-aimed shot being fired. When mastered, the first-round kill becomes a certainty.

3-98. Follow-through is a continued mental and physical application of the fundamentals after each round is fired. It is the act of continuing to apply all of the sniper marksmanship fundamentals as the weapon fires and immediately after it fires. Follow-through consists of—

- Keeping the head in firm contact with the stock (stock weld).
- Keeping the finger on the trigger all the way to the rear.
- Continuing to look through the rear aperture or scope tube.
- Concentrating on the front sight or crosshairs.
- Keeping muscles relaxed.
- Avoiding reaction to recoil and noise.
- Releasing the trigger only after the recoil has stopped.

3-99. Good follow-through ensures that the weapon is allowed to fire and recoil naturally. The sniper and rifle combination reacts as a single unit to such actions. From a training viewpoint, follow-through may allow the observer to observe the strike of the bullet in relation to the sniper's point of aim and to help him rapidly correct and adjust his sights for a second shot. Also, a good follow-through will indicate to the sniper the quality of his natural POA. The weapon should settle back on target. If it does not, then muscles were used to get the weapon on target.

CALLING THE SHOT

3-100. Calling the shot is being able to tell where the round should impact on the target. Because live targets invariably move when hit, the sniper will find it almost impossible to use his telescope to locate the target after the round is fired. Using iron sights, the sniper will find that searching for a downrange hit is beyond his capabilities. He must be able to accurately call his shots. Proper follow-through will aid in calling the shot. However, the dominant factor in shot calling is where the reticle or post is located when the weapon discharges. The sniper refers to this location as his **final focus point**.

3-101. With iron sights, the final focus point should be on the top edge of the front sight blade. The blade is the only part of the sight picture that is moving (in the wobble area). Focusing on the blade aids in calling the shot and detecting any errors in sight alignment or sight picture. Of course, lining up the sights and the target initially requires the sniper to shift his focus from the target to the blade and back until he is satisfied that he is properly aligned with the target. This shifting exposes two more facts about eye focus. The eye can instantly shift focus from near objects (the blade) to far objects (the target). The eye cannot, however, be focused so that two objects at greatly different ranges (again the blade and target) are both in sharp focus. After years of experience, many snipers find that they no longer hold final focus on the front sight blade. Their focus is somewhere between the blade and the target. This act has been related to many things, from personal preference to failing eyesight. Regardless, inexperienced snipers are still advised to use the blade as a final focus point. With iron sights the final check before shooting will be sight alignment, as misalignment will cause a miss.

3-102. The sniper can easily place the final focus point with telescopic sights because of the sight's optical qualities. Properly focused, a scope should present both the field of view and the reticle in sharp detail. Final focus should then be on the reticle. While focusing on the reticle, the sniper moves his head slightly from side to side. The reticle may seem to move across the target face, even though the rifle and scope are motionless. Parallax is present when the target image is not correctly focused onto the reticule's focal plane. Therefore, the target image and the reticle appear to be in two separate positions inside the scope, causing the effect of reticle movement across the target. A certain amount of parallax is unavoidable throughout the range of the ART series of scopes. The M3A on the M24 has a focus/parallax adjustment that eliminates parallax. The sniper should adjust this knob until the target's image is on the same focal plane as the reticle. To determine if the target's image appears at the ideal location, the sniper should move his head slightly left and right to see if the reticle appears to move. If it does not move, the focus is properly adjusted and no parallax is present. The sniper will focus and concentrate on the reticle for the final shot, not the target.

3-103. In calling the shot, the sniper predicts where the shot will hit the target. The sniper calls the shot while dry firing and actual firing by noting the position of the sights in relation to the aiming point the instant the round is fired. If his shot is not on call, the sniper must review the fundamentals to isolate his problem or make a sight change as indicated to move his shot to his POA. Unless he can accurately call his shots, the sniper will not be able to effectively zero his rifle.

DETECTION AND CORRECTION OF ERRORS

3-104. During the process of teaching or using the fundamentals of marksmanship, it will become evident that errors may plague any sniper. When an error is detected, it must be corrected. Sometimes errors are not obvious to the sniper. Therefore, a coach or instructor will be invaluable. The procedure for correcting errors is to pinpoint or isolate the error, prove to the sniper that he is making this error, and convince him that through his own efforts and concentration he can correct his error. Knowing what to look for through analyzing the shot groups, observing the sniper, questioning the sniper, and reviewing the fundamentals of training exercises will assist the coach in this process. Even during sustainment a trained sniper will use detection and correction to ensure bad habits have not been developed.

TARGET ANALYSIS

3-105. Target or shot-group analysis is an important step in processing the detection and correction of errors. When analyzing a target, the coach should correlate errors in performance to loose groups, the shape of groups, and the size of groups. With some snipers, especially the experienced, this analysis cannot be done readily. However, the coach must be able to discuss the probable error. A bad shot group is seldom caused by only one error. Remember, in the initial analysis of groups, the coach must take into consideration the capabilities of the sniper, the weapon, and the ammunition.

OBSERVATION

3-106. When the coach or instructor has an indication that the sniper is committing one or more errors, it will usually be necessary for him to observe the sniper while he is in the act of firing to pinpoint his errors. If the instructor has no indication of the sniper's probable errors, the initial emphasis should be on his firing position and breath control. Next, the instructor should look for the most common errors—anticipation of the shot and improper trigger control. If observing the sniper fails to pinpoint his errors, the instructor must then question him.

QUESTIONING

3-107. The coach or instructor should ask the sniper if he could detect his errors. He should have the sniper explain the firing procedure, to include position, aiming, breath control, trigger control, and follow-through. If questioning does not reveal all of the errors, the instructor should talk the sniper through the procedures listed in Figure 3-22.

1. Set the sights.	7. Obtain a sight picture.
2. Build the position.	8. Focus on the front sight.
3. Align the sight.	9. Control the trigger.
4. Check the natural POA.	10. Follow through.
5. Adjust the natural POA.	11. Call the shot.
6. Control the breath.	

Figure 3-22. Fundamental Procedures for Firing One Round

NOTE: If errors still occur, there are several training exercises that can help to pinpoint them.

TRAINING EXERCISES

3-108. The instructor can use the following training exercises or devices at any time to supplement the detection procedure:

- Trigger exercise.
- Metal disk exercise.
- Ball and dummy exercise.
- Blank target-firing exercise.
- M2 aiming device.
- Air rifles.

3-109. When the sniper leaves the firing line, he compares weather conditions to the information needed to hit the POA or POI. Since he fires in all types of weather conditions, he must be aware of temperature, light, mirage, and wind. Other major tasks that the sniper must complete are as follows:

- Compare sight settings with previous firing sessions. If the sniper always has to fine-tune for windage or elevation, there is a chance he needs a sight change (slip a scale).
- Compare ammunition by lot number for the best rifle and ammunition combination.
- Compare all groups fired under each condition. Check the low and high shots and those to the left and right of the main group—the less dispersion, the better. If groups are tight, they are easily moved to the center of the target; if loose, there is a problem. Check the telescope focus and make sure the rifle is cleaned correctly. Remarks in the data book will also help.
- Make corrections. Record corrections in the data book, such as position and sight adjustment information, to ensure retention.
- Analyze a group on a target. These results are important for marksmanship training. The sniper may not notice errors during firing, but errors become apparent when analyzing a group. This study can only be done if the data book has been used correctly.

3-110. As the stability of a firing position decreases, the wobble area increases. The larger the wobble area, the harder it is to fire the shot without reacting to it. This reaction occurs when the sniper—

- Anticipates recoil.
- Jerks the trigger.
- Flinches.
- Avoids recoil.

APPLICATION OF FIRE

3-111. Following the Austrian-Prussian War of 1866, the Prussian Army began a systematic study of the effectiveness and control of small-arms fire. The result of this study, conducted over a 6-year period, was the introduction of the science of musketry, a misnomer as all major armies were by then equipped with rifles. Musketry is the science of small-arms fire under field conditions, as opposed to range conditions, and is concerned entirely with firing at unknown distances; thus the importance of musketry to the sniper. The material presented is merely an overview of the fundamentals of musketry. At the peak of the study of musketry as a martial science, musketry schools often extended their courses to six weeks. Only the introduction of machine guns and automatic small arms precipitated the doctrine of its study, although various aspects of musketry were retained as separate subjects, such as judging distances and issuing fire control orders. This study ties together the scattered remnants of the study of musketry as it pertains to sniping.

MINUTE OF ANGLE

3-112. Most weapon sights are constructed with a means of adjustment. Although the technicalities of adjustment may vary with weapon type or means of sighting, generally the weapon sight will be correctable for windage and elevation. The specific method by which adjustment is accomplished is angular displacement of the sight in relation to the bore of the rifle. This angular displacement is measured in MOAs, and establishes the angle of departure in relationship to LOS.

3-113. An MOA is the unit of angular measure that equals 1/60 of 1 degree of arc. With few exceptions the universal method of weapon sight adjustment is in fractions or multiples of MOAs. An MOA equals a distance of 1.0472 inches at 100 yards and 2.9 centimeters at 100 meters. Since an MOA is an angular unit of measure, the arc established by an MOA increases proportionately with distance (Figure 3-23, page 3-38).

3-114. Fractions are difficult to work with when making mental calculations. For this reason, snipers should assume that 1 MOA is the equivalent of 1 inch at 100 yards or 3 centimeters at 100 meters. By rounding off the angular displacement of the MOA in this manner, only 1/2 inch of accuracy at 1,000 yards and 1 centimeter at 1,000 meters are lost. This manual presents data in both the English and the metric system (Table 3-1, pages 3-38 and 3-39), allowing the sniper to use whichever one he is most comfortable with.

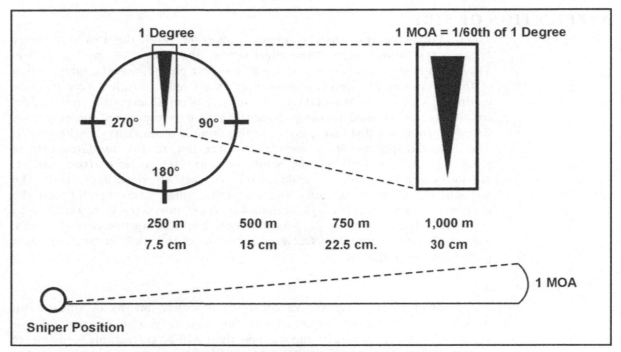

Figure 3-23. An MOA Measurement

Table 3-1. Metric and English Equivalents Used to Measure MOAs

Metric 1 MOA (cm)	Yards	Meters ←	Yards →	Meters	English 1 MOA (inches)
3	109	100		91	1
4.5	164	150		137	1.5
6	219	200		183	2
7.5	273	250		228	2.5
9	328	300		274	3
10.5	383	350		320	3.5
12	437	400		365	4
13.5	492	450		411	4.5
15	546	500		457	5
16.5	602	550		503	5.5
18	656	600		548	6
19.5	711	650		594	6.5
21	766	700		640	7
22.5	820	750		686	7.5
24	875	800		731	8

Table 3-1. Metric and English Equivalents Used to Measure MOAs (Continued)

Metric 1 MOA (cm)	Yards	Meters ←	Yards →	Meters	English 1 MOA (inches)
25.5	929	850		777	8.5
27	984	900		823	9
28.5	1,039	950		869	9.5
30	1,094	1,000		914	10
31.5	1,148	1,050		960	10.5
33	1,203	1,100		1,005	11

SIGHT CORRECTIONS

3-115. With the knowledge of how much the displacement of 1 MOA at a given distance is, snipers can calculate sight corrections. All that the sniper needs to know is how many MOAs, or fractions of an MOA, each sight graduation (known as a "click") equals. This amount depends on the type of sight used.

3-116. To determine the amount of correction required in MOAs for the English system, the error in inches is divided by the range expressed in whole numbers. The correction formula follows:

$$\text{Minutes} = \frac{\text{Error (inches)}}{\text{Range (expressed in whole numbers)}}$$

3-117. To determine the amount of correction required in MOAs using the metric system, the error in centimeters is divided by the range expressed in whole numbers, then the result is divided by 3. The correction formula follows:

$$\text{Minutes} = \frac{\text{Error (centimeters)}}{\text{Range (expressed in whole numbers)} \div 3}$$

3-118. There will be times when the impact of a shot is observed, but there is no accurate indication of how much the error is in inches or centimeters. Such occasions may occur when there is a great distance between the aiming point and the impact point or when there is a lack of an accurate reference. It is possible to determine the distance of the impact point from the POA in mils, then to convert the mils to MOAs. The conversion factor follows:

1 mil = 3.439 MOA (This is rounded to 3.5 for field use.)

EXAMPLE: When a round is fired, the observer sees the impact of the round to be several feet to the right of the target. He notes the impact point and determines it to be 2 mils to the right of the aiming point: 3.5 x 2 = 7 Minutes.

3-119. Table 3-2, page 3-40, gives the inch equivalents of mils at the given ranges of 91 meters to 1,000 meters and 100 yards to 1,000 yards. This data will aid the sniper in computing his sight change in mils for a given distance to the target with a given miss in estimated inches. For example, a miss of 28 inches left at 400 yards would be a 2-mil hold to the right.

Table 3-2. Inch Equivalents of Mils

Range (Meters/Yards)	Inches	Range (Meters/Yards)	Inches
91/100	3.6	549/600	22.0
100 m	4.0	600 m	24.0
183/200	7.0	640/740	25.0
200 m	8.0	700 m	27.5
274/300	11.0	731/800	29.0
300 m	12.0	800 m	31.5
365/400	14.0	823/900	32.5
400 m	15.75	900 m	35.5
457/500	18.0	914/1,000	36.0
500 m	20.0	1,000 m	39.0

BALLISTICS

3-120. As applied to sniper marksmanship, ballistics may be defined as the study of the firing, flight, and effect of ammunition. To fully understand ballistics, the sniper should be familiar with the terms listed in Table 3-3, page 3-41. Proper execution of marksmanship fundamentals and a thorough knowledge of ballistics ensure the successful completion of the mission. Tables and formulas in this section should be used only as guidelines since every rifle performs differently. Maintaining extensive ballistics data eventually results in a well-kept data book and provides the sniper with actual knowledge gained through experience. Appendix H provides additional ballistics data.

APPLIED BALLISTICS

3-121. Ballistics can be broken down into three major areas. Interior or internal ballistics deals with the bullet in the rifle from primer detonation until it leaves the muzzle of the weapon. Exterior and external ballistics picks up after the bullet leaves the muzzle of the weapon and extends through the trajectory until the bullet impacts on the target or POA. Terminal ballistics is the study of what the bullet does upon impact with the target. The effectiveness of the terminal ballistics depends upon—

- Terminal velocity.
- Location of the hit.
- Bullet design and construction.

Table 3-3. Ballistics Terminology

Muzzle Velocity	The speed of a bullet as it leaves the rifle barrel, measured in fps. It varies according to various factors, such as ammunition type and lot number, temperature, and humidity.
Line of Sight	A straight line from the eye through the aiming devices to the POA.
Line of Departure	The line defined by the bore of the rifle or the path the bullet would take without gravity.
Trajectory	The path of the bullet as it flies to the target.
Midrange Trajectory	The high point the bullet reaches half way to the target. This point must be known to engage a target that requires firing underneath an overhead obstacle, such as a bridge or a tree. Inattention to midrange trajectory may cause the sniper to hit the obstacle instead of the target.
Maximum Ordinate	The highest point of elevation that a bullet reaches during its time of flight for a given distance.
Bullet Drop	How far the bullet drops from the line of departure to the POI.
Time of Flight	The amount of time it takes for the bullet to exit the rifle and reach the target.
Retained Velocity	The speed of the bullet when it reaches the target. Due to drag, the velocity will be reduced.

TARGET MATERIAL OR CONSTRUCTION

3-122. When it is fired, a bullet travels a straight path in the bore of the rifle as long as the bullet is confined in the barrel. As soon as the bullet is free of this constraint (exits the barrel), it immediately begins to fall due to the effects of gravity, and its motion is retarded due to air resistance. The path of the bullet through the air is called the bullet's trajectory.

3-123. If the barrel is horizontal, the forward motion imparted to the bullet by the detonation of the cartridge will cause it to travel in the direction of point A, but air resistance and the pull of gravity will cause it to strike point B (Figure 3-24). As soon as the bullet is free from the constraint of the barrel, it begins to pull from the horizontal.

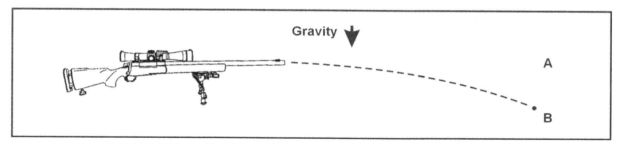

Figure 3-24. Bullet's Trajectory When the Sniper Fires the Rifle Horizontally

3-124. For point A to be struck, the barrel of the rifle must be elevated to some predetermined angle (Figure 3-25, page 3-42). The bullet's initial impulse will be in the direction of point C. However, because of the initial angle, the bullet will fall to point A, due again to air resistance and gravity. This initial angle is known as the angle of departure.

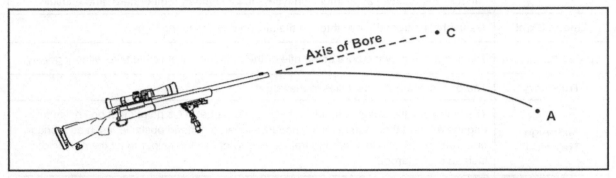

Figure 3-25. Bullet's Trajectory When the Sniper Fires the Rifle at an Elevated Angle

3-125. The angle of departure is set by the sights and establishes the shape of the trajectory. The trajectory varies with the range to the target. For any given range, the angle of departure varies with the determining factors of the trajectory. The form of the trajectory is influenced by—

- The initial velocity (muzzle velocity).
- The angle of departure.
- Gravity.
- Air resistance.
- The rotation of the projectile (bullet) about its axis.

3-126. The relationship between initial velocity and air resistance is that the greater the amount of air resistance the bullet must overcome, the faster the bullet slows down as it travels through the air. A bullet with a lower initial velocity will be retarded less by air resistance and will retain a greater proportion of its initial velocity over a given distance. This relationship is important in that a light projectile with a higher initial (or short range) velocity will have a "flatter" initial trajectory but will have less initial and retained energy with which to incorporate the target, will be deflected more by wind, and will have a steeper trajectory at longer ranges. A comparatively heavy projectile will have a lower initial velocity and a steeper initial trajectory, will retain its energy over a great distance (retained energy is proportional to the mass of the projectile), will be deflected less by wind, and will have a "flatter" long-range trajectory.

3-127. Angle of departure is the angle to which the muzzle of the rifle must be elevated above the horizontal line in order for the bullet to strike a distant point. When the bullet departs the muzzle of the rifle, it immediately begins to fall to earth in relation to the angle of departure, due to the constant pull of gravity. The angle of departure increases the height the bullet must fall before it reaches the ground. If a rifle barrel were set horizontally in a vacuum, a bullet fired from the barrel would reach the ground at a distant point at the same moment that a bullet merely dropped from the same height as the barrel would reach the ground. Despite the horizontal motion of the bullet, its velocity in the vertical plane is constant (due to the constant effect of gravity). However, angle of departure in the air is directly related to the time of flight of the projectile in that medium. The greater the angle at which the projectile departs the muzzle, the longer it will remain in the air and the further it will travel before it strikes the ground. However, the effect of gravity causes the bullet to begin to lose distance at the 33-degree point.

3-128. The angle of departure is not constant. Although the angle of departure may remain fixed, a number of variables will influence the angle of departure in a series of shots fixed at the same given distance. The differences in the internal ballistics of a given lot of ammunition will have an effect. A muzzle velocity, within a proven lot, will often vary as much as 60 feet per second between shots. Imperfections in the human eye will cause the angles of departure of successive shots to be inconsistent. Imperfections in the weapon, such as faulty bedding, worn bore, or worn sights, are variables. Errors in the way the rifle is held or canted will affect the angle of departure. These are just a few factors that cause differences in the angle of departure and are the main reasons why successive shots under seemingly identical conditions do not hit at the same point on the target.

3-129. Gravity's influence on the shape of a bullet's trajectory is a constant force. It neither increases nor decreases over time or distance. It is present, but given the variable dynamics influencing the flight of a bullet, it is unimportant. Given that both air resistance and gravity influence the motion of a projectile, the initial velocity of the projectile and the air resistance are interdependent and directly influence the shape of the trajectory.

3-130. The single most important variable affecting the flight of a bullet is air resistance. It is not gravity that determines the shape of a bullet's trajectory. If gravity alone were the determining factor, the trajectory would have the shape of a parabola, where the angle of fall would be the same (or very nearly so) as the angle of departure. However, the result of air resistance is that the shape of the trajectory is an ellipse, where the angle of fall is steeper than the angle of departure.

3-131. The lands and grooves in the bore of the rifle impart a rotational motion to the bullet about its own axis. This rotational motion causes the projectile (as it travels through the air) to shift in the direction of rotation (in almost all cases to the right). This motion causes a drift that is caused by air resistance. A spinning projectile behaves precisely like a gyroscope. Pressure applied to the front of the projectile (air resistance) retards its forward motion but does not significantly upset its stability. However, upward pressure applied to the underside of the projectile (due to its downward travel caused by gravity) causes it to drift in the direction of spin. This drift is relatively insignificant at all but the greatest ranges (more than 1,000 yards).

3-132. Due to the combined influences just discussed, the trajectory of the bullet first crosses the LOS with a scarcely perceptible curve. The trajectory continues to rise to a point a little more than halfway to the target, called the maximum ordinate, beyond which it curves downward with a constantly increasing curve (possibly recrossing the LOS) until it hits the target (or ground). The point where the LOS meets the target is the POA. The point where the bullet (trajectory) strikes the target is the POI. Theoretically, the POA and the POI should coincide. In practical terms, because of one or more of the influences discussed, they rarely do. The greater the skill of the sniper and the more perfect the rifle and ammunition, the more often these two points will coincide (Figure 3-26).

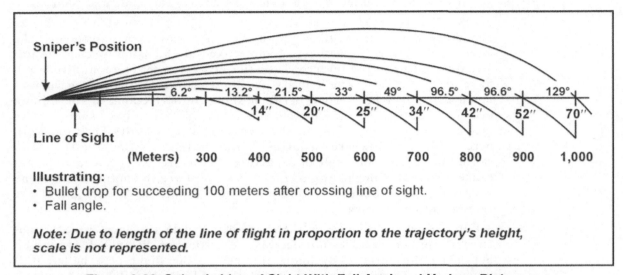

Figure 3-26. Sniper's Line of Sight With Fall Angles at Various Distances

3-133. The part of the trajectory between the muzzle and the maximum ordinate is called the rising branch of the trajectory; the part beyond the maximum ordinate is called the falling branch of the trajectory. Snipers are most concerned with the falling branch because this part of the trajectory contains the target and the ground in its vicinity. In computing the height of the trajectory, assuming the LOS is horizontal and at regular intervals (usually 100 yards), the sniper measures and records the height of the trajectory as the ordinate. The distance from the muzzle to the ordinate is known as the abscissa. The distance in front of the muzzle, within which the bullet does not rise higher than the target, is called the danger space of the rising branch of the trajectory. The falling branch of the trajectory also has a danger space. The danger space of the falling branch is the point where the bullet falls into the height of the target and continues to the ground.

3-134. Assuming that the POA is taken at the center of the target, the extent of the danger space depends on the following:

- Height of the sniper—whether he is standing, kneeling, or prone.
- Height of the target—whether he is standing, kneeling, or prone.
- "Flatness" of the trajectory—the ballistic properties of the cartridge used.
- Angle of the LOS—above or below the horizontal.
- Slope of the ground—where the target resides.

3-135. The POA also has a significant influence on the extent of the danger space. If the sniper takes the POA at the top of the target, the total danger space will lie entirely behind the target. If he takes the POA at the foot of the target, the total danger space will lie entirely in front of the target. Thus, the extent of the total danger space, including the target, will be determined by where the POA is taken on the target. Only when the POA is at the center of the target will the total danger space (in relative terms) extend an equal distance in front of and behind the target.

EFFECTS ON TRAJECTORY

3-136. Mastery of marksmanship fundamentals and field skills are not the only requirements for being a sniper. Some of the factors that have an influence on the trajectory include the following:

- *Gravity.* The sniper would not have a maximum range without gravity; a fired bullet would continue to move much the same as items floating in space. As soon as the bullet exits the muzzle of the weapon, gravity begins to pull it down, requiring the sniper to use his elevation adjustment. At extended ranges, the sniper actually aims the muzzle of his rifle above his LOS and lets gravity pull the bullet down into the target. Gravity is always present, so the sniper must compensate for it through elevation adjustments or holdover techniques.

- *Drag.* It is the slowing effect the atmosphere has on the bullet. This effect either increases or decreases according to the air—that is, the less dense the air, the less drag and vice versa. Factors affecting drag and air density are—

    - *Temperature.* The higher the temperature, the less dense the air. If the sniper zeroes at 60 degrees F and he fires at 80 degrees F, the air is less dense, thereby causing an increase in muzzle velocity and a higher impact. A 20-degree change equals a 1-minute elevation change on the rifle. This generally applies for a 7.62-mm weapon.

    - *Altitude/barometric pressure.* Since the air pressure is less at higher altitudes, the air is less dense and there is less drag. Therefore, the bullet is more efficient and impacts higher. Table 3-4, page 3-46, shows the appropriate effect of change of impact from sea level to 10,000 feet if the rifle is zeroed at sea level. Impact will be the POA at sea level. For example, a rifle zeroed at sea level and fired at 700 meters at 5,000 feet will hit 1.6 minutes high.

    - *Humidity.* Humidity varies along with the altitude and temperature. Problems can occur if extreme humidity changes exist in the area of operations. When humidity goes up, impact goes down and vice versa. Keeping a good data book during training and acquiring experience are the best teachers.

    - *Bullet efficiency.* This term refers to a bullet's ballistic coefficient. The imaginary perfect bullet is rated as being 1.00. Match bullets range from .500 to about .600. The M118 173-grain match bullet is rated at .515. Table 3-5, page 3-46, lists other ammunition, bullet types, ballistics, and the velocity for each.

    - *Wind.* The effects of wind are discussed later in this chapter.

Table 3-4. Point of Impact Rise at New Elevation (Minutes)

Range (Meters)	2,500 Feet *	5,000 Feet *	10,000 Feet *
100	0.05	0.08	0.13
200	0.1	0.2	0.34
300	0.2	0.4	0.6
400	0.4	0.5	0.9
500	0.5	0.9	1.4
600	0.6	1.0	1.8
700	1.0	1.6	2.4
800	1.3	1.9	3.3
900	1.6	2.8	4.8
1,000	1.8	3.7	6.0

* Above Sea Level

Table 3-5. Selected Ballistics Information

Ammunition	Bullet Type	Ballistic Coefficient	Muzzle Velocity
M193	55 FMJBT	.260	3,200 fps
M180	147 FMJBT	.400	2,808 fps
M118	173 FMJBT	.515	2,610 fps
M852	168 HPBT	.475	2,675 fps
M72	173 FMJBT	.515	2,640 fps

SHOT GROUPS

3-137. If a rifle is fired many times under uniform conditions, the bullets striking the target will group themselves about a central point called the center of impact and will form a circular or elliptical group. The dimensions and shape of this shot group will vary depending on the distance of the target from the sniper. The circle or ellipse formed by these shots constantly increases in size with the range. The line connecting the centers of impact of all shots at all ranges measured is called the mean trajectory, and the core containing the circumferences of all the circles would mark the limits of the sheaf. The mean trajectory is the average trajectory. All ordinates are compared to it, and angles of departure and fall refer only to it.

3-138. The pattern on the target made by all of the bullets is called the shot group. If the shot group is received on a vertical target, it is called a vertical shot group and is circular. If the group is received on a horizontal target, it is called a horizontal shot group and is elliptical A large number of shots will form a shot group having the general shape of an ellipse, with its major axis vertical. The shots will be symmetrically grouped about the center of impact, not necessarily about the POA. They will be grouped more densely near the center of impact than at the edges, and half of all the shots will be found in a strip approximately 1/4 the size of the whole group. The width of this strip is called the mean vertical (or the 50 percent dispersion) if measured vertically or the mean lateral if measured laterally.

3-139. When considering the horizontal shot group, the mean lateral dispersion retains its same significance, but what is called the mean vertical dispersion on a vertical target is known as the mean longitudinal dispersion on a horizontal target. There is a significant relationship between the size or dimensions of a shot group and the size or dimensions of the target fired at. With a shot group of fixed dimensions, when the target is made sufficiently large, all shots fired will strike the target. Conversely, with a very small target, only a portion of the shots fired will strike the target. The rest of the shots will pass over, under, or to the sides of the target.

3-140. It is evident that the practical application of exterior ballistics—hitting a target of variable dimensions at unknown distances—is one probability of a shot group of fixed dimensions (the sniper's grouping ability) conforming to the dimensions of a given target. Added to this probability is the ability of the sniper to compensate for environmental conditions and maintain an accurate zero.

3-141. One of the greatest paradoxes of sniping is that an average marksman has a slightly higher chance of hitting targets at unknown distances than a good marksman, if their respective abilities to judge distances, determine effects of environmental conditions, and maintain an accurate zero are equal. (The classification of **good marksman** and **average marksman** refers only to the sniper's grouping ability.) A good marksman who has miscalculated wind or who is not accurately zeroed would expect to miss the target entirely. An average marksman, under identical conditions, would expect to obtain at least a few hits on the target; or if only one shot was fired, would have a slight chance of obtaining a first-round hit. The above statement does not mean that average marksmen make better snipers. It does mean that the better the individual shoots, the more precise his ability to judge distance, calculate wind, and maintain his zero must be.

3-142. Practical exterior ballistics is the state of applying a shot group or a sheaf of shots over an estimated distance against a target of unknown or estimated dimensions. It also includes estimating the probability of obtaining a hit with a single shot contained within the sheaf of shots previously determined through shot-group practices.

INFLUENCE OF GROUND ON THE SHOT GROUP (SHEAF OF SHOTS)

3-143. When firing at targets of unknown distances under field conditions, the sniper must take into consideration the lay of the ground and how it will affect his probable chances of hitting the target. Generally, the ground a sniper fires over will—

- Be level.
- Slope upward.
- Slope downward.

3-144. The extent of the danger space depends on the—
- Relationship between the trajectory and the LOS, angle of fall, and the range curvature of the trajectory.
- Height of the target.
- Point of aim.
- Point of impact.

NOTE: The longer the range, the shorter the danger space, due to the increasing curvature of the trajectory.

3-145. The displacing of the center of impact from the center of the target is a factor that the sniper must also consider. It will often be the controlling factor. The danger space at ranges under 700 yards is affected by the position of the sniper (height of the muzzle above the ground). The danger space increases as the height of the muzzle decreases. At longer ranges, no material effect is felt from different positions of the sniper.

3-146. The influence of the ground on computing hit probability on a target at unknown distances results in the necessity of distinguishing between danger space and swept space (which are functions of the mean trajectory), and between these (danger space and swept space) and the dangerous zone (which is a function of the whole or a part of the cone of fire). For a given height of target and POA, the danger space is of fixed dimensions. The swept space varies in relation with the slope of the ground. Swept space is shorter on rising ground and longer on falling ground than the danger space. All the functions of the danger zone, such as the density of the group at a given distance from the center of impact, are correspondingly modified.

SNIPER DATA BOOK

3-147. The sniper data book contains a collection of data cards. The sniper uses these cards to record firing results and all elements that have an effect on firing the weapon. This information could include weather conditions or even the sniper's attitude on a particular day. The sniper can refer to this information later to understand his weapon, the weather effects, and his firing ability on a given day. One of the most important items of information he will record is the cold barrel zero of his weapon. A cold barrel zero refers to the first round fired from the weapon. It is critical that the sniper know this by firing the first round at 200 meters. When the barrel warms up, later shots may begin to group 1 or 2 minutes higher or lower, depending on rifle specifics. Figure 3-27, page 3-49, shows a sample sniper data card.

3-148. When used properly, the data card will provide the necessary information for initial sight settings at each distance or range. It also provides a basis for analyzing the performance of the sniper and his rifle and is a valuable aid in making bold and accurate sight changes. The most competent sniper would not be able to consistently hit the center of the target if he were unable to analyze his performance or if he had no record of his performance or conditions affecting his firing.

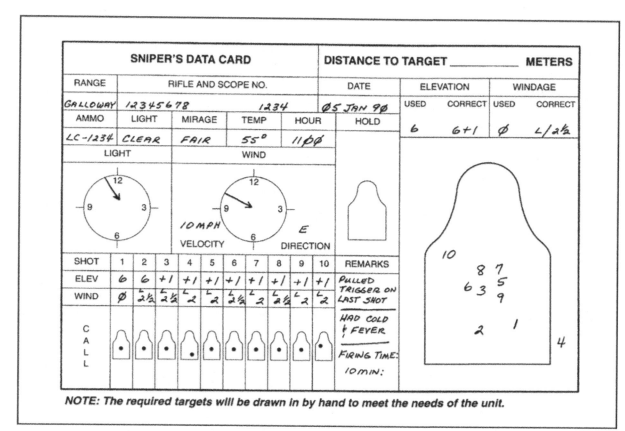

Figure 3-27. Sample of a Sniper's Data Card

ENTRIES

3-149. The three phases in writing information on the data card are *before firing*, *during firing*, and *after firing*. Each phase requires specific data and provides an excellent learning tool for future training. Each sniper should complete the following information for each phase and analyze his performance to stay proficient at all times.

Phase I–Before Firing

3-150. Before the sniper fires, he should record the following data:

- *Range*. The distance to the target.
- *Rifle and telescope number*. The serial numbers of the rifle and telescope.
- *Date*. Date of firing.
- *Ammunition*. Type and lot number of ammunition.
- *Light*. Amount of light (overcast, clear).
- *Mirage*. Whether a mirage can be seen or not (bad, fair, good).
- *Temperature*. Temperature on the range.
- *Hour*. Time of firing.

- *Light (diagram).* He draws an arrow in the direction the light is shining.
- *Wind.* He draws an arrow in the direction that the wind is blowing and records its average velocity and cardinal direction (N, NE, S, SW).

Phase II–During Firing

3-151. The sniper should also record specific data during firing. This information includes the following:

- *Elevation.* Elevation setting used and any correction needed. (For example: The target distance is 600 meters; the sniper sets the elevation dial to 6. The sniper fires and the round hits the target 6 inches low of center. He then adds 1 minute [one click] of elevation.)
- *Windage.* Windage setting used and any correction needed. (For example: The sniper fires at a 600 meter target with a windage setting on 0; the round impacts 15 inches right of center. He will then add 2 1/2 minutes left to the windage dial [L/2 1/2].)
- *Shot.* The column of information about a particular shot. (For example: Column 1 is for the first round; column 10 is for the tenth round.)
- *Wind.* Windage used. (For example: L/2, 1/2, O, R/1/2.) This is for iron sights or compensation for spin drift. Mil holds are used for the scope.
- *Call.* Where the aiming point was when the weapon fired.
- *Large silhouette or target representation.* Used to record the exact impact of the round on the target. This is recorded by writing the shot's number on the large silhouette that is in the same place it hit the target.

Phase III–After Firing

3-152. The sniper also records data after firing that will enable him to better understand his results and to improve his performance. This data includes:

- Comments about the weapon, firing conditions (time allowed for fire), or his condition (nervous, felt bad, felt good).
- Corrected no-wind zero. Show the elevation and windage in minutes and clicks that was correct for this position and distance under no-wind conditions.
- Remarks. Note any equipment, performance, weather conditions, or range conditions that had a good or bad effect on the firing results.

ANALYSIS

3-153. When the sniper leaves the firing line, he compares weather conditions to the information needed to hit the POA or POI. Since he fires in all types of weather conditions, he must be aware of temperature, light, mirage, and wind. He must also consider the following possibilities:

- Compare sight settings with previous firing sessions. If the sniper always has to fine-tune for windage or elevation, there is a chance he needs a sight change (slip a scale).
- Compare the ammunition by lot number for the best rifle and ammunition combination.

- Compare all groups fired under each condition. Check the low and high shots as well as those to the left and the right of the main group. Of course, less dispersion is desired. If groups are tight, they are easily moved to the center of the target; if scattered, there is a problem. Check the telescope focus and ensure that the rifle is cleaned correctly. Remarks in the data book will also help.
- Make corrections. Record corrections in the data book, such as position and sight adjustment information, to ensure retention. The sniper should compare hits to calls. If they agree, the result is an indication that the zero is correct and that any compensation for the effects of the weather is correct. If the calls and hits are consistently out of the target, sight adjustment or more position and trigger control work is necessary.

3-154. The sniper should compare the weather conditions and location of the groups on the latest data sheet to previous data sheets to determine how much and in which direction the sights should be moved to compensate for the weather conditions. If better results are obtained with a different sight picture under an unusual light condition, he should use this sight picture whenever firing under that particular light condition. A different sight picture may necessitate adjusting the sights. After establishing how much to compensate for the effects of weather or which sight picture works best under various light conditions, the sniper should commit this information to memory.

3-155. The sniper should keep the training and zeroing data sheets for future reference. Rather than carry the firing data sheets during sniper training exercises or combat, he can carry or tape on his weapon stock a list of the elevation and windage zeros at various ranges.

ZEROING THE RIFLE

3-156. A zero is the alignment of the sights with the bore of the rifle so that the bullet will impact on the target at the desired POA. However, the aiming point, the sight, and the bore will coincide at two points. These points are called the zero.

3-157. Depending upon the situation, a sniper may have to deliver an effective shot at ranges up to 1,000 meters or more. This need requires the sniper to zero his rifle (with telescopic and iron sights) at most of the ranges that he may be expected to fire. When using telescopic sights, he needs only zero for elevation at 300 meters (100 meters for windage) and confirms at the more distant ranges. His success depends on a "one round, one hit" philosophy. He may not get a second shot. Therefore, he must accurately zero his rifle so that when applying the fundamentals he can be assured of an accurate hit.

CHARACTERISTICS OF THE SNIPER RIFLE IRON SIGHTS

3-158. The iron sights of the M24 are adjustable for both windage and elevation. While these sights are a backup to the telescope and used only under extraordinary circumstances, it is in the sniper's best interest to be fully capable with them. Iron sights are excellent for developing marksmanship skills. They force the sniper to concentrate on sight alignment, sight picture, and follow-through.

3-159. The M24 has a hooded front sight that has interchangeable inserts. These various-sized inserts range from circular discs to posts. The sniper should use the post front sight to develop the sight picture that is consistent with the majority of U.S. systems. The rear sight is the Palma match sight and has elevation and windage adjustments in 1/4 MOA. The elevation knob is on the top of the sight and the windage knob is on the right side of the sight. Turning the elevation knob in the direction marked "UP" raises the POI, and turning the windage knob in the direction marked "R" moves the POI to the right.

ADJUSTMENT OF THE REAR SIGHT

3-160. The sniper determines mechanical windage zero by aligning the sight base index line with the centerline of the windage gauge. The location of the movable index line indicates the windage used or the windage zero of the rifle. For example, if the index line is to the left of the centerline of the gauge, this point is a left reading. The sniper determines windage zero by simply counting the number of clicks back to the mechanical zero. He determines the elevation of any range by counting the number of clicks down to mechanical elevation zero.

3-161. Sight adjustment or manipulation is a very important aspect of training that must be thoroughly learned by the sniper. He can accomplish this goal best through explanation and practical work in manipulating the sights.

3-162. The sniper must move the rear sight in the direction that the shot or shot group is to be moved. To move the rear sight or a shot group to the right, he turns the windage knob clockwise. The rule to remember is *push left—pull right*. To raise the elevation or a shot group, he turns the elevation knob clockwise. To lower it, he turns counterclockwise.

ZEROING THE SNIPER RIFLE USING THE IRON SIGHTS

3-163. The most precise method of zeroing a sniper rifle with the iron sights is to fire the rifle and adjust the sights to hit a given point at a specific range. The rifle is zeroed in 100-meter increments from 100 to 900 meters. The targets are placed at each range, then the sniper fires one or more five-round shot groups at each aiming point. He must adjust the rear sight until the center of the shot group and the aiming point coincide at each range. The initial zeroing for each range should be accomplished from the prone supported position. The sniper can then zero from those positions and ranges that are most practical. There is no need to zero from the least steady positions at longer ranges.

3-164. The sniper should use the following zeroing procedure for M24 iron sights:

- *Elevation knob adjustments.* Turning the elevation knob located on the top of the rear sight in the UP direction raises the POI; turning the knob downward lowers the POI. Each click of adjustment equals 0.25 MOA.

- *Windage knob adjustments.* Turning the windage knob located on the right side of the rear sight in the R direction moves the impact of the round to the right; turning the knob in the opposite direction moves the POI to the left. Each click of adjustment equals 0.25 MOA. Windage should be zeroed at 100 meters to negate the effects of wind.

- *Calibrating rear sight.* After zeroing the sights to the rifle, the sniper loosens the elevation and windage indicator plate screws with the wrench provided. He should align the "0" on the plate with the "0" on the sight body, then retightens the plate screws. Next, he loosens the setscrews in each knob and aligns the "0" of the knob with the reference line on the sight. He presses the sight and tightens the setscrews. The sniper then sharpens or softens the click to preference by loosening or tightening the spring screws equally on the knob. He must now count down the number of clicks to the bottom of the sight. He then records this number and uses it as a reference whenever he believes there has been a problem with his rear sight. He only needs to bottom out the sight and count up the number of clicks required to the desired zero. Windage and elevation corrections can now be made, and the sniper can return quickly to the zero standard. Elevation should be zeroed at 200 meters to increase the accuracy of the zero.

- *Graduations.* There are 12 divisions or 3 MOA adjustments in each knob revolution. Total elevation adjustment is 60 MOAs and total windage adjustment is 36 MOAs. Adjustment scales are of the "vernier" type. Each graduation on the scale plate equals 3 MOAs. Each graduation on the sight base scale equals 1 MOA.

3-165. To use the scales, the sniper—

- Notes the point at which graduations on both scales are aligned (Figure 3-28, page 3-54).
- Counts the number of full 3 MOA graduations from "0" on the scale plate to "0" on the sight base scale.
- Adds this figure to the number of MOAs from "0" on the bottom scale to the point where the two graduations are aligned.

NOTE: The Redfield Palma sight is the issued sight of the SWS and no longer available commercially.

CHARACTERISTICS OF THE SNIPER RIFLE TELESCOPIC SIGHT

3-166. Sniper telescopic sights have turret assemblies for the adjustment of elevation and windage. The upper assembly is the elevation and the assembly on the right is for windage. These assemblies have knobs that are marked for corrections of a given value in the direction indicated by the arrow. The M3A and the ART series use a similar system for zeroing. The sniper moves the knobs in the direction that he wants the shot group to move on the target.

3-167. The M3A is graduated to provide 1 MOA of adjustment for each click of its elevation knob and 1/2 MOA of adjustment for each click of its windage knob. This sight is designed to provide audible and tactile clicks. The elevation turret knob is marked in 100-meter increments from 100 to 500, and 50-meter increments from 500 to 1,000.

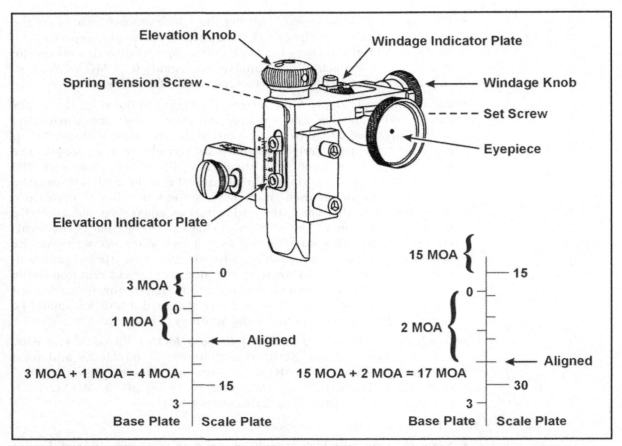

Figure 3-28. Adjusting the Elevation and Windage on the Rear Sight Assembly

ZEROING THE SNIPER WEAPON SYSTEM WITH THE TELESCOPIC SIGHT

3-168. The most precise method of zeroing the sniper rifle for elevation using the scope sight is to fire and adjust the sight to hit a given point at 200 meters. For windage, the scope should be zeroed at 100 meters. This point rules out as much wind effect as possible. After zeroing at 100 meters, the sniper should confirm his zero out to 900 meters in 100-meter increments. The bull's-eye-type target (200-yard targets, NSN SR1-6920-00-900-8204) can be used for zeroing. Another choice is a blank paper with black pastees forming a 1-MOA aim point.

3-169. The sniper should use the following zeroing procedures for a telescopic sight. He should—

- Properly mount the scope to the rifle.
- Select or prepare a distinct target (aiming cross) at 200 meters for elevation or 100 meters for windage. If 200 meters is used for windage, then impact must be compensated for a no-wind effect.
- Assume the prone supported position.
- Focus the reticle to his eye.
- Set parallax for target range.

- Boresight scope to ensure round on paper, considering that the M3A is a fixed 10x scope.
- Fire a single shot and determine its location and distance from the aiming cross.
- Using the elevation and windage rule, determine the number of clicks necessary to move the center of the group to the center of the aiming cross.
- Remove the elevation and windage turret caps and make the necessary sight adjustments. Then replace the turret caps. In making sight adjustments, the sniper must turn the adjusting screws in the direction that he wants to move the strike of the bullet or group.
- Fire 5-round groups as necessary to ensure that the center of the shot group coincides with the POA at 200 meters for elevation and compensated for wind.
- Zero the elevation and windage scales and replace the turret caps.

3-170. The rifle is now zeroed for 200 meters with a no-wind zero.

3-171. To engage targets at other ranges, set the range on the elevation turret. To engage targets at undetermined ranges, use the mil dots in the scope, determine the range to the target, and then manually set the elevation turret.

NOTE: Elevation and windage turrets should not be forced past the natural stops as damage may occur.

AN/PVS-2 Night Vision Device

3-172. The AN/PVS-2 may be zeroed during daylight hours or during hours of darkness. However, the operator may experience some difficulty in attempting to zero just before darkness (dusk). The light level is too low at dusk to permit the operator to resolve his zero target with the lens cap cover in place, but the light level at dusk is still intense enough to cause the sight to automatically cut off unless the lens cap cover is in position over the objective lens. The sniper will normally zero the sight for the maximum practical range that he can be expected to observe and fire, depending on the level of illumination.

3-173. The sniper should zero the sight in the following manner. He should—

- Place or select a distinct target at the desired zeroing range. A steel target provides the easiest target to spot because bullet splash is indicated by a spark as the bullet strikes the steel. He should assume the prone supported position, supporting the weapon and night vision sight combination with sandbags or other available equipment that will afford maximum stability.
- Boresight the sight to the rifle. The sniper places the iron sight windage and elevation zero on the rifle for the zeroing range and adjusts the weapon position until the correct sight picture is obtained on the aiming point at the zeroing range. He moves the eye to the night vision sight and observes the location of the reticle pattern in relation to the reference aiming point. If the reference aiming point on the

FM 3-05.222

target and the reference POA of the reticle pattern do not coincide, move the elevation and azimuth adjustment knobs until these aiming points coincide.

- Place the reference POA of the reticle pattern (Figure 3-29) on the center of mass of the target or on a distinct aiming point on the target. Then fire enough rounds to obtain a good shot group. Check the target to determine the center of the shot group in relation to the reticle POA.
- Adjust the night vision sight to move the reticle aiming reference point to the center of the shot group. When making adjustments for errors in elevation or azimuth, move the sight in the direction of the error. For example, if the shot group is high and to the left of the reticle POA, compensate for the error by moving the sight to the left and up.

NOTE: Each click of the azimuth or elevation knob will move the strike of the round 2 inches for each 100 meters of range.

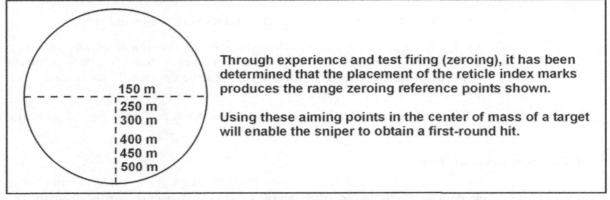

Figure 3-29. The Range References and POAs for the AN/PVS-2 Black Line Reticle Pattern

3-174. To engage targets at ranges other than the zero range, apply hold-off to compensate for the rise and fall in the trajectory of the round.

AN/PVS-4 Night Vision Device

3-175. Zeroing the AN/PVS-4 is similar to zeroing with standard optical sights because (unlike the AN/PVS-2) the AN/PVS-4 mounts over the bore of the weapons system and has internal windage and elevation adjustments (Figure 3-30, page 3-57).

Periodic Checking

3-176. A sniper cannot expect his zero to remain absolutely constant. Periodic checking of the zero is required after disassembly of the sniper rifle for maintenance and cleaning, for changes in ammunition lots, as a result of severe weather changes, and to ensure first-shot hits. The rifle must be zeroed by the individual who will use it. Individual differences in stock weld, eye relief, position, and trigger control usually result in each sniper having a different zero with the same rifle or a change in zero after moving from one position to another.

3-56

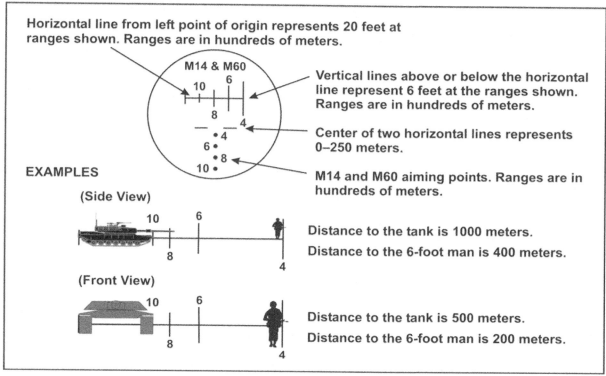

Figure 3-30. Using the M14 and M60 Reticle of the AN/PVS-4 for Range Estimation and POA

Confirming Zero

3-177. After a rifle has been zeroed and it becomes necessary to confirm this zero for any reason, the rifle can be zeroed again by firing at a known distance with the sight set on the old zero. If a sight adjustment is necessary to hit the aiming point, this zero change will remain constant at all ranges. For example, if a sniper is firing at a distance of 500 meters with the old zero and it becomes necessary to raise the elevation three clicks to hit the aiming point, he should raise the elevation zero three clicks at all ranges.

Changing Zero

3-178. Before changing the zero, windage, or elevation, the sniper must consider the effects of weather. Extreme changes of humidity or temperature can warp the stock or affect the ammunition. Wear, abuse, or repairs can also cause a sniper rifle's zero set to change.

Field-Expedient Zeroing

3-179. The sniper should use the boresight to confirm zero retention in a denied area. The sniper may need to confirm his zero in a field environment. Dropping a weapon or taking it through excessive climatic changes (by deploying worldwide) are good reasons for confirming the SWS's zero. The sniper may also use this method when the time or situation does not permit the use of a known distance range. This technique works best when confirming old zeros.

3-180. The sniper will need an observer equipped with binoculars or a spotting telescope to assist him. The sniper and observer pick out an aiming point in the center of an area; for example, a hillside, brick house, or any surface where the strike of the bullet can be observed. The team can determine the range to this point by using the ranging device on the telescope, by laser range finder, by map survey, by the range card of another weapon, or by ground measurement.

3-181. Once the sniper has assumed a stable position, the observer must position himself close and to the rear of the sniper. The observer's binoculars or telescope should be positioned approximately 18 to 24 inches above the weapon and as close in line with the axis of the bore as possible. With his optics in this position, the observer can see the trace of the bullet as it moves downrange. The trace or shock wave of the bullet sets up an air turbulence sufficient enough to be observed in the form of a vapor trail. The trace of the bullet enables the observer to follow the path of the bullet in its trajectory toward its impact area. The trace will disappear prior to impact and make it appear to the inexperienced observer that the bullet struck above or beyond its actual impact point. For example, at 300 meters the trace will disappear approximately 5 inches above the impact point. At 500 meters the trace will disappear approximately 25 inches above the impact point.

3-182. Wind causes lateral movement of the bullet. This lateral movement will appear as a drifting of the trace in the direction that the wind is blowing. This movement must be considered when determining windage zero. The observer must be careful to observe the trace at its head and not be misled by the bending tail of the trace in a stout crosswind. Before firing the first round, the sniper must set his sights so that he will hit on or near his aiming point. This setting is based on the old zero or an estimate. The sniper fires a shot and gives a call to the observer. If the strike of the bullet could not be observed, the observer gives a sight adjustment based on the trace of the bullet.

3-183. If the first shots do not hit the target, and the observer did not detect trace, the sniper may fire at the four corners of the target. One of the rounds will hit the target and the sniper can use this hit to make an adjustment to start the zeroing process. Once the strike of the bullet can be observed in the desired impact area, the observer compares the strike with the call and gives sight adjustments until the bullet impact coincides with the aiming point.

Firing at Targets With No Definite Zero Established

3-184. The sniper should use the 100-meter zero when firing on targets at a range of 100 meters or less. The difference between the impact of the bullet and the aiming point increases as the range increases if the sights are not moved. If the sniper's zero is 9+2 at 900 meters and 8+1 at 800 meters, and he establishes the range of the target at 850 meters, he should use a sight setting of 850+1 rather than using his 800- or 900-meter zero or the hold-off method. At any range, moving the sights is preferred over the hold-off method.

Firing the 25-Meter Range

3-185. The sniper should dial the telescope to 300 meters for elevation and to zero for windage. He then aims and fires at a target that is 25 yards away. He adjusts the telescope until rounds are impacting 1 inch above the POA. For the sniper to confirm, he fires the SWS on a known distance range out to its maximum effective range.

3-186. For iron sights, the sniper may fire on a 25-meter range to obtain a battle-sight zero. He then subtracts 1 minute (four clicks) of elevation from the battle-sight zero to get a 200-meter zero. The sniper may then use the following measures to determine the necessary increases in elevation to engage targets out to 600 meters:

- 200 to 300 meters—2 minutes.
- 300 to 400 meters—3 minutes.
- 400 to 500 meters—4 minutes.
- 500 to 600 meters—5 minutes.

NOTE: These measures are based on the average change of several sniper rifles. While the changes may not result in an "exact" POA or POI zero, the sniper should not miss his target.

ENVIRONMENTAL EFFECTS

3-187. For the highly trained sniper, the effects of weather are the main cause of error in the strike of the bullet. Wind, mirage, light, temperature, and humidity all have some effect on the bullet and the sniper. Some effects are insignificant, depending on average conditions of sniper employment.

3-188. It must be noted that all of the "rules of thumb" given here are for the 7.62-mm bullet (168 to 175 grains) at 2600 feet per second (fps). If the sniper is using a caliber 5.56, 300 Win Mag, .338 Lapua Mag, or any other round then these rules do not apply. It is incumbent on the sniper to find the specific wind constants and other "effects" for the round he will be using.

WIND

3-189. The condition that constantly presents the greatest problem to the sniper is the wind. The wind has a considerable effect on the bullet, and the effect increases with the range. This result is due mainly to the slowing of the bullet's velocity combined with a longer flight time. This slowing allows the wind to have a greater effect on the bullet as distances increase. The result is a loss of stability. Wind also has a considerable effect on the sniper. The stronger the wind, the more difficult it is for the sniper to hold the rifle steady. The effect on the sniper can be partially offset with good training, conditioning, and the use of supported positions.

Classification

3-190. Since the sniper must know how much effect the wind will have on the bullet, he must be able to classify the wind. The best method to use is the clock system (Figure 3-31, page 3-60). With the sniper at the center of the clock and the target at 12 o'clock, the wind is assigned the following three values:

- *Full value* means that the force of the wind will have a full effect on the flight of the bullet. These winds come from 3 and 9 o'clock.
- *Half value* means that a wind at the same speed, but from 1, 2, 4, 5, 7, 8, 10, and 11 o'clock will move the bullet only half as much as a full-value wind. While this half-value definition is generally accepted,

it is not accurate. Applying basic math will illustrate that the actual half-value winds are from 1, 5, 7, and 11 on the clock. Winds from 2, 4, 8, and 10 have values of 86 percent.

NOTE: To determine the exact effect of the wind on the bullet when the wind is between full and no-value positions, multiply the wind speed by the following constants: 90-degree, full; 75 degree, 0.96; 60 degree, 0.86; 45 degree, 0.70; 30 degree, 0.50; 15 degree, 0.25.

- *No value* means that a wind from 6 or 12 o'clock will have little or no effect on the flight of the bullet at close ranges. The no-value wind has a definite effect on the bullet at long ranges (beyond 600 meters) if it is not blowing directly from 6 or 12 o'clock. This wind is the most difficult to fire in due to its switching or fishtail effect, which requires frequent sight changes. Depending on the velocity of this type of wind, it will have an effect on the vertical displacement of the bullet.

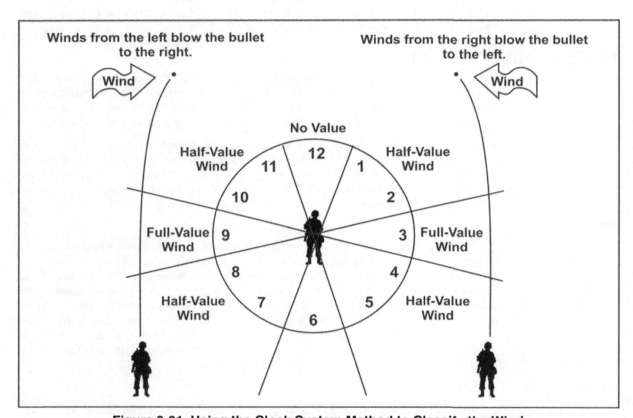

Figure 3-31. Using the Clock System Method to Classify the Wind

Velocity

3-191. Before adjusting the sight to compensate for wind, the sniper must determine wind direction and velocity. He may use certain indicators to make this determination. These indicators include range flags, smoke, trees, grass, rain, and the sense of feel. In **most** cases, wind direction can be determined simply by observing the indicators. However, the preferred method of determining wind direction and velocity is reading mirage.

3-192. A method of estimating the velocity of the wind during training is to watch the range flag (Figure 3-32). The sniper determines the angle in degrees between the flag and pole, then divides by the constant number 4. The result gives the approximate velocity in miles per hour (mph). This amount is based on the use of the heavier cotton range flags, not nylon flags, which are now used on most ranges.

NOTE: Nylon flags are not a reliable indicator for determining wind speed because of their susceptibility to minor wind speed variations.

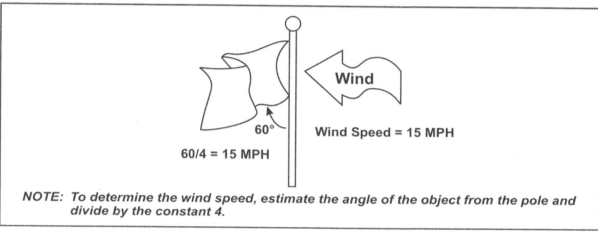

Figure 3-32. Estimating Wind Velocity Using the Range Flag

3-193. If no flag is visible, the sniper holds a piece of paper, grass, cotton, or some other light material at shoulder level, then drops it. He then points directly at the spot where it lands and divides the angle between his body and arm by the constant number 4. This number gives him the approximate wind velocity in mph (Figure 3-33).

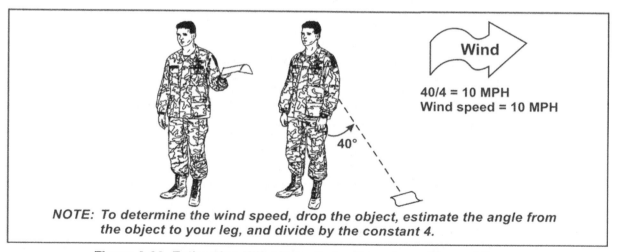

Figure 3-33. Estimating Wind Velocity by Dropping a Piece of Paper

3-194. If the sniper is unable to use these methods, he can apply the information in Table 3-6 to determine velocity.

Table 3-6. Determining Velocity

Wind Velocity (mph)	Effect
0 – 3	The wind can barely be felt but may be seen by mirage or smoke drifts.
3 – 5	The wind can be felt on the face. Grass begins to move.
5 – 8	The leaves in the trees and long grass are in constant motion.
8 – 12	The wind raises dust and loose paper and moves small branches in trees.
12 – 15	The wind causes trees to sway.

MIRAGE

3-195. A mirage is a reflection of the heat through layers of air at different temperatures and densities as seen on a warm, bright day. With the telescope, the sniper can see a mirage as long as there is a difference in ground and air temperatures. Proper reading of the mirage enables the sniper to estimate wind direction with a high degree of accuracy. The sniper uses the spotting scope to read the mirage. Since the wind nearest to midrange has the greatest effect on the bullet, he should try to determine velocity at that point. He can determine the amount in one of two ways:

- Focus on an object at midrange, then place the telescope back on to the target without readjusting the focus.
- Focus on the target, then back off the focus one-quarter turn counterclockwise. This movement makes the target appear fuzzy, but the mirage will be clear.

3-196. As observed through the telescope, the mirage appears to move with the same velocity as the wind, except when blowing straight into or away from the telescope. Then the mirage gives the appearance of moving straight upward with no lateral movement. It is then called a boiling mirage. A boiling mirage may also be seen when the wind is constantly changing direction; for example, a full-value wind blowing from 9 to 3 o'clock suddenly changes direction. The mirage will appear to stop moving from left to right and will present a boiling appearance. When this image occurs, the inexperienced observer may direct the sniper to fire with the "0" wind. As the sniper fires, the wind begins blowing from 3 to 9 o'clock and causes the bullet to miss the target. Therefore, firing in a "boil" can hamper shot placement. Unless there is a no-value wind, the sniper must wait until the boil disappears. In general, changes in the velocity of the wind, up to about 12 mph, can readily be determined by observing the mirage. Beyond that speed, the movement of the mirage is too fast for detection of minor changes. In general, when the waves of the mirage are shallow, its velocity and resultant wind speed are fast. Mirage will disappear at wind speeds above 15 mph.

3-197. The sniper can determine the true direction of the wind by traversing the telescope until the heat waves appear to move straight up with no lateral motion (a boiling mirage).

3-198. A mirage is particularly valuable in reading no-value winds. If the mirage is boiling, the effective wind velocity is zero. If there is any lateral movement of the mirage, it is necessary to make windage adjustments.

3-199. Another important effect of mirage is the light diffraction caused by the uneven air densities, which are characteristic of heat waves. Depending on atmospheric conditions, this diffraction will cause a displacement of the target image in the direction of the movement of the mirage. Thus if a mirage is moving from left to right, the target will appear to be slightly to the right of its actual location. Since the sniper can only aim at the image received by his eye, he will actually aim at a point that is offset slightly from the center of the target. This error will be in addition to the displacement of the bullet caused by the wind. Since the total effect of the visible mirage (effective wind plus target displacement) will vary considerably with atmospheric conditions and light intensity, it is impossible to predict the amount of error produced at any given place and time. It is only through considerable experience in reading mirage that the sniper will develop proficiency as a "wind doper."

3-200. Before firing, the sniper should check the mirage and make the necessary sight adjustments or hold-off to compensate for any wind. Immediately after firing, but before plotting the call in the scorebook, he again checks the mirage. If any changes are noted, they must be considered in relating the strike of the bullet to the call. The above procedure should be used for each shot.

CONVERSION OF WIND VELOCITY TO MINUTES OF ANGLE

3-201. All telescopic sights have windage adjustments that are graduated in MOAs or fractions thereof. An MOA is 1/60th of a degree. This number equals about 1 inch (1.0472 inches) for every 100 yards and 3 centimeters (2.97 centimeters) for every meter.

Example: 1 MOA = 2 inches at 200 yards
1 MOA = 15 centimeters at 500 meters

3-202. Snipers use MOAs to determine and adjust the elevation and windage needed on the telescope. After finding the wind direction and velocity in mph, the sniper must then convert it into MOAs using the wind formula as a rule of thumb only. The wind formula is as follows:

$$\frac{\textbf{Range (hundreds)} \times \textbf{Velocity (mph)}}{\textbf{Given Variable}} = \textbf{Minutes Full-Value Wind}$$

3-203. The given variable (GV) for M80 ball depends on the target's range (R) and is due to bullet velocity loss:

- 100 to 500 GV = 15
- 600 GV = 14
- 700 to 800 V = 13
- 900 GV = 12
- 1,000 GV = 11

3-204. The variable for M118, M118LR, and M852 is 10 at all ranges.

3-205. If the target is 700 meters away and the wind velocity is 10 mph, the formula is as follows:

$$\frac{7 \times 10}{10} = 7 \text{ MOA}$$

NOTE: This formula determines the number of minutes for a full-value wind. For a half-value (1/2V) wind, the 7 MOA would be divided in half, resulting in 3.5 MOA.

3-206. The observer makes his own adjustment estimations and then compares them to the wind conversion table, which can be a valuable training tool. He must not rely on this table. If it is lost, his ability to perform the mission could be severely hampered. Until the observer gains skill in estimating wind speed (WS) and computing sight changes, he may refer to the wind conversion table (Table 3-7). The observer will give the sniper a sight adjustment for iron sights or a mil hold off for the scope.

Table 3-7. Wind Conversion Table in Mils

WS / R (m)	2	4	6	8	10	12	14	16	18	20
100	0.00	0.00	0.25	0.25	0.25	0.25	0.50	0.50	0.50	0.50
1/2V	0.00	0.00	0.00	0.00	0.00	0.00	0.25	0.25	0.25	0.25
200	0.00	0.25	0.25	0.50	0.50	0.75	0.75	1.00	1.00	1.25
1/2V	0.00	0.00	0.00	0.25	0.25	0.50	0.50	0.50	0.50	0.75
300	0.25	0.25	0.50	0.75	1.00	1.00	1.25	1.50	1.50	1.75
1/2V	0.00	0.00	0.25	0.50	0.50	0.50	0.75	0.75	0.75	1.00
400	0.25	0.50	0.75	1.00	1.25	1.50	1.50	2.00	2.25	2.25
1/2V	0.00	0.25	0.50	0.50	0.75	0.75	0.75	1.00	1.25	1.25
500	0.25	0.50	1.00	1.25	1.50	1.75	2.00	2.25	2.50	3.00
1/2V	0.00	0.25	0.50	0.75	0.75	1.00	1.00	1.25	1.25	1.50
600	0.25	0.75	1.00	1.50	1.75	2.25	2.50	2.75	3.25	3.50
1/2V	0.00	0.50	0.50	0.75	1.00	1.00	1.25	1.25	1.50	1.75
700	0.50	0.75	1.25	1.50	2.00	2.50	3.00	3.25	3.75	4.25
1/2V	0.25	0.50	0.75	0.75	1.00	1.25	1.50	1.75	2.00	2.25
800	0.50	1.00	1.50	2.00	2.25	2.75	3.25	3.75	4.25	4.75
1/2V	0.25	0.50	0.75	1.00	1.25	1.50	1.75	1.75	2.25	2.50
900	5.00	1.00	1.50	2.00	2.50	3.25	3.75	4.25	4.75	5.25
1/2V	2.50	0.50	0.75	1.00	1.25	1.50	1.75	2.25	2.50	2.50
1,000	5.00	1.25	1.75	2.50	3.00	3.50	4.25	4.75	5.25	6.00
1/2V	2.50	0.63	0.75	1.25	1.50	1.75	2.25	2.50	2.50	3.00

LIGHT

3-207. Light does not affect the trajectory of the bullet. However, it may affect the way the sniper sees the target through the telescope. Light affects different people in different ways. The sniper generally fires high on a dull, cloudy day and low on a bright, clear day. Extreme light conditions from the left or the right may have an effect on the horizontal impact of a shot group.

3-208. This effect can be compared to the refraction (bending) of light through a medium, such as a prism or a fish bowl. The same effect can be observed on a day with high humidity and with sunlight from high angles. To solve the problem of light and its effects, the sniper must accurately record the light conditions under which he is firing. Through experience and study, he will eventually determine the effect of light on his zero. Light may also affect firing of unknown distance ranges since it affects range determination capabilities, by elongating the target.

TEMPERATURE

3-209. Temperature has a definite effect on the elevation setting required to hit the center of the target. This effect is caused by the fact that an increase in temperature of 20 degrees Fahrenheit (F) will increase the muzzle velocity by approximately 50 fps. When ammunition sits in direct sunlight, the burn rate of powder is increased. The greatest effect of temperature is on the density of the air. As the temperature rises, the air density is lowered. Since there is less resistance, velocity decreases at a slower rate and the impact rises. This increase is in relation to the temperature in which the rifle was zeroed. If the sniper zeroes at 50 degrees and he is now firing at 90 degrees, the impact rises considerably. How high it rises is best determined by past firing recorded in the data book. The general rule is that a 20-degree increase from zero temperature will raise the impact by 1 minute; conversely, a 20-degree decrease will drop impact by 1 minute from 100 to 500 meters, 15 degrees will affect the strike by 1 MOA from 600 to 900 meters, and 10 degrees over 900 meters will affect the strike by 1 MOA.

ELEVATION

3-210. Elevation above sea level can have an important effect on bullet trajectory. At higher elevations, air density, temperature, and air drag on the bullet decrease. The basic rule of thumb is that the bullet strike will vary by 1 MOA for every 5,000 feet of elevation. This amount will roughly correspond to the same barometric rule for changes in round strike.

BAROMETRIC PRESSURE

3-211. The effects of barometric pressure are that the higher the pressure, the denser the air. Thus the higher the pressure the lower the bullet will strike. As the pressure goes up, the sights go up. The basic rule is that from 100 to 500 meters, 1 inch in barometric pressure will affect the strike by 0.25 MOA; from 600 to 800 meters, a 1-inch change will affect the strike by 0.75 MOA, and from 900 to 1000 meters, a 1-inch change will effect the strike by 1.5 MOA.

HUMIDITY

3-212. Humidity varies along with the altitude and temperature. The sniper can encounter problems if drastic humidity changes occur in his area of operation. If humidity goes up, impact goes up; if humidity goes down, impact goes down. As a rule of thumb, a 20-percent change will equal about 1 minute affecting the impact. The sniper should keep a good data book during training and refer to his own record.

3-213. To understand the effects of humidity on the strike of the bullet, the sniper must realize that the higher the humidity, the thinner the air; thus there is less resistance to the flight of the bullet. This will tend to slow the bullet at a slower rate, and, as a result, the sniper must lower his elevation to compensate for these factors. The effect of humidity at short ranges is not as noticeable as at longer ranges. The sniper's experience and his analysis of hits and groups under varied conditions will determine the effect of humidity on his zero.

3-214. Some snipers fail to note all of the factors of weather. Certain combinations of weather will have different effects on the bullet. For this reason, a sniper may fire two successive days in the same location and under what appears to be the same conditions and yet use two different sight settings. For example, a 30-percent rise in humidity cannot always be determined readily. This rise in humidity makes the air less dense. If this thinner air is present with a 10-mile-per-hour wind, less elevation will be required to hit the same location than on a day when the humidity is 30 percent lower.

3-215. By not considering all the effects of weather, some snipers may overemphasize certain effects and therefore make bad shots from time to time. Snipers normally fire for a certain period of time under average conditions. As a result, they zero their rifles and (with the exception of minor displacements of shots and groups) have little difficulty except for the wind. However, a sniper can travel to a different location and fire again and find a change in his zero. Proper recording and study based on experience are all important in determining the effects of weather. Probably one of the most difficult things to impress upon a sniper is the evidence of a probable change in his zero. If a change is indicated, it should be applied to all ranges.

SLOPE FIRING

3-216. The sniper team conducts most firing practices by using the military range facilities, which are relatively flat. However, snipers may deploy to other regions of the world and have to operate in a mountainous or urban environment. This type of mission would require target engagements at higher and lower elevations. Unless the sniper takes corrective action, bullet impact will be above the POA. How high the bullet hits is determined by the range and angle to the target (Table 3-8, page 3-67). The amount of elevation change applied to the telescope of the rifle for angle firing is known as slope dope.

Table 3-8. Bullet Rise at Given Angle and Range in Minutes

Range (Meters)	Slant Degrees											
	5	10	15	20	25	30	35	40	45	50	55	60
100	0.01	0.04	0.09	0.16	0.25	0.36	0.49	0.63	0.79	0.97	1.2	1.4
200	0.03	0.09	0.2	0.34	0.53	0.76	1.0	1.3	1.7	2.0	2.4	2.9
300	0.03	0.1	0.3	0.5	0.9	1.2	1.6	2.1	2.7	3.2	3.9	4.5
400	0.05	0.19	0.43	0.76	1.2	1.7	2.3	2.9	3.7	4.5	5.4	6.3
500	0.06	0.26	0.57	1.0	1.6	2.3	3.0	3.9	4.9	6.0	7.2	8.4
600	0.08	0.31	0.73	1.3	2.0	2.9	3.9	5.0	6.3	7.7	9.2	10.7
700	0.1	0.4	0.9	1.6	2.5	3.6	4.9	6.3	7.9	9.6	11.5	13.4
800	0.13	0.5	1.0	2.0	3.0	4.4	5.9	7.7	9.6	11.7	14.0	16.4
900	0.15	0.6	1.3	2.4	3.7	5.3	7.2	9.3	11.6	14.1	16.9	19.8
1,000	0.2	0.7	1.6	2.8	4.5	6.4	8.6	11.0	13.9	16.9	20.2	23.7

NOTE: Range given is slant range (meters), not map distance.

3-217. The following is a list of compensation factors to use in setting the sights of the SWS when firing from any of the following angles. To use Table 3-9, pages 3-68 and 3-69, the sniper finds the angle at which he must fire and then multiplies the estimated range by the decimal figure shown to the right. For example, if the estimated range is 500 meters and the angle of fire is 35 degrees, the zero of the weapon should be set for 410 meters.

Example: 500 x .82 = 410 meters

3-218. As can be seen, the steeper the angle, the shorter the range will be set on the scope or sights for a first-round hit. Also, the steeper the angle, the more precise the sniper must be in estimating or measuring the angle. Interpolation is necessary for angles between tens and fives.

Example: Find the compensation factor for 72 degrees.
70 degrees = 0.34; 75 degrees = 0.26;
72 is 40 percent between 70 and 75 degrees.
0.34 - 0.26 = 0.08; 0.08 x 40 percent
(0.40) = 0.03; 0.34 - 0.03 = 0.31

NOTE: Table 3-9B and C, pages 3-68 and 3-69, are additional means of determining where to set the sights on the SWS to fire from a given angle. These tables are excellent references to reproduce and make into small cards for quick and easy access when conducting a mission. The data is for ranging only; the actual distance is used when determining the effects of the environment.

Table 3-9. Compensation Factors Used When Firing From a Given Angle

A

Percent of Slope Angle Up or Down (Degrees)	Multiply Range By
5	.99
10	.98
15	.96
20	.94
25	.91
30	.87
35	.82
40	.77
45	.70
50	.64
55	.57
60	.50
65	.42
70	.34
75	.26
80	.17
85	.09
90	.00

NOTE: This chart can also be used to compensate for apparent size of a target when miling for distance; for example, a 5-meter target at 30 degrees to the viewer is .87 x 5 = 4.35, which is the apparent size of a 5-meter target.

B

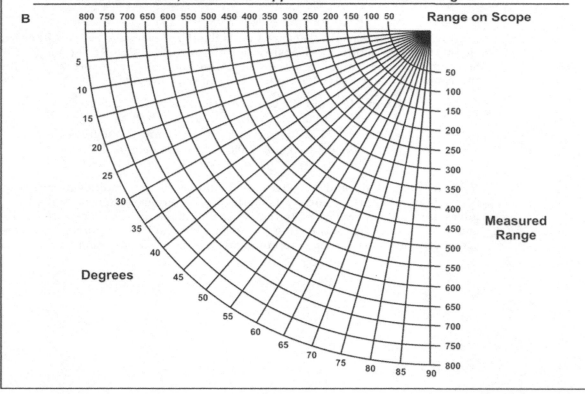

Table 3-9. Compensation Factors Used When Firing From a Given Angle (Continued)

C Range (m)	Angular Degree																	
	5	10	15	20	25	30	35	40	45	50	55	60	65	70	75	80	85	90
50	49.5	49	48	47	45.5	43.5	41	38.5	35	32	28.5	25	21	17	13	8.5	4.5	0
100	99	98	96	94	91	87	82	77	70	64	57	50	42	34	26	17	9	0
150	149	147	144	141	137	131	123	116	105	96	85.5	75	63	51	39	25.5	13.5	0
200	198	196	192	188	182	174	164	154	140	128	114	100	84	68	52	34	18	0
250	248	245	240	235	228	218	205	193	175	160	143	125	105	85	65	42.5	22.5	0
300	297	294	288	282	273	261	246	231	210	192	171	150	126	102	78	51	27	0
350	347	343	336	329	319	305	287	270	245	224	200	175	147	119	91	59.5	31.5	0
400	396	392	384	376	364	348	328	308	280	256	228	200	168	136	104	68	36	0
450	446	441	432	423	410	392	369	347	315	288	257	225	189	153	117	76.5	40.5	0
500	595	490	480	470	455	435	410	385	350	320	285	250	210	170	130	85	45	0
550	545	539	528	517	501	479	451	424	385	352	314	275	231	187	143	93.5	49.5	0
600	594	588	576	564	546	522	492	462	420	384	342	300	252	204	156	102	54	0
650	644	637	624	611	592	566	533	501	455	416	371	325	273	221	169	111	58.5	0
700	693	686	672	658	637	609	574	539	490	448	399	350	294	238	182	119	63	0
750	743	735	720	705	683	653	615	578	525	480	428	375	315	255	195	128	67.5	0
800	792	784	768	752	728	696	656	616	560	512	456	400	336	272	208	136	72	0
850	842	833	816	799	774	740	697	655	595	544	485	425	357	289	221	145	76.5	0
900	891	882	864	846	819	783	738	693	630	576	513	450	378	306	234	153	81	0
950	941	931	912	893	865	827	779	732	665	608	542	475	399	323	247	162	85.5	0
1,000	990	980	960	940	910	870	820	770	700	640	570	500	420	340	260	170	90	0

HOLD-OFF

3-219. Hold-off is shifting the POA to achieve a desired POI. Certain situations such as multiple targets at varying ranges do not allow proper elevation adjustments. Therefore, familiarization and practice of elevation hold-off techniques prepare the sniper to meet these situations. Windage is almost always held off by the sniper and will be practiced each range session.

ELEVATION

3-220. The sniper uses this technique only when he does not have time to change his sight setting. He rarely achieves pinpoint accuracy when holding off because a minor error in range determination or a lack of a precise aiming point might cause the bullet to miss the desired point. The sniper uses hold-off with the telescope only if several targets appear at various ranges and time does not permit adjusting the scope for each target.

3-221. The sniper uses hold-off to hit a target at ranges other than the range for which the rifle is presently adjusted. When he aims directly at a target at ranges greater than the set range, his bullet will hit below the POA. At closer distances, his bullet will hit higher than the POA. If the sniper understands this point and the effect of trajectory and bullet drop, he will be able to hit the target at ranges other than that for which the rifle was adjusted. For example, the sniper adjusts the rifle for a target located 500 meters downrange but another target appears at a range of 600 meters. The hold-off would be 25 inches; that is, the sniper should hold off 25 inches above the center of visible mass to hit the center of mass of that particular target (Figure 3-34). If another target were to appear at 400 meters, the sniper would aim 15 inches below the center of visible mass to hit the center of mass.

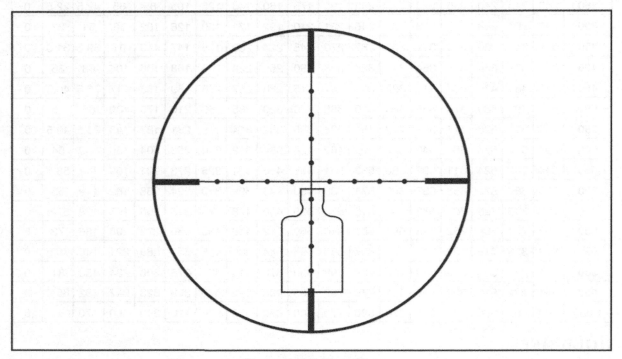

Figure 3-34. Elevation

3-222. The vertical mil dots on the M3A's reticle can be used as aiming points when using elevation hold-offs (Figure 3-35, page 3-71). For example, if the sniper has to engage a target at 500 meters and the scope is set at 400 meters, he would place the first mil dot 5 inches below the vertical line on the target's center mass. This setting gives the sniper a 15-inch hold-off at 500 meters.

3-223. For a 500-meter zero, the following measures apply:
- 100 and 400 meters, the waist or beltline.
- 200 and 300 meters, the groin.
- 500 meters, the chest.
- 600 meters, the top of the head.

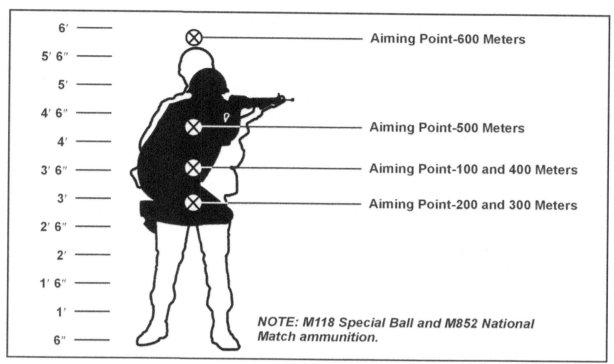

Figure 3-35. Correct Holds for Various Ranges With Sights Set for 500 Meters

WINDAGE

3-224. The sniper can use a hold-off to compensate for the effects of wind. When using the M3A scope, the sniper uses the horizontal mil dots on the reticle to hold off for wind. The space between each mil dot equals 3.375 MOAs, and a very accurate hold can be determined with the mil dots. For example, if the sniper has a target at 500 meters that requires a 10-inch hold-off, he would place the target's center mass halfway between the crosshairs and the first mil dot (1/2 mil) (Figure 3-36).

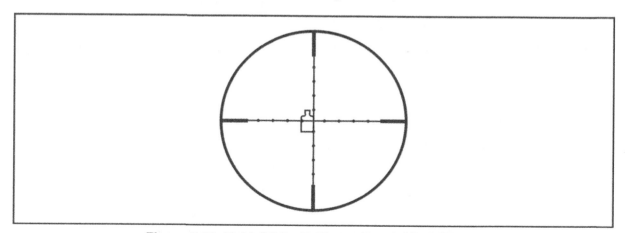

Figure 3-36. Hold-Off for 7.62-mm Special Ball (M118)

3-225. When holding off, the sniper aims into the wind. If the wind is moving from the right to left, his POA is to the right. If it is moving from left to right, his POA is to the left. Constant practice in wind estimation can bring about proficiency in making sight adjustments or learning to apply hold-off correctly. If the sniper misses the target and the impact of the round is observed, he notes the lateral distance of his error and refires, holding off that distance in the opposite direction. The formula used to find the hold-off distance is as follows:

$$\frac{\text{MOA (from wind formula)}}{3.5} = \text{Hold-off in mils}$$

NOTE: The wind formula must be computed first to find the MOA.

Example: Range to a target is 400 yards; wind is from 3 o'clock at 8 mph. Find the hold-off required to hit the target (M118).

$$R \times V = \text{MOA} \qquad \frac{4 \times 8}{10} = \frac{32}{10} = 3.2 \text{ MOA}$$

$$\frac{\text{MOA}}{3.5} = \text{Hold-off in mils} \qquad \frac{3.2}{3.5} = .91 = \text{right 1 mil}$$

For a half-value wind, divide mils by 2 for the hold-off.

ENGAGEMENT OF MOVING TARGETS

3-226. Moving targets are generally classified as walking or running and are the most difficult to hit. When engaging a target that is moving laterally across the LOS, the sniper must concentrate on moving his weapon with the target while aiming at a point some distance ahead. He must hold the lead, fire, and follow through after the shot. To engage moving targets, the sniper uses one of the techniques discussed below.

LEADING

3-227. A quarterback throwing a pass to his receiver can demonstrate the best example of a lead. He has to throw the ball at some point downfield in front of the receiver; the receiver will then run to that point. The same principle applies to firing at moving targets. Engaging moving targets requires the sniper to place the crosshairs ahead of the target's movement. The distance the crosshairs are placed in front of the target's movement is called a lead. The sniper uses the following four factors in determining leads.

Speed of the Target

3-228. Target speed will be a significant factor in determining the lead of the target. Running targets will require a greater lead than walking targets. Once target speed is determined, the sniper estimates the proper lead for the target at that specific range. Simultaneously, he applies the angle value to his lead estimation for the target (full-lead, half-lead).

3-229. For example, a target walking at a 45-degree angle toward the sniper at an average of 300 meters would require a 6-inch lead. This amount is determined by using the full-value lead of a walking target 300 meters away (a 12-inch lead) and dividing it in half for a half-value lead (as the target is moving at a 45-degree angle toward the sniper). Wind must also be considered, as it will affect the lead used. For a target moving with the wind, the sniper subtracts the wind value from the lead. For a target moving against the wind, he adds to the lead.

3-230. Double leads are sometimes necessary for a sniper who uses the swing-through method on a target that is moving toward his firing side. The double lead is necessary because of the difficulty that a person has in swinging his weapon smoothly toward his firing side. Practice on a known-distance range and meticulous record keeping are required to hone a sniper's moving target engagement skill.

Angle of Target Movement

3-231. A target moving perpendicular to the bullet's flight path moves a greater lateral distance during its flight time than a target moving at an angle away from or toward the bullet's path. A method of estimating the angle of movement of a target moving across the sniper's front follows (Figure 3-37):

- *Full-value lead target.* When only one arm and one side of the target are visible, the target is moving at or near a 90-degree angle and requires a full-value lead.
- *Half-value lead target.* When one arm and two-thirds of the front or back of the target are visible, the target is moving at approximately a 45-degree angle and requires a one-half value lead.
- *No-lead target.* When both arms and the entire front or back are visible, the target is moving directly toward or away from the sniper and requires no lead.

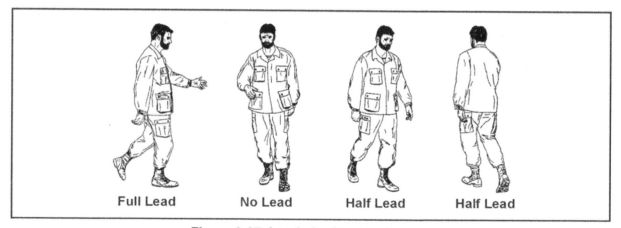

Figure 3-37. Leads for Moving Targets

Range to the Target

3-232. The farther away a target is, the longer it takes for the bullet to reach it. Therefore, the lead must be increased as the distance to the target increases.

FM 3-05.222

Wind Effects

3-233. The sniper must consider how the wind will affect the trajectory of the round. A wind blowing opposite to the target's direction requires more of a lead than a wind blowing in the same direction as the target's movement. When the target is moving against the wind, the wind effect is added to the lead. When he is moving with the wind, the wind effect is subtracted from the lead. Thus, "against add, with subtract."

3-234. Once the required lead has been determined (Table 3-9, pages 3-68 and 3-69), the sniper should use the mil scale in the telescope for precise hold-off. The mil scale can be mentally sectioned into 1/4-mil increments for leads. The chosen point on the mil scale becomes the sniper's point of concentration, just as the crosshairs are for stationary targets. The sniper concentrates on the lead point and fires the weapon when the target is at this point.

Lead Values

3-235. Tables 3-10 through 3-12, pages 3-74 and 3-75, list the recommended leads for movers at various ranges and speeds. Snipers should usually not engage movers beyond 400 yards due to the excessive lead required and low probability of a hit. If a mover is engaged at distances beyond 400 yards, an immediate follow-up shot must be ready.

3-236. The classification of a walker, fast walker, and a runner is based on a walker moving at 2 mph, a fast walker at 3 1/2 mph, and a runner at 5 mph.

3-237. These are starting point leads and are only guides. Each individual will have his own leads based on how he perceives movement and his reaction time to it.

Table 3-10. Recommended Leads in Mils for Movers

Range (Meters)	Walkers	Fast Walkers	Runners
100	Leading Edge	7/8	1 3/4
200	7/8	1 1/4	1 3/4
300	1 1/8	1 3/4	2 1/4
400	1 1/4	1 3/4	2 1/2
500	1 1/2	1 3/4	2 1/2
600	1 1/2	2 1/4	3
700	1 1/2	2 1/4	3
800	1 1/2	2 1/2	3
900	1 3/4	2 1/2	3 1/2
1,000	1 3/4	2 1/2	3 1/2

Table 3-11. Recommended Leads in Minutes of Angle for Movers

Range (Meters)	Walkers	Fast Walkers	Runners
100	Leading Edge	3	6
200	3	4.5	6
300	4	6	8
400	4.5	6	9
500	4.5	6	9
600	5	7.5	10
700	5	7.5	10
800	5.5	8	11
900	5.5	8	11
1,000	5.5	8	11

Table 3-12. Recommended Leads in Feet for Movers

Range (Meters)	Walkers	Fast Walkers	Runners
100	Leading Edge	0.25	0.5
200	0.5	0.75	1
300	1	1.5	2
400	1.5	2.25	3
500	2	3	4
600	2.5	3.75	5
700	3	4.5	6
800	3.5	5.25	7
900	4	6	8
1,000	4.5	7.25	9

TRACKING

3-238. Tracking requires the sniper to establish an aiming point ahead of the target's movement and to maintain it as the weapon is fired. This technique requires the weapon and body position to be moved while following the target and firing. This method is preferred and needs to be perfected after the basics are mastered.

TRAPPING OR AMBUSHING

3-239. Trapping or ambushing is the sniper's alternate method of engaging moving targets. The sniper must establish an aiming point ahead of the target that is the correct lead for speed and distance. As the target reaches this point, the sniper fires his weapon. This method allows the sniper's weapon and body position to remain motionless. With practice, a sniper can determine exact leads and aiming points using the horizontal stadia lines in the ART scopes or the mil dots in the M3A. The sniper must remember to concentrate on the crosshairs and not on the target. He must also not jerk the trigger. However, he must make the weapon go off the lead. The sniper can use a combination of tracking and ambushing to aid in determining target speed and direction. This technique is best suited for sentries who walk a set pattern.

TRACKING AND HOLDING

3-240. The sniper uses this technique to engage an erratically moving target. While the target is moving, the sniper keeps his crosshairs centered as much as possible and adjusts his position with the target. When the target stops, the sniper quickly perfects his hold and fires. This technique requires concentration and discipline to keep from firing before the target comes to a complete halt.

FIRING A SNAPSHOT

3-241. A sniper may often attempt to engage a target that only presents itself briefly, then resumes cover. Once he establishes a pattern, he can aim in the vicinity of the target's expected appearance and fire a snapshot at the moment of exposure.

COMMON ERRORS WITH MOVING TARGETS

3-242. When engaging moving targets, the sniper makes common errors because he is under greater stress than with a stationary target. There are more considerations, such as retaining a steady position and the correct aiming point, how fast the target is moving, and how far away it is. The more practice a sniper has firing at moving targets the better he will become. Some common mistakes that a sniper makes are when he—

- Watches his target instead of his aiming point. He must force himself to watch his lead point.
- Jerks or flinches at the moment his weapon fires because he thinks he must fire **NOW**. This reflex can be overcome through practice on a live-fire range.
- Hurries and thus forgets to adjust for wind speed and direction as needed. Windage must be calculated for moving targets just as for stationary targets. Failing to estimate when acquiring a lead will result in a miss.

3-243. Engaging moving targets requires the sniper to determine target distance and wind effects on the round, the lateral speed of the target, the round's time of flight, and the placement of a proper lead to compensate for both. These added variables increase the chance of a miss. Therefore, the sniper should engage moving targets when it is the only option.

ENGAGEMENT OF SNAP TARGETS

3-244. Many times the sniper will see a target that shows itself for only a brief moment, especially in urban and countersniper environments. Under these circumstances it is very important to concentrate on trigger control. Trigger control is modified to a very rapid pull of the finger directly to the rear without disturbing the lay of the weapon, similar to the moving target trigger control.

3-245. Another valuable skill for the sniper to learn is the quick-kill firing technique. He is most vulnerable during movement. Not only is he compromised because of the heavier equipment requirement, but also because of his large, optically sighted sniper rifle. Using the quick-kill technique, the sniper or observer can engage a target very rapidly at close range. This method is very useful for chance encounters with the enemy and when security is threatened. The sniper carries the rifle pointed toward the front, with the muzzle always pointing where he is looking and not at port arms. When the rifle is raised to fire, the eye is looking at the target. As the sniper looks at his target, the weapon lines on the target and he fires in the same movement. This technique must be practiced to obtain proficiency. It is not "wild firing," but a learned technique. A close analogy could be made to a skeet shooter who points his shotgun as opposed to sighting it.

FIRING THROUGH OBSTACLES AND BARRIERS

3-246. Another variable the sniper may encounter is the effect that glass penetration has on exterior and terminal ballistics. Firing through glass is unpredictable, and unless the target is close to the glass, more than one shot may be required. The sniper should never shoot through glass if it is close to his position. He is better off opening the window or having someone else break the window for him. The U.S. Army conducted a penetration test by firing through a glass plate from a distance of 1 yard at a silhouette target 100 yards away. Of the 14 test shots through various types of glass, only 2 shots hit the target.

GLASS PENETRATION

3-247. The United States Marine Corps (USMC) conducted a test by firing at an 8- by 9-inch pane of safety glass at 90- and 45-degree angles with the following results:

- Regardless of the angle, the path of the test bullet core was not greatly affected up to 5 feet beyond the point of initial impact; further from the glass, the apparent deflection became more pronounced.
- At an angle, glass fragments were always blown perpendicular to the glass plate.
- The M118 173 grain bullet's copper jacket fragments upon impact. All of the bullet fragments followed an erratic path both in height and width. Each of the main cores (lead) began to tumble about 2 feet from the initial impact point.
- Due to the lamination of safety glass with a sheet plastic, large fragments of plastic were embedded in the target 1 foot from the POI. These fragments were large enough to cause severe wounds.
- Glass fragments did not penetrate targets farther than 1 foot from the POI.
- It can be concluded that anyone near the glass would be injured.

3-248. Therefore, as indicated by both the USMC and U.S. Army tests, snipers should try to avoid engaging targets requiring glass penetration.

PENETRATION PERFORMANCE OF M118 SPECIAL BALL

3-249. To support the M24 SWS program, two tests were conducted with the M118 Special Ball ammunition at a range of 800 meters. The first test used a test sample of ballistic Kevlar, and the second test used a 10-gauge, mild steel plate. Testing personnel positioned a witness plate behind each of these targets. Witness plates consist of a 0.5-mm sheet of 2024T3 aluminum to measure residual velocity or energy. To pass the test, the bullet had to penetrate both the target and witness plate. Results of these tests follow:

- *M118 versus ballistic Kevlar.* When 10 rounds were fired at 13 layers of ballistic Kevlar (equivalent to the U.S. personal armor system ground troop vest), full penetration was achieved of both the test sample and the aluminum witness plate.
- *M118 versus mild steel plate.* When 20 rounds were fired at a 3.42-mm thick (10-gauge) SAE 1010 or 1020 steel plate (Rockwell hardness of B55 to B70), 16 achieved full penetration of both the test sample and aluminum witness plate. The 4 failing rounds penetrated the steel plate but only dented the witness plate. These 4 rounds were considered to have insufficient terminal energy to be effective.

COLD BORE FIRST-SHOT HIT

3-250. On a mission, a sniper will rarely get a second shot at the intended target. Therefore, he must be 98 percent sure that he will hit his target with the first shot. This requirement places a great deal of importance on the maintenance of a sniper's logbook. Whenever the sniper conducts a live-fire exercise, he should develop a database on his SWS and its cold bore zero. The sniper uses the integrated act of firing one round to hone his sniping skills. By maintaining a detailed logbook, he develops confidence in his system's ability to provide the "one shot–one kill" goal of every sniper. The sniper must pay close attention to the maintenance and cleanliness of his rifle. He must also maintain proficiency in the marksmanship fundamentals. He should attempt to obtain his cold bore data at all ranges and climatic conditions. The bore and chamber must be completely dry and free of all lubricants. The exact POI of the bullet should be annotated in the logbook. Also keeping a file of the actual paper targets used is even better. This data will help detect trends that can be used to improve the sniper's performance. This exercise also develops the teamwork required for the sniper pair to accomplish the mission. A sniper going on a mission will foul his bore with 5 shots to preclude problems with the so-called cold bore shot.

LIMITED VISIBILITY FIRING

3-251. The U.S. Army currently fields the AN/PVS 10 as the night vision sight for its SWS. If unavailable, then the sniper can compromise by using issued equipment to mount a PVS-4 onto an M4 or M16. This NVD should be kept permanently mounted to avoid zeroing problems. This system is adequate because the rifle's effective range matches that of the NVD's ability to distinguish target details. The M24 can be used during limited visibility operations if the conditions are favorable. Moonlight, artificial illumination, and terrain will determine the potential effectiveness. The sniper will find that the reticle will fade out during limited visibility. Rather than trying to strain his eyes to make out the reticle, he should use the entire field of view of the telescope as the aiming device. Live-fire exercises will help the sniper determine his own maximum effective range. The sniper needs to use off-center vision in the rifle scope to see the heavy crosshair post and target.

3-252. The PVS-10 and the Universal Night Sight (UNS) are now standard issue. The NAD 750 and KN 200/250 (mentioned in Chapter 2), as well as the other NVDs, are not yet a standard-issue item. These sights are available for contingency operations and are in contingency stocks. Every effort should be made to acquire these sights for training. It is the best sight currently available for precision firing during limited visibility. It has a longer effective range for discriminating targets and does not need to be mounted and dismounted from the rifle.

3-253. Another consideration during limited visibility firing is that of muzzle flash. Both the M4 and the M16 are equipped with an excellent flash suppressor. The enemy would have to be very close or using NVDs to pinpoint a couple of muzzle flashes. The M24 has a flash suppressor that attaches to the front sight block and is locked in place by a ring. To minimize the compromising effects of muzzle flash, the sniper should carefully select hide sites and ammunition lots. However, at a range of greater than 100 meters, the muzzle flash is not noticeable, and even with NVDs the flash is barely noticeable. Snipers should not use flash hiders because they increase the muzzle blast signature of the weapon and increase the likelihood of detection. They also change the barrel harmonics and are detrimental to the weapon's accuracy.

NUCLEAR, BIOLOGICAL, AND CHEMICAL FIRING

3-254. Performance of long-range precision fire is difficult at best during nuclear, biological, and chemical (NBC) conditions. Enemy NBC warfare creates new problems for the sniper. Not only must the sniper properly execute the fundamentals of marksmanship and contend with the forces of nature, he must overcome obstacles presented by protective equipment.

3-255. Firing in mission-oriented protective posture (MOPP) has a significant effect on the ability to deliver precision fire. The following problems and solutions have been identified:

- *Eye relief.* Special emphasis must be made in maintaining proper eye relief and the absence of scope shadow. It is a must to maintain consistent stock weld. However, care must be taken not to break the mask's seal.
- *Trigger control.* Problems encountered with trigger control consist of the following:
  - *Sense of touch.* When gloves are worn, the sniper cannot determine the amount of pressure he is applying to the trigger. This point is particularly important if the sniper has his trigger

adjusted for a light pull. Training with a glove will be beneficial; however, the trigger should be adjusted to allow the sniper to feel the trigger without accidental discharge.

- *Stock drag*. While training, the sniper should have his observer watch his trigger finger to ensure that the finger and glove are not touching any part of the rifle but the trigger. The glove or finger resting on the trigger guard moves the rifle as the trigger is pulled to the rear. The sniper must wear a well-fitted glove.

• *Vertical sight picture*. The sniper naturally cants the rifle into the cheek of the face while firing with a protective mask. Using the crosshair of the reticle as a reference mark, he keeps the weapon in a vertical position. Failure to stay upright will cause shots to hit low and in the direction of the cant. Also, windage and elevation corrections will not be true.

• *Sniper/observer communications*. The absence of a voice-emitter on the M25-series protective mask creates an obstacle in relaying information. The team either speaks louder or uses written messages. A system of foot taps, finger taps, or hand signals may be devised. Communication is a must; training should include the development and practice of communications at different MOPP levels.

3-256. The easiest solution to NBC firing with the M24 SWS is to use the Harris bipod. The bipod helps stabilize the rifle and allows the sniper to maintain a solid position behind the rifle as he cants his head to achieve a proper sight picture. The sniper can also try tilting his head down so he is looking up through the telescope. NBC firing must be incorporated into live-fire ranges so that the most comfortable and effective position can be developed. Also, a detailed logbook should be developed that addresses the effects of NBC firing.

Chapter 4
Field Skills

The sniper's primary mission is to interdict selected enemy targets with long-range precision fire. How well he accomplishes his mission depends on the knowledge, understanding, and application of various field techniques and skills that allow him to move, hide, observe, and detect targets (Appendix I). This chapter discusses those techniques and skills that the sniper must learn before employment in support of combat operations. The sniper's application of these skills will affect his survival on the battlefield.

CAMOUFLAGE

4-1. Camouflage is one of the basic weapons of war. To the sniper team, it can mean the difference between life and death. Camouflage measures are important since the team cannot afford to be detected at any time while moving alone, as part of another element, or while operating from a firing position. Marksmanship training teaches the sniper to hit a target. Knowing how and when to camouflage can enable the sniper to escape becoming a target. He must be camouflage-conscious from the time he departs on a mission until he returns. Paying attention to camouflage fundamentals is a mark of a well-trained sniper. FM 20-3, *Camouflage, Concealment, and Decoys*, provides more details.

FUNDAMENTALS

4-2. The sniper must pay careful attention when using camouflage clothing and equipment (artificial and natural). He should apply the following fundamental rules when determining his camouflage needs:

- Take advantage of all available natural concealment such as trees, bushes, grass, earth, man-made structures, and shadows.
- Alter the form, shadow, texture, and color of objects.
- Camouflage against ground and air observation.
- Camouflage a sniper post as it is prepared.
    - Study the terrain and vegetation in the area. Arrange grass, leaves, brush, and other natural camouflage to conform to the area.
    - Use only as much material as is needed. Excessive use of material (natural or artificial) can reveal a sniper's position.
    - Obtain natural material over a wide area. Do not strip an area, as this may attract the enemy's attention.

- Dispose of excess soil by covering it with leaves and grass or by dumping it under bushes, into streams, or into ravines. Piles of fresh dirt indicate that an area is occupied and reduce the effectiveness of camouflage.

4-3. The sniper and his equipment must blend with the natural background. Remember that vegetation changes color many times in an area.

VARIOUS GEOGRAPHICAL AREAS

4-4. A sniper cannot use one type of camouflage in all types of terrain and geographic areas. Before operations in an area, a sniper should study the terrain, the vegetation, and the lay of the land to determine the best possible type of personal camouflage.

4-5. In areas with heavy snow or in wooded areas with snow-covered brush, the sniper should use a full, white camouflage suit with gray shading. With snow on the ground and the brush not covered, he should wear white trousers and green-brown tops. A hood or veil in snow areas is very effective, and equipment should be striped or totally covered in white. In snow regions, visibility during a bright night is nearly as good as during the day. This advantage gives the sniper full-time capabilities, but he must move along carefully concealed routes.

4-6. In sandy and desert areas that have minimal vegetation, textured camouflage is normally not necessary. Still, proper coloring of a suit that breaks up the sniper's human outline is needed. Blending tan and brown colors is most effective. A bulky-type smock of light material with a hood works well. The sniper must be sure his hands, face, and all equipment blend into a solid pattern that corresponds with the terrain. The sniper must make full use of the terrain by using properly selected and concealed routes of movement.

4-7. When deployed with regular troops in an urban area, the sniper should be dressed like the troops in the area. When the sniper is in position, he should be camouflaged to match his area of operations. He can use a bulky, shapeless, gray camouflage suit that has been colored to match rubble and debris. He should make sure some type of hood breaks up the outline of the head. Movement during daylight hours should be extremely slow and careful, if at all, because of the unlimited amount of possible enemy sniper positions.

4-8. In jungle areas, the sniper can use foliage, artificial camouflage, and camouflage paint in a contrasting pattern that will blend with the texture of the terrain. In a very hot and humid area, he should wear only a light camouflage suit. A heavy suit will cause a loss of too much body fluid. The vegetation is usually very thick in jungle areas, so the sniper can rely more on the natural foliage for concealment.

DISCIPLINE

4-9. The sniper must always practice camouflage discipline. The sniper will change his camouflage to match the terrain patterns and foliage as he moves and as it dries or wilts. He ensures his camouflage presents a natural appearance at all times.

CONFIGURATION

4-10. The sniper must constantly observe the terrain and vegetation changes to pick the most concealed routes of advance and be certain he is camouflaged properly. He should use shadows caused by vegetation, terrain features, and man-made features to remain undetected. He must master the techniques of hiding, blending, and deceiving.

Hiding

4-11. This technique enables the sniper to completely conceal his body from observation by lying in thick vegetation, lying under leaves, or even by digging a shallow trench and covering up in it. Hiding may be used if the sniper stumbles upon an enemy patrol and immediate concealment is needed or if he wishes to stay out of sight during daylight hours to await darkness. However, the sniper should not use the hiding technique in the final firing position (FFP), as he would be unable to see his target.

Blending

4-12. A sniper should use this technique since it is not possible to completely camouflage in such a way as to be indistinguishable from the surrounding area. Camouflage needs to be so nearly perfect that the sniper cannot be recognized through optical gear or with the human eye. He must be able to be looked at directly and not be seen. This trait takes much practice and experience. The ghillie suit is a form of blending.

NOTE: A sniper should not attempt to use disguising as a camouflage technique. This requires him to change his appearance to look like another object.

Deceiving

4-13. In this method, the sniper tricks the enemy into a false conclusion regarding his location, intentions, or movement. By planting objects such as ammunition cans, food cartons, or something intriguing, the sniper decoys the enemy into the open where he can be brought under fire. Cutting enemy communications wire and waiting for the repair personnel is another technique. After a unit has left a bivouac area, a sniper can stay behind to watch for enemy scouts that may search the area. The unit can also use mannequins to lure the enemy sniper into firing, thereby revealing his position.

TARGET INDICATORS

4-14. A target indicator is anything a sniper does or fails to do that will reveal his position to an enemy. A sniper must know these indicators if he is to locate the enemy and prevent the enemy from locating him. There are four general areas: olfactory, tactile, auditory, and visual.

Olfactory

4-15. The enemy can smell these target indicators. Cooking food, fires, cigarettes, aftershave lotion, soap, and insect repellents are examples. Most of these indicators are caused by the sniper's bodily functions. The sniper

FM 3-05.222

usually can eliminate this target indicator by washing the body, burying body wastes, and eliminating the cause. The indicator only gives a sign that the sniper is in the area.

Tactile

4-16. The sniper can touch these indicators; for example, trip wire, phone wire, and hide positions. He uses them mainly at night. Tactile indicators are defeated through the proper construction of sniper hides, and awareness of the altered vegetation he has left behind while constructing his hide.

Auditory

4-17. This indicator is a sound that the sniper might make by moving, rattling equipment, or talking, and is most noticeable during hours of darkness. The enemy may dismiss small noises as natural, but when they hear someone speak, they know for certain that others are near. The sniper should silence all equipment before a mission so that he will make no sound while running or walking. He can defeat auditory indicators through noise discipline and proper equipment preparation.

Visual

4-18. This factor is the most important target indicator. The main reason a sniper is detected is because the enemy sees him. Being familiar with subcategories of visual target indicators can help the sniper locate the enemy and prevent him from being detected. The sniper can overcome the following visual indicators by properly using the principles of concealment.

4-19. **Why Things are Seen.** The proper understanding and application of the principles of concealment used with the proper camouflage techniques can protect the sniper from enemy observation. The following principles explain why things are seen:

- *Siting*. This detection involves anything that is out of place or in a location that it does not belong. It includes wrong foliage or items in an area that the sniper is occupying. Siting is dependent upon—
  - Mission.
  - Dispersion (more than one sniper team per objective).
  - Terrain patterns (rural, urban, wooded, barren).
- *Shape*. Military equipment and personnel have familiar outlines and specific shapes that are easily recognizable. A sniper must alter or disguise these revealing shapes and outlines. Geometric shapes are manmade.
- *Shadows*. If used correctly, shadows can be very effective in hiding a sniper's position. They can be found under most conditions of day and night. However, the sniper can cast a shadow that can give him away.
- *Silhouettes*. They can easily be seen in the daytime as well as at night. A sniper must break up the outline of his body and his equipment so it blends with the background to reduce the chance of his silhouette being recognized.

- *Surface.* Reflections of light on shiny surfaces can instantly attract attention and can be seen for great distances. The sniper must camouflage all objects that have a distinguishable surface, such as hats, gloves, and shirtsleeves. He must also consider the texture of the surface he is camouflaging.

- *Spacing.* This factor is normally more important when two or more sniper teams are deployed together. Teams should coordinate their locations so that one does not compromise another. Teams should also coordinate their movements so that only one team is moving near the objective at one time. Spacing is also a factor when dealing with one sniper team. A sniper team must consider the distance between team members when moving to, and when at, the objective or firing position. Team members may need to move forward into the firing position individually so as not to compromise the firing position. This movement will normally depend on the terrain and the enemy situation.

- *Color.* Changing seasons cause vegetation to change. A sniper must be aware of the color of vegetation so that he does not contrast with it. The sniper must never use points of color, as the eye will notice any movement in the color.

- *Movement.* The main reason a sniper's position is revealed to the enemy is due to movement. Even if all other indicators are absent, movement can give a sniper's position away. Rapid or jerky movement is very noticeable; while slow movement may be seen, it is not as noticeable nor will it attract the eye as readily. The sniper must also remember that animal and foliage movements can give him away.

4-20. **Effects of Terrain Patterns and Weather Conditions.** The sniper must consider the weather conditions throughout the mission because they can constantly change. He must also consider terrain patterns because the patterns at the objective may be quite different from the ones en route to and from the objective.

TYPES OF CAMOUFLAGE

4-21. The two types of camouflage that the sniper team can use to camouflage itself and its equipment are natural and artificial. Each type has specific effects that can help the sniper remain undetected.

4-22. Natural camouflage is vegetation or materials that are native to the given area. The sniper team should always augment its appearance by using some natural camouflage. Natural foliage, properly applied, is preferred to artificial material, but the sniper must be aware of wilting.

4-23. Artificial camouflage is any manmade material or substance that the sniper uses for coloring or covering something to conceal it. He can use camouflage sticks or face paints to cover all exposed areas of skin such as face, hands, and the back of the neck. He should darken the parts of the face that form shadows. The sniper team uses the following types of camouflage patterns:

- *Striping.* Used when in heavily wooded areas, and leafy vegetation is scarce.

FM 3-05.222

- *Blotching.* Used when an area is thick with leafy vegetation.
- *Combination.* Used when moving through changing terrain. It is normally the best all-round pattern.

MATERIALS

4-24. There are many types of camouflage materials. The sniper can use any of the following items to cover exposed skin:

- Artificial materials (or manufactured materials).
- Army-issued camouflage paint sticks:
    - Loam and light green—used for light-skinned personnel in all but snow regions.
    - Sand and light green—used for dark-skinned personnel in all but snow regions.
    - Loam and white—used for all personnel in snow-covered terrain.

NOTE: The use of camouflage in a cold weather environment will make detecting cold weather injuries more difficult.

- Commercial hunter's paint. There are many different colors.
- Stage makeup.
- Bear grease.
- Natural materials (or self-made materials):
    - Burnt cork.
    - Charcoal.
    - Lampblack (carbide).
    - Mud.

CAUTION

Dyes or paints should not be used, as they do not come off. Mud may contain dangerous parasites.

CLOTHING

4-25. The sniper can wear many types of clothing to conceal himself from the enemy. Battle dress uniforms (BDUs) have a camouflage pattern but often require additional camouflaging, especially in operations that occur very close to the enemy. The sniper can wear any of the following:

- U.S. Army uniforms:
    - Camouflage fatigues.
    - BDUs.
    - Desert BDUs.
    - Overwhites.
    - Desert night camouflage uniforms.

- Nonstandard uniforms with other camouflage patterns may help blend into the surrounding population.
- Gloves or mittens.
- Head masks:
    - Balaclavas.
    - Veils.
    - Head covers.
    - Kaffiyehs.
    - Ghillie or sniper hats.

GHILLIE SUIT

4-26. The term "ghillie suit" originated in Scotland during the 1800s. ("Ghillie" is a Scottish and Irish term for a fishing and hunting guide.) Scottish game wardens made special camouflage suits to catch poachers. Today the ghillie suit is a specially made camouflage uniform that is covered with irregular patterns of garnish or netting (Figure 4-1, page 4-8).

4-27. The sniper can make a ghillie suit from BDUs or one-piece aviator-type uniforms. Turning the uniform inside out places the pockets inside the suit and protects items in the pockets from damage caused by crawling on the ground. The sniper should cover the front of the ghillie suit with canvas or some type of heavy cloth to reinforce it. He should cover the knees and elbows with two layers of canvas, and reinforce the seam of the crotch with heavy nylon thread since these areas are prone to wear out more often. Shoo-goo is excellent for attaching the canvas to the uniform.

4-28. The next step is to make a garnish or net cover. The sniper should make sure the garnish or netting covers the shoulders and reaches down to the elbows on the sleeves. The garnish applied to the back of the suit should be long enough to cover the sides of the sniper when he is in the prone position. A bush hat is also covered with garnish or netting. The garnish should be long enough to break up the outline of the sniper's neck, but should not be so long in front to obscure his vision or hinder movement. A cut-up hammock makes an excellent foundation for the garnish.

4-29. A veil can be made from a net or pieces of cloth covered with garnish or netting. It covers the weapon and the sniper's head when he is in a firing position. The sniper can sew the veil into the ghillie suit or a boonie hat, or he can carry it separately. He must remember that a ghillie suit does not make him invisible but is only a camouflage base. The sniper can add natural vegetation to help blend with the surroundings, at a rate of 60 to 70 percent natural to 30 to 40 percent man-made.

NOTE: The ghillie suit is made to meet the sniper's need. However, he must take great care to ensure that he does not place an excessive amount of material on the netting. Doing so may form a new outline that can be seen by the enemy, or create a suit that will overheat him.

NOTE: It may be to the advantage of the sniper to use only a veil, as a full ghillie suit will be very bulky and difficult to pack and transport.

FM 3-05.222

Figure 4-1. Construction of the Ghillie Suit

> **CAUTION**
>
> **If using camouflage netting as a base, remove the radar scattering rings. Also remember the plastic camouflage shines when wet and the netting may catch on foliage when the sniper is crawling.**

CAMOUFLAGE FOR EQUIPMENT

4-30. The sniper must camouflage all the equipment that he will use. However, he must ensure that the camouflage does not interfere with or hinder the operation of the equipment. Equipment that the sniper should camouflage is as follows:

- *Rifles*. The SWS and the M4/M16/M203 should also be camouflaged to break up their outlines. The SWS can be carried in a "drag bag" (Figure 4-2, page 4-9), which is a rifle case made of canvas and covered with garnish similar to the ghillie suit. However, the rifle will not be combat ready while it is in the drag bag. The drag bag can become a liability in many circumstances.

NOTE: The sniper should use drag bags carefully because they grab and snag on foliage during movement, but are beneficial when climbing buildings.

- *Optics.* The sniper must also camouflage optics to break up the outline and to reduce the possibility of light reflecting off the lenses. He can cover the lenses with mesh-type webbing or nylon hose material. He can also use a cover cutout that changes the circular appearance of the optic's objective lens.
- *ALICE Packs.* If the sniper uses the ALICE pack while wearing the ghillie suit, he must camouflage the pack the same as the suit. He can use paints, dyes, netting, and garnish. However, the sniper should avoid wearing the ALICE pack with the ghillie suit.

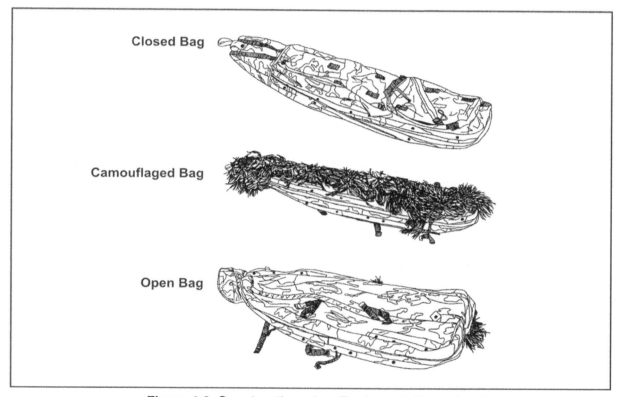

Figure 4-2. Construction of an Equipment "Drag Bag"

FACIAL CAMOUFLAGE PATTERNS

4-31. Facial patterns can vary from irregular stripes across the face to bold splotching. The best pattern, perhaps, is a combination of both strips and blotches. The sniper should avoid wild types of designs and colors that stand out from the background. He should cover all exposed skin, to include the—

- Hands and forearms.
- Neck, front and back.

- Ears, as well as behind the ears.
- Face:
    - Forehead–darkened.
    - Cheekbones–darkened.
    - Nose–darkened.
    - Chin–darkened.
    - Under eyes–lightened.
    - Under nose–lightened.
    - Under chin–lightened.

USING REMOVABLE CAMOUFLAGE SPRAY PAINT ON THE SWS AND EQUIPMENT

4-32. The sniper should paint his weapon with a removable paint (such as Bow Flage) so that he can change the colors to suit different vegetation and changing seasons. Bow Flage spray paint will not affect the accuracy or performance of the weapon. However, the sniper must take care when applying this paint. Bow Flage should not make contact with the lens of optical equipment, the bore of the weapon, the chamber, the face of the bolt, the trigger area, or the adjustment knobs of the telescope. It will not damage the weapon to be stored with the paint on it, but it is easily removed with Bow Flage remover or Shooter's Choice cleaning solvent.

FIELD-EXPEDIENT CAMOUFLAGE

4-33. The sniper may have to use field-expedient camouflage if other methods are not available. Instead of camouflage sticks or face paint, he may use charcoal, walnut stain, mud, or whatever works. He should not use oil or grease due to the strong odor. The sniper can attach natural vegetation to the body using boot bands or rubber bands, or by cutting holes in the uniform.

COVER AND CONCEALMENT

4-34. Properly understanding and applying the principles of cover and concealment, along with proper camouflage techniques, protects the sniper from enemy observation.

COVER

4-35. Cover is natural or artificial protection from the fire of enemy weapons. Natural (ravines, hollows, reverse slopes) and artificial (fighting positions, trenches, walls) cover protect the sniper from flat trajectory fires and partly protect him from high-angle fires and the effects of nuclear explosions. Even a 6-inch depression (if properly used) or fold in the ground may provide enough cover to save the sniper under fire. He must always look for and take advantage of all cover the terrain offers. By combining this habit with proper movement techniques, he can protect himself from enemy fire. To get protection from enemy fire when moving, the sniper should use routes that put cover between himself and the places where the enemy is known or thought to be. He should use natural and artificial cover to keep the enemy from seeing him and firing at him.

CONCEALMENT

4-36. Concealment is natural or artificial protection from enemy observation. The surroundings may provide natural concealment that needs no change before use (bushes, grass, and shadows). The sniper can create artificial concealment from materials such as burlap and camouflage nets, or he can move natural materials (bushes, leaves, and grass) from their original location. He must consider the effects of the change of seasons on the concealment provided by both natural and artificial materials.

4-37. The principles of concealment include the following:

- *Avoid Unnecessary Movement.* Remain still; movement attracts attention. The sniper's position may be concealed when he remains still, yet easily detected if he moves. This movement against a stationary background will make the sniper stand out. When he must change positions, he should move carefully over a concealed route to the new position, preferably during limited visibility. He should move inches at a time, slowly and cautiously, always scanning ahead for the next position.

- *Use All Available Concealment.* Background is important; the sniper must blend in to avoid detection. The trees, bushes, grass, earth, and man-made structures that form the background vary in color and appearance. This feature makes it possible for the sniper to blend in with them. The sniper should select trees or bushes to blend with the uniform and to absorb the figure outline. He must always assume that his area is under observation. The sniper in the open stands out clearly, but the sniper in the shadows is difficult to see. Shadows exist under most conditions, day and night. A sniper should never fire from the edge of a woodline; he should fire from a position inside the woodline (in the shade or shadows provided by the treetops).

- *Stay Low to Observe.* A low silhouette makes it difficult for the enemy to see a sniper. Therefore, he should observe from a crouch, a squat, or a prone position.

- *Expose Nothing That Shines.* Reflection of light on a shiny surface instantly attracts attention and can be seen from great distances. The sniper should uncover his rifle scope only when indexing and reducing a target. He should then use optics cautiously in bright sunshine because of the reflections they cause.

- *Avoid Skylining.* Figures on the skyline can be seen from a great distance, even at night, because a dark outline stands out against the lighter sky. The silhouette formed by the body makes a good target.

- *Alter Familiar Outlines.* Military equipment and the human body are familiar outlines to the enemy. The sniper should alter or disguise these revealing shapes by using a ghillie suit or outer smock that is covered with irregular patterns of garnish. He must alter his outline from his head to the soles of his boots.

- *Keep Quiet.* Noise, such as talking, can be picked up by enemy patrols or observation posts. The sniper should silence gear before a mission so that it makes no sound when he walks or runs.

INDIVIDUAL AND TEAM MOVEMENT

4-38. In many cases the success of a sniper's mission will depend upon his being able to close the range to his target, engage or observe the target, and withdraw without being detected. To succeed, he must be able to move silently through different types of terrain.

PREPARATION FOR MOVEMENT

4-39. As with any mission, the sniper must make preparations before moving. He must make a detailed study of large-scale maps and aerial photographs of the area, interview inhabitants and people who have been through the areas before, and review any other intelligence available about the area. He may construct sand tables of the area of operations (AO) to assist in forming and rehearsing the plan. The sniper must select camouflage to suit the area. He must also allow enough time for the selection of the proper camouflage, which should match the type of terrain the team will be moving through. Before moving, personnel should make sure that all shiny equipment is toned down and all gear is silenced. The sniper must ensure that only mission-essential gear is taken along.

Route Selection

4-40. A sniper should try to avoid known enemy positions and obstacles, open areas, and areas believed to be under enemy observation. He should select routes that make maximum use of cover and concealment and should never use trails. A sniper should try to take advantage of the more difficult terrain such as swamps or dense woods.

Movement

4-41. The sniper team cannot afford to be seen at any time by anyone. Therefore, its movement will be slow and deliberate. The movement over any given distance will be considerably slower than infantry units. Stealth is a sniper's security.

Rules of Movement

4-42. When moving, the sniper should always remember the following rules:

- Always assume that the area is under enemy observation.
- Move slowly; progress by feet and inches.
- Do not cause the overhead movement of trees, bushes, or tall grasses by rubbing against them.
- Plan every movement and traverse the route in segments.
- Stop, look, and listen often.
- Move during disturbances such as gunfire, explosions, aircraft noise, wind, or anything that will distract the enemy's attention or conceal the team's movement.

TYPES OF MOVEMENT

4-43. The sniper team will always move with caution. It will use various methods of walking and crawling based upon the enemy threat and the speed of movement required.

Walking

4-44. Walking is the fastest, easiest, and most useful way to move when extreme silence is desired. It is used when threat is low and speed is important. The sniper walks in a crouch to maintain a low profile with shadows and bushes so as not to be silhouetted. To ensure solid footing, he keeps his weight on one foot as he raises the other, being sure to clear all brush. He then gently sets the moving foot down, toes first, and then the heel. He takes short steps to maintain balance and carries the weapon in-line with the body by grasping the forward sling swivel (muzzle pointed down). At night, he holds the weapon close to his body to free his other hand to feel for obstacles. The sniper should use this walking technique when near the enemy; otherwise, he would use the standard patrol walk.

Hands and Knees Crawl

4-45. The sniper uses this self-explanatory crawl when cover is adequate or silence is necessary (Figure 4-3). The sniper holds the rifle in one hand close to the chest and in-line with the body, or places it on the ground alongside the body. The weight of the upper body is supported by the opposite arm. While supporting the rifle in one hand, the sniper picks a point ahead to position the opposite hand and slowly and quietly moves the hand into position. When moving the hand into position, the sniper can support the weight of his upper body on the opposite elbow. The sniper then alternately moves his hands forward, being careful not to make any noise. Leaves, twigs, and pebbles can be moved out of the way with the hand if absolute silence is required.

Figure 4-3. Hands and Knees Crawl

High Crawl

4-46. When cover is more prevalent or when speed is required, the sniper uses this movement (Figure 4-4, page 4-14). The body is kept free of the ground and the weight rests on the forearms and the lower legs (shins). The rifle can either be carried, as in the low crawl, or cradled in the arms. Movement is made by alternately pulling with each arm and pushing with one leg. The sniper can alternate legs for pushing when cover is adequate. An alternate method is to pull with both arms and push with one leg. The sniper should **always** keep in mind that the head and buttocks cannot be raised too high and the legs must not be allowed to make excessive noise when being dragged over brush and debris. Both heels must remain in contact with the ground. This is the standard Army high crawl.

Figure 4-4. High Crawl

Medium Crawl

4-47. The medium crawl allows the sniper to move in fairly low cover because it is faster and less tiring to the body (Figure 4-5, page 4-15). This movement is similar to the low crawl, except that one leg is cocked forward to push with. One leg is used until tired, then the other leg is used. However, the sniper must not alternate legs, as this causes the lower portion of the body to rise into the air. This is the standard Army low crawl and is conducted in the same manner.

Low Crawl

4-48. The sniper uses the low crawl when an enemy is near, when vegetation is sparse, or when moving in or out of position to fire or to observe (Figure 4-6, page 4-15). To low crawl, he lies face down on the ground, legs together, feet flat on the ground, and arms to the front and flat on the ground. To carry the rifle, he grasps the upper portion of the sling and lays the stock on the back of his hand or wrist, with the rifle lying on the inside of his body under one arm. He can push the rifle forward as he moves. However, care must be taken to ensure that the muzzle does not protrude into the air or stick into the ground. To move forward, the sniper extends his arms and pulls with his arms while pushing with his toes, being careful not to raise his heels or head. This movement is extremely slow and requires practice to keep from using quick or jerky movements. The head is maintained down one side of the face.

Figure 4-5. Medium Crawl

Figure 4-6. Low Crawl

Turning While Crawling

4-49. It may be necessary to change direction or turn completely around while crawling. To execute a right turn, the sniper moves his upper body as far to the right as possible and then moves his left leg to the left as far as possible. He then closes the right leg to the left leg. This turn will create a pivot-type movement (Figure 4-7, page 4-16). Left turns are done in the opposite fashion.

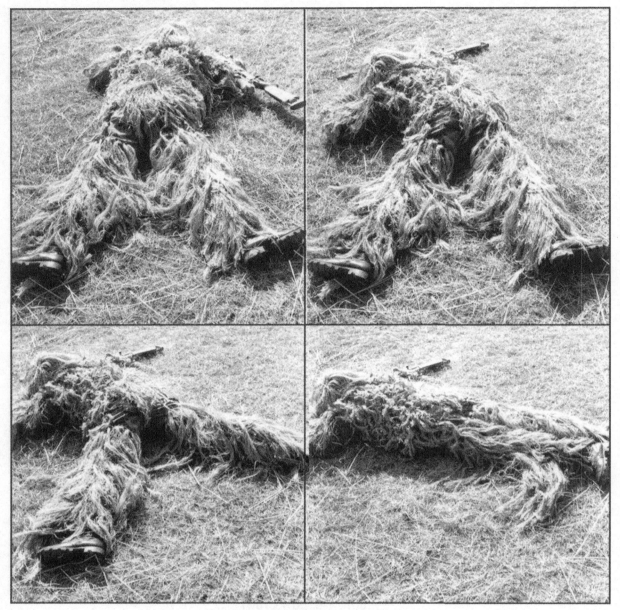

Figure 4-7. Turning While Crawling

Backward Movement

4-50. The sniper moves backward by reversing the crawling movement.

Assuming the Prone Position

4-51. The sniper assumes the prone position from a walk by stopping, tucking his rifle under his arm, and crouching slowly. Simultaneously, he feels the ground with the free hand for a clear spot. He then lowers his knees, one at a time, to the ground. He shifts his weight to one knee and lifts and extends the free leg to the rear. The sniper uses his toes to feel for a clear spot. Rolling onto that side, he then lowers the rest of his body into position.

Night Movement

4-52. Movement at night is basically the same as during the day except it is slower and more deliberate because of the limited visibility. The sniper has to rely on the senses of touch and hearing to a greater extent. If at all possible, the sniper should move under the cover of darkness, fog, haze, rain, or high winds to conceal his movement. This is a safety factor; however, it makes the enemy harder to spot and specific positions or landmarks harder to locate.

STALKING

4-53. Stalking is the sniper's art of moving unseen into a firing position within a range that will ensure a first-round kill and then withdrawing undetected. Stalking incorporates all aspects of fieldcraft and can only be effectively learned by repeated practice over various types of ground.

Reconnaissance

4-54. The sniper should conduct a complete reconnaissance before his mission. Seldom will he have an opportunity to view the ground. He must rely on maps and aerial photographs for his information. The sniper should address the following before stalking:

- Location, position, or target to be stalked.
- Cover and concealment.
- Best possible firing position to engage targets.
- Best line of advance to stalk.
- Obstacles, whether natural or artificial.
- Observation points along the route.
- Known or suspected enemy locations.
- Method of movement throughout the mission.
- Withdrawal route (to include method of movement).

Conduct of the Stalk

4-55. A sniper may lose his sense of direction while stalking, particularly if he has to crawl a great distance. Losing direction can be reduced if the sniper—

- Uses a compass, map, and aerial photograph, and thoroughly and accurately plans the route, direction, and distance to various checkpoints.
- Memorizes a distinct landmark or two, or even a series.
- Notes the direction of the wind and sun. However, he must bear in mind that over a long period of time the wind direction can change and the sun will change position.
- Has the ability to use terrain association.

4-56. The sniper must be alert at all times. Any relaxation on a stalk can lead to carelessness, resulting in an unsuccessful mission and even death. He should also conduct an observation at periodic intervals. If the sniper is surprised or exposed during the stalk, immediate reaction is necessary. The sniper must decide whether to freeze or move quickly to the nearest cover and hide.

4-57. Disturbed animals or birds can draw attention to the area of approach. If animals are alarmed, the sniper should stop, wait, and listen. Their flight may indicate someone's approach or call attention to his position. However, advantage should be taken of any local disturbances or distractions that could enable him to move more quickly than would otherwise be possible. It should be emphasized that such movement includes a degree of risk, and when the enemy is close, risks should be avoided.

4-58. While halted, the sniper identifies his next position and the position after that position. If he is moving through tall grass, he should occasionally make a slight change of direction to keep the grass from waving in an unnatural motion. If crossing roads or trails, he should look for a low spot or cross on the leading edge of a curve and always avoid cleared areas, steep slopes, and loose rocks. The sniper should never skyline himself. He should also be aware of any changes in local cover, since such changes will usually require an alteration to his personal camouflage.

4-59. During route selection, the sniper must always plan one or two points ahead of his next point. Doing so prevents the sniper from crawling into a dead-end position.

Night Stalking

4-60. A sniper is less adapted to stalking at night than during the day. He must use slower, more deliberate movement to occupy an observation post or a firing position. The principal differences between day and night stalking are that at night—

- There is a degree of protection offered by the darkness against aimed enemy fire. However, a false sense of security may compromise the sniper.
- The sniper should use NVDs to aid in movement.
- While observation is still important, much more use is made of hearing, making silence vital.
- Cover is less important than background. The sniper should particularly avoid crests and skylines against which he may be silhouetted. He should use lunar shadows to hide in to help defeat NVDs.
- Maintaining direction is much more difficult to achieve, which places greater emphasis on a thorough reconnaissance. A compass or knowledge of the stars may help.

Silent Movement Techniques

4-61. Stealthful movement is critical to a sniper. Survival and mission success require the sniper to learn the skills of memorizing the ground and the surrounding terrain, applying silent and stealth movement, moving over different terrain, and using various noise obstacles. The sniper must memorize the terrain, select a route, move, communicate using touch signals, and avoid or negotiate obstacles using stealth techniques. The sniper can accomplish his mission by—

- Using binoculars to observe the terrain to the front, simultaneously selecting a route of advance and memorizing the terrain.
- Specifying signals with his team partner for different obstacles. Considerations include—
  - Finding the obstacles.
  - Identifying the obstacles (barbed wire, explosives, mines).
  - Negotiating the obstacles (Should the team go around, over, or under the obstacles?).
  - Clearing the obstacles (or getting caught in the obstacle).
  - Signaling partner (a signal must be relayed to the sniper's partner).
- Using stealth and silent movement techniques. They include—
  - Cautious and deliberate movement.
  - Frequent halts to listen and observe.
  - No unnecessary movement.
  - Silent movement. All equipment is taped and padded.
  - Looking where the next move is going to be made.
  - Clearing foliage or debris from the next position.
  - Constant awareness of the natural habitat of birds and animals in your area.
- Obtaining a safe passage of obstacles. Factors include—
  - Avoiding or bypassing noise obstacles.
  - If noise obstacles must be moved through, checking the debris and clearing loose noise obstacles from the path.
  - Memorizing locations of obstacles for night movement.
- Using the basic elements of walking stealthily. They are—
  - Maintaining balance.
  - Shifting weight gradually from the rear foot to the front foot.
  - Moving the rear foot to the front, taking care to clear brush. The moving foot may be placed either heel first, toe first, edge of foot first, or flat on the ground.
- Knowing how to move through rubble and debris. The sniper must—
  - Test the debris with his hand.
  - Remove debris that will break.
  - Put his feet down flat-footed. This way will reduce noise.
- Avoiding movement through mud and muck. If it cannot be avoided, the boots should be wrapped with burlap rags or socks.

- Crossing in the sand. Movement is noiseless and can be fairly fast.
- Keeping a low silhouette when moving over an obstacle. Trying not to brush or scrape against the obstacle, he should lower himself silently on the other side and move away at a medium-slow pace.
- Always maintaining positive control of his weapon.
- Never pulling or tugging at snagged equipment to free it; he should untangle or cut it free.

Detection Devices

4-62. The sniper must be constantly vigilant in his movements and acts to defeat enemy detection. He should be able to use the following devices:

- *Passive and Active Light Intensification Devices.* The sniper must be aware of enemy detection devices and remember that he could unknowingly be under observation. Where there is the possibility that NVDs are being used, the sniper can combat them by moving very slowly and staying very low to the ground. This way his dark silhouette will be broken up by vegetation. Preferably, he will move in dark shadows or tree lines that will obscure the enemy's vision. Also, moving in defilade through ground haze, fog, or rain will greatly benefit the sniper by helping him to remain undetected. Using the new IR reflecting material (used in equipment netting) as a base for the ghillie suit will limit the enemy's IR viewing capabilities. This should be used with caution, and the sniper must experiment with the correct balance.
- *Sensors.* Sensors are remote monitoring devices with seismic sensors, magnetic sensors, motion sensors, IR sensors, or thermal sensors planted in the ground along likely avenues of advance or perimeters. These devices normally vary in sensitivity. They are triggered by vibration of the ground, metal, movement, breaking a beam of light, or heat within their area of influence. The sniper can move past these devices undetected only by using the slowest, most careful, and errorless movement. He can help combat the effects of seismic devices by moving when other actions that will activate the devices, such as artillery fire, low-flying aircraft, rain, snow, or even a heavy wind, are in progress or, in some instances, moving without rhythm. The sniper can defeat most other sensors if he knows their limitations and capabilities.
- *Ground Surveillance Radars.* Ground surveillance radars can detect troop or vehicle movement at an extended range, but only along its line of sight and only if the object is moving at a given speed or faster. It takes a well-trained individual to properly monitor the device. A sniper can combat the use of ground surveillance radars by moving in defilade, out of the direct line of sight of the equipment, or slower than the radar can detect. He should move extremely slow and low to the ground, using natural objects and vegetation to mask the movement. The more laterally to the radar the sniper moves, the easier it is for the radar to detect the sniper's movement.
- *Thermal Imagers.* Thermal imagers are infrared heat detectors that locate body heat. The difference between heat sources is what is

registered. These devices could locate even a motionless and camouflaged sniper. One possible way to confuse such a detector would be to attach a space blanket (Mylar) to the inside of the camouflage suit. The blanket would reflect the body heat inward and could possibly keep the sniper from being distinguished from the heat pattern of the surrounding terrain. This method would work best when the temperature is warm and the greatest amount of radiant heat is rising from the ground. Active infrared spotlights and metascopes may be used against the sniper. The sniper must always avoid the IR light or he will be detected.

> **CAUTION**
>
> **By trapping the body heat and not allowing it to dissipate, the sniper increases the chance of becoming a heat casualty.**

Selecting Lines of Advance

4-63. Part of the sniper's mission will be to analyze the terrain, select a good route to the target, use obstacles (man-made and natural) and terrain to their best advantage, and determine the best method of movement to arrive at his target. Once at the target site, he must be able to select firing positions and plan a stalk.

4-64. On the ground, the sniper looks for a route that will provide the best cover and concealment. He should fully use low ground, dead space, and shadows and avoid open areas. He looks for a route that will provide easy movement, yet will allow quiet movement at night. The sniper selects the route, then chooses the movement techniques that will allow undetected movement over that specific terrain.

4-65. Position selection is also critical to mission success. The sniper should not select a position that looks obvious and ideal; it will appear that way to the enemy. He should select a position away from prominent terrain features of contrasting background. When possible, he selects an area that has an obstacle (natural or man-made) between him and the target.

4-66. Stalk planning involves map and ground reconnaissance, selection of a route to the objective, selection of the type of movement, notation of known or suspected enemy locations, and selection of a route of withdrawal. Sniper teams must not be detected or even suspected by the enemy. To maintain efficiency, each sniper must master *individual* movement techniques and ensure team effort is kept at the highest possible level.

Sniper Team Movement and Navigation

4-67. Normally, the sniper carries the SWS, the observer carries an M4/M16/M203, and both have sidearms. Due to the number of personnel and firepower, the sniper team cannot afford to be detected by the enemy nor can it successfully meet the enemy in sustained engagements. Another technique is for the sniper to carry the M24 bagged and on his back, while carrying an M4 at the ready. This gives the team greater firepower.

4-68. When possible, the sniper team should have a security element (squad/platoon) attached. The security element allows the team to reach its area of operations quicker and safer. Plus, it provides the team a reaction force should the team be detected.

4-69. Snipers use the following guidelines when attaching a security element:

- The security element leader is in charge of the team while it is attached.
- Sniper teams always appear as an integral part of the element.
- Sniper teams wear the same uniform as the element members.
- Sniper teams maintain proper intervals and positions in all formations.
- The SWS is carried in-line and close to the body, hiding its outline and barrel length, or it is bagged and the shooter carries an M4.
- All equipment that is unique to sniper teams is concealed from view (optics, ghillie suits).
- Once in the area of operations, the sniper team separates from the security element and operates alone.

4-70. Two examples of sniper teams separating from security elements follow:

- The security element provides security while the team prepares for its operation. The team—
  - Dons the ghillie suits and camouflages itself and its equipment (if mission requires).
  - Ensures that all equipment is secure and caches any nonessential equipment (if mission requires).
  - Once it is prepared, assumes a concealed position, and the security element departs the area.
  - Once the security element has departed, waits in position long enough to ensure neither it nor the security element have been compromised. The team then moves to its tentative position.
- The security element conducts a short security halt at the separation point. The snipers halt, ensuring they have good available concealment and know each other's location. The security element then proceeds, leaving the sniper team in place. The sniper team remains in position until the security element is clear of the area. The team then organizes itself as required by the mission and moves on to its tentative position. This type of separation also works well in military operations in urban terrain (MOUT) situations.

4-71. When selecting routes, the sniper team must remember its strengths and weaknesses. The following guidelines should be used when selecting routes:

- Avoid known enemy positions and obstacles.
- Seek terrain that offers the best cover and concealment.

- Take advantage of difficult terrain (swamps, dense woods).
- Avoid natural lines of drift.
- Do not use trails, roads, or footpaths.
- Avoid built-up or populated areas.
- Avoid areas of heavy enemy guerrilla activity.
- Avoid areas between opposing forces in contact with each other.

4-72. When the sniper team moves, it must always assume its area is under enemy observation. Because of this threat and the small amount of firepower that the team has, it can use only one type of formation—the sniper movement formation. Characteristics are as follows:

- The observer is the point man; the sniper follows.
- The observer's sector of security is 9 o'clock to 3 o'clock; the sniper's sector of security is 3 o'clock to 9 o'clock (overlapping each other).
- Team members maintain visual contact, even when lying on the ground.
- Team members maintain an interval of no more than 2 meters.
- The sniper reacts to the point man's actions.
- Team leader designates the movement techniques and routes used.
- Team leader designates rally points.
- During the stalk, team moves by using individual bounding techniques. It can move by successive bounds or alternating bounds.
- Team crosses linear danger areas by moving together across the danger area after a security or listing halt.

Sniper Team Immediate Action Drills

4-73. A sniper team must never become decisively engaged with the enemy. It must rehearse immediate action drills so they become a natural and immediate reaction should it make unexpected contact with the enemy. Examples of such actions are as follows:

- *Visual Contact*. If the sniper team sees the enemy and the enemy does not see the team, it freezes. If the team has time, it will do the following:
    - Assume the best covered and concealed position.
    - Remain in position until the enemy has passed.

NOTE: The team will not initiate contact.

- *Ambush*. The sniper team's objective is to break contact immediately during an ambush. One example of this technique involves performing the following:
    - The observer delivers rapid fire on the enemy and the team immediately moves out of the area.
    - The team moves to a location where the enemy cannot observe or place direct fire on it.
    - If contact cannot be broken, the sniper calls for indirect fire or security element (if attached).

- If team members get separated, they should either link up at the objective rally point (ORP) or move to the next designated rally point. This move will depend upon the team SOP.
- *Indirect Fire.* Indirect fire can cause the team to move out of the area as quickly as possible and may result in its exact location and direction being pinpointed. Therefore, the team must not only react to indirect fire but also take the following actions to conceal its movement once it is out of the impact area:
  - The team leader moves the team out of the impact area using the quickest route by giving the direction and distance (clock method).
  - Both members move out of the impact area the designated distance and direction.
  - The team leader then moves the team farther away from the impact area by using the most direct concealed route. They continue the mission using an alternate route.
  - If the team members get separated, they should either link up at the ORP or move next designated rally point.
- *Air Attack.* If the sniper team finds itself caught in an air attack or its position is about to be destroyed, it should react as follows:
  - Assume the best available covered and concealed positions.
  - Between passes of aircraft, move to a position that offers better cover and concealment.
  - Do not engage the aircraft.
  - Remain in position until the attacking aircraft departs.
  - Link up at the ORP or move to the next designated rally point if the members get separated.

Navigational Aids

4-74. To aid the sniper team in navigation, it should memorize the route by studying maps, aerial photos, or sketches. The team notes distinctive features (hills, streams, and roads) and its location in relation to the route. It plans an alternate route in case the primary route cannot be used. It plans an offset to circumvent known obstacles to movement. The team uses terrain countdown, which involves memorizing terrain features from the start to the objective, to maintain the route. During the mission, the sniper team mentally counts each terrain feature, thus ensuring it maintains the proper route. The team designates all en route rally points along the routes.

4-75. The sniper team maintains orientation at all times. As it moves, it observes the terrain carefully and mentally checks off the distinctive features noted in the planning and study of the route. The team must be aware of the map terrain interval to prevent counting low terrain features not represented on a map.

4-76. The following aids are available to ensure orientation:
- Global positioning system (GPS).
- The location and direction of flow of principal streams.

- Hills, valleys, roads, and other peculiar terrain features.
- Railroad tracks, power lines, and other man-made objects.

TRACKING AND COUNTERTRACKING

4-77. Tracking is the art of being able to follow a person or an animal by the signs that they leave during their movement. It is nearly impossible to move cross-country and not leave signs of one's passage. These signs, no matter how small, can be detected by a trained and experienced tracker. However, a person who is trained in tracking techniques can use deception drills that can minimize telltale signs and throw off or confuse trackers who are not well trained or who do not have the experience to spot the signs of a deception.

4-78. As a tracker follows a trail, he builds a picture of the enemy in his mind by asking himself these questions: How many persons am I following? What is their state of training? How are they equipped? Are they healthy? What is their state of morale? Do they know they are being followed? To answer these questions, the tracker uses available indicators—that is, signs that tell an action occurred at a specific time and place (Figure 4-8). By comparing indicators, the tracker obtains answers to his questions.

NOTE: Throughout this section, the terms tracker and sniper are used interchangeably.

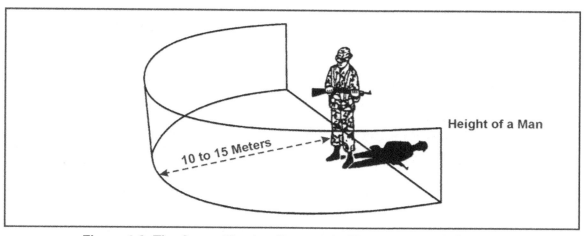

Figure 4-8. The Area a Tracker Surveys to Find Tracking Indicators

TRACKING SIGNS

4-79. Signs are visible marks left by individuals or animals as they pass through an area. The sniper must know the following categories of signs:
- *Ground Signs*. These are signs left below the knees. All ground signs are further divided as follows:
    - *Large signs* are caused by the movement of ten or more individuals through the area.
    - *Small signs* are caused by the movement of one to nine individuals through the area.

- *High Signs (also known as top signs).* These are signs left above the knees. They are also divided into large and small top signs.
- *Temporary Signs.* These signs will eventually fade with time (for example, a footprint).
- *Permanent Signs.* These signs require weeks to fade or will leave a mark forever (for example, broken branches or chipped bark).

TRACKING INDICATORS

4-80. Any sign the tracker discovers can be defined by one of six tracking indicators. They include displacement, stains, weathering, litter, camouflage, and immediate-use intelligence.

Displacement

4-81. Displacement takes place when anything is moved from its original position. A well-defined footprint in soft, moist ground is a good example of displacement. The footgear or bare feet of the person who left the print displaced the soil by compression, leaving an indentation in the ground. The tracker can study this sign and determine several important facts. For example, a print left by worn footgear or by bare feet may indicate lack of proper equipment. Displacement can also result from clearing a trail by breaking or cutting through heavy vegetation with a machete; these trails are obvious to the most inexperienced tracker. Individuals may unconsciously break more branches as they move behind someone who is cutting the path. Displacement indicators can also be made by persons carrying heavy loads who stop to rest; prints made by box edges can help to identify the load. When loads are set down at a rest halt or campsite, they usually crush grass and twigs. A reclining man can also flatten the vegetation.

4-82. **Analyzing Footprints.** Footprints can indicate direction, rate of movement, number, sex, and whether the individual knows he is being tracked. Figures 4-9 through 4-12 show different appearances of tracks made during various activities and countertracking techniques. The footprint can be the whole print but is usually only the "heel dig" and "toe push" footprint. They may also be found on the underside of large leaves that have not dried out and are lying on the ground.

4-83. If footprints are deep and the pace is long, rapid movement is apparent. Extremely long strides and deep prints with toe prints deeper than heel prints indicate running (Figure 4-9).

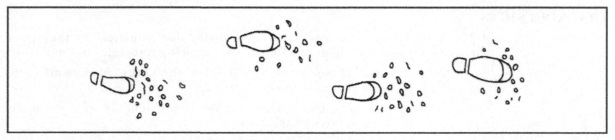

Figure 4-9. Running

4-84. Prints that are deep, have a short stride, are narrowly spaced, and show signs of shuffling indicate the person who left the print is carrying a heavy load (Figure 4-10).

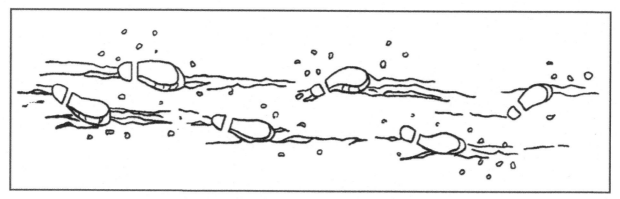

Figure 4-10. Carrying a Heavy Load

4-85. If the party members realize they are being followed, they may try to hide their tracks. Persons walking backward have a short, irregular stride (Figure 4-11). The prints have an unnaturally deep toe, and soil is displaced in the direction of movement. These types of prints are characterized by "toe digs" and "heel push" as opposed to the normal footprint.

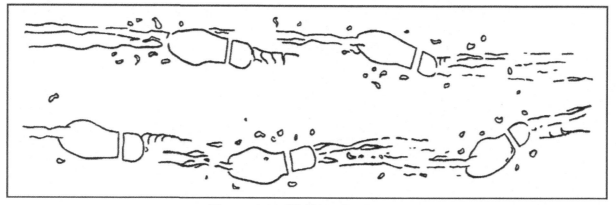

Figure 4-11. Walking Backward

4-86. To determine the sex of a member of the party being followed, the tracker should study the size and position of the footprints (Figure 4-12, page 4-28). Women tend to be pigeon-toed; men walk with their feet straight ahead or pointed slightly to the outside. Prints left by women are usually smaller and the stride is usually shorter than that taken by men.

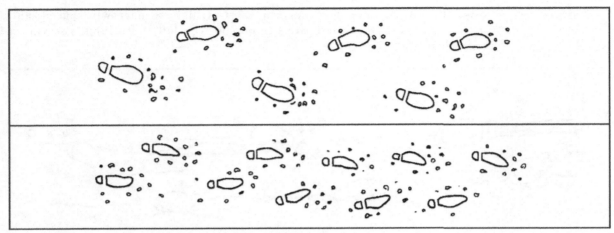

Figure 4-12. Man Versus Woman

4-87. **Determining Key Prints.** Normally, the last man in the file leaves the clearest footprints; these should be the key prints. The tracker cuts a stick to match the length of the prints and notches it to show the length and widest part of the sole. He can then study the angle of the key prints in relation to the direction of march. He looks for an identifying mark or feature, such as worn or frayed footgear, to identify the key prints. If the trail becomes vague, erased, or merges with another, the tracker can employ his stick-measuring device and identify the key prints with close study. This method helps him to stay on the trail. By using the box method, he can count up to 18 persons. The tracker can—

- Use the stride as a unit of measure when determining key prints (Figure 4-13). He uses these prints and the edges of the road or trail to box in an area to analyze.

- Also use the 36-inch box method if key prints are not evident (Figure 4-14, page 4-29). To use this method, the tracker uses the edges of the road or trail as the sides of the box. He measures a cross section of the area 36 inches long, counting each indentation in the box and dividing by two. This method gives a close estimate of the number of individuals who made the prints; however, this system is not as accurate as the stride measurement.

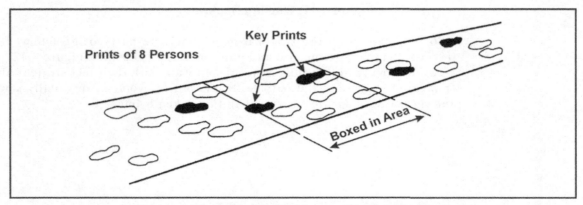

Figure 4-13. Using the Stride as a Unit of Measure

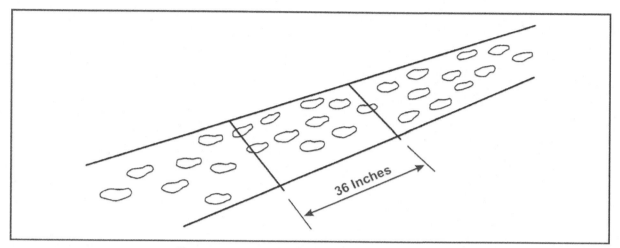

Figure 4-14. Using the 36-Inch Box Method

4-88. **Recognizing Other Signs of Displacement.** Foliage, moss, vines, sticks, or rocks that are scuffed or snapped from their original position form valuable indicators. Broken dirt seals around rocks, mud or dirt moved to rocks or other natural debris, and water moved onto the banks of a stream are also good indicators (Figure 4-15). Vines may be dragged, dew droplets displaced, or stones and sticks overturned to show a different color underneath. Grass or other vegetation may be bent or broken in the direction of movement (Figure 4-16).

Figure 4-15. Turned Over Rocks and Sticks

Figure 4-16. Crushed or Disturbed Vegetation

4-89. The tracker inspects all areas for bits of clothing, threads, or dirt from torn footgear or can fall and be left on thorns, snags, or the ground.

4-90. Flushed from their natural habitat, wild animals and birds are another example of displacement. Cries of birds excited by unnatural movement are an indicator; moving tops of tall grass or brush on a windless day indicate that something is moving the vegetation.

4-91. Changes in the normal life of insects and spiders may indicate that someone has recently passed. Valuable clues are disturbed bees, ant holes covered by someone moving over them, or torn spider webs. Spiders often spin webs across open areas, trails, or roads to trap flying insects. If the tracked person does not avoid these webs, he leaves an indicator to an observant tracker.

4-92. If the person being followed tries to use a stream to cover his trail, the tracker can still follow successfully. Algae and other water plants can be displaced by lost footing or by careless walking. Rocks can be displaced from their original position or overturned to indicate a lighter or darker color on the opposite side. The person entering or exiting a stream creates slide marks or footprints, or scuffs the bark on roots or sticks (Figure 4-17). Normally, a person or animal seeks the path of least resistance; therefore, when searching the stream for an indication of departures, trackers will find signs in open areas along the banks.

Figure 4-17. Slip Marks and Waterfilled Footprints on Stream Banks

Stains

4-93. A stain occurs when any substance from one organism or article is smeared or deposited on something else. The best example of staining is blood from a profusely bleeding wound. Bloodstains often appear as spatters or drops and are not always on the ground; they also appear smeared on leaves or twigs of trees and bushes. The tracker can determine the seriousness of the wound and how far the wounded person can move unassisted. This process may lead the tracker to enemy bodies or indicate where they have been carried.

4-94. By studying bloodstains, the tracker can determine the wound's location as follows:

- If the blood seems to be dripping steadily, it probably came from a wound on the trunk.

- If the blood appears to be slung toward the front, rear, or sides, the wound is probably in the extremity.
- Arterial wounds appear to pour blood at regular intervals as if poured from a pitcher. If the wound is veinous, the blood pours steadily.
- A lung wound deposits pink, bubbly, and frothy bloodstains.
- A bloodstain from a head wound appears heavy, wet, and slimy.
- Abdominal wounds often mix blood with digestive juices so the deposit has an odor and is light in color.

4-95. Any body fluids such as urine or feces deposited on the ground, trees, bushes, or rocks will leave a stain.

4-96. On a calm, clear day, leaves of bushes and small trees are generally turned so that the dark top side shows. However, when a man passes through an area and disturbs the leaves, he will generally cause the lighter side of the leaf to show. This movement is also true with some varieties of grass. Moving causes an unnatural discoloration of the area, which is called "shine." Grass or leaves that have been stepped on will have a bruise on the lighter side.

4-97. Staining can also occur when muddy footgear is dragged over grass, stones, and shrubs. Thus, staining and displacement combine to indicate movement and direction. Crushed leaves may stain rocky ground that is too hard to show footprints. Roots, stones, and vines may be stained where leaves or berries are crushed by moving feet.

4-98. The tracker may have difficulty determining the difference between staining and displacement since both terms can be applied to some indicators. For example, muddied water may indicate recent movement; displaced mud also stains the water. Muddy footgear can stain stones in streams, and algae can be displaced from stones in streams and can stain other stones or the bank. Muddy water collects in new footprints in swampy ground; however, the mud settles and the water clears with time. The tracker can use this information to indicate time. Normally, the mud clears in about one hour, although time varies with the terrain. Since muddied water travels with the current, it is usually best to move downstream.

Weathering

4-99. Weathering either aids or hinders the tracker. It also affects indicators in certain ways so that the tracker can determine their relative ages. However, wind, snow, rain, or sunlight can erase indicators entirely and hinder the tracker. The tracker should know how weathering affects soil, vegetation, and other indicators in his area. He cannot properly determine the age of indicators until he understands the effects that weathering has on trail signs.

4-100. For example, when bloodstains are fresh, they are bright red. Air and sunlight first change blood to a deep ruby-red color, then to a dark brown crust when the moisture evaporates. Scuff marks on trees or bushes darken with time. Sap oozes on trees and then hardens when it makes contact with the air.

4-101. Weather greatly affects footprints (Figure 4-18). By carefully studying this weathering process, the tracker can estimate the age of the print. If particles of soil are just beginning to fall into the print, this is a sign that the print is very recent. At this point, the tracker should then focus on becoming a stalker. If the edges of the print are dried and crusty, the prints are probably about 1 hour old. This process varies with terrain and is only a guide.

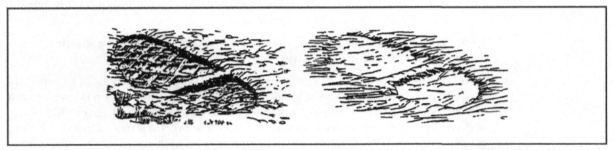

Figure 4-18. Effects of Weather on the Clarity of Footprints

4-102. A light rain may round the edges of the print. By remembering when the last rain occurred, the tracker can place the print into a time frame. A heavy rain may erase all signs.

4-103. Trails exiting streams may appear weathered by rain due to water running from clothing or equipment into the tracks. This trait is especially true if the party exits the stream single file. Then, each person deposits water into the tracks. The existence of a wet, weathered trail slowly fading into a dry trail indicates the trail is fresh.

4-104. Wind dries out tracks and blows litter, sticks, or leaves into prints. By recalling wind activity, the tracker may estimate the age of the tracks. For example, the tracker may reason "the wind is calm at the present but blew hard about an hour ago. These tracks have litter blown into them, so they must be over an hour old." However, he must be sure that the litter was blown into the prints and not crushed into them when the prints were made.

4-105. Wind affects sound and odors. If the wind is blowing down the trail (toward the tracker), sounds and odors may be carried to him; conversely, if the wind is blowing up the trail (away from the tracker), he must be extremely cautious since wind also carries sounds toward the enemy. The tracker can determine wind direction by dropping a handful of dust or dried grass from shoulder height. By pointing in the same direction the wind is blowing, the tracker can localize sounds by cupping his hands behind his ears and turning slowly. When sounds are loudest, the tracker is facing the origin.

4-106. In calm weather (no wind), air currents that may be too light to detect can carry sounds to the tracker. Air cools in the evening and moves downhill toward the valleys. If the tracker is moving uphill late in the day or night, air currents will probably be moving toward him if no other wind is blowing. As the morning sun warms the air in the valleys, it moves uphill. The tracker considers these factors when plotting patrol routes or other operations. If he keeps the wind in his face, sounds and odors will be carried to him from his objective or from the party being tracked.

4-107. The tracker should also consider the sun. It is difficult to fire directly into the sun, but if the tracker has the sun at his back and the wind in his face, he has a slight advantage.

Litter

4-108. Litter consists of anything not indigenous to the area that is left on the ground. A poorly trained or poorly disciplined unit moving over terrain is apt to leave a trail of litter. Unmistakable signs of recent movement are gum or candy wrappers, ration cans, cigarette butts, remains of fires, urine, and bloody bandages. Rain flattens or washes litter away and turns paper into pulp. Exposure to weather can cause ration cans to rust at the opened edge; then, the rust moves toward the center. The tracker must consider weather conditions when estimating the age of litter. He can use the last rain or strong wind as the basis for a time frame.

4-109. The sniper should also know the wildlife in the area. Even sumps, regardless of how well camouflaged, are a potential source of litter. The best policy you can follow is to take out with you everything you bring in.

Camouflage

4-110. Camouflage applies to tracking when the followed party uses techniques to baffle or slow the tracker—that is, walking backward to leave confusing prints, brushing out trails, and moving over rocky ground or through streams. Camouflaged movement indicates a trained adversary.

Immediate-Use Intelligence

4-111. The tracker combines all indicators and interprets what he has seen to form a composite picture for on-the-spot intelligence. For example, indicators may show contact is imminent and require extreme stealth.

4-112. The tracker avoids reporting his interpretations as facts. He reports what he has seen, rather than stating these things exist. There are many ways a tracker can interpret the sex and size of the party, the load, and the type of equipment. Time frames can be determined by weathering effects on indicators.

4-113. Immediate-use intelligence is information about the enemy that can be used to gain surprise, to keep him off balance, or to keep him from escaping the area entirely. The commander may have many sources of intelligence such as reports, documents, or prisoners of war. These sources can be combined to form indicators of the enemy's last location, future plans, and destination.

4-114. However, tracking gives the commander definite information on which to act immediately. For example, a unit may report there are no men of military age in a village. This information is of value only if it is combined with other information to make a composite enemy picture in the area. Therefore, a tracker who interprets trail signs and reports that he is 30 minutes behind a known enemy unit, moving north, and located at a specific location, gives the commander information on which he can act.

FM 3-05.222

DOG-TRACKER TEAMS

4-115. The three types of tracker dogs are as follows:
- *Visual dogs* rely upon their acute vision. They usually are the final part of tracking before shifting over to the attack mode.
- *Search dogs* are allowed to run free and search using airborne scents.
- *Tracker dogs* run on leashes and use ground scents.

4-116. Many myths surround the abilities and limitations of canine trackers. The first and perhaps greatest myth is that tracking involves only the dog's sense of smell. Canine tracking involves a team—a merging of man and dog. Dogs use both their eyes and ears; the tracker uses his eyes and knowledge of the quarry. Together, they create an effective team that maximizes their strengths and minimizes their weaknesses. The sniper team is not only trying to evade and outwit "just" a dog but also the dog's handler. The most common breed of dog used is the German shepherd. These dogs are trained to respond independently to a variety of situations and threats. Good tracking dogs are a rare and difficult-to-replace asset.

4-117. A visual tracker assists the dog handlers in finding a track if the dog loses the trail. He can radio ahead to another tracker and give him an oral account of the track picture. A visual tracker is slower than dogs because he must always use his powers of observation, which creates fatigue. His effectiveness is limited at night.

4-118. Tracker dogs smell microbes in the earth that are released from disturbed soil. The trail has no innate smell of a specific quarry, although trails do vary depending on the size and number of the quarry. For example, a scent is like the wake a ship leaves in the ocean, but no part of the ship is left in the wake. It is the white, foamy, disturbed water that is the trail. The result is entirely different from a point smell of the quarry such as sweat, urine, or cigarette smoke. The same training that makes tracking dogs adept at tracking a scent trail applies to finding a point smell.

4-119. Smelling is a highly complex process and many variables affect it. The most important element in tracking is the actual ground such as earth and grass. It contains living microbes that are always disturbed by the quarry's passage. Artificial surfaces (concrete and macadam) and mainly inorganic surfaces (stone) provide little or no living microbes to form a scent track.

4-120. A search or a scent-discrimination dog builds a scent picture of the person that he is tracking. Scent may be short-lived and its life span is dependent upon the weather and the area that the person last passed through. The sun and the wind, as well as time, destroy the scent. There are both airborne and ground scents. Airborne scents can be blown away within minutes or a few hours. Ground scents can last longer than 48 hours under ideal conditions. Bloodhounds have been known to successfully track a scent that was left behind 7 days before.

4-121. Wind and moisture are other major variables that affect tracking. Foggy and drizzly weather that keeps the ground moist is best. Too much rain can wash a trail away; depending on the strength of the trail, it takes persistent, hard rain to erase a scent trail. Usually, the scent is not washed away but only sealed beneath a layer of ground water. A short, violent rainfall could deposit enough water to seal the scent track, but after the rain stops and the water layer evaporates, the microbe trail would again be detectable by dogs. Hard, dry ground releases the fewest microbes and is the most difficult terrain for dogs to track on. A dog may also have difficulty following a trail on a beach or dusty path, but his human tracker could easily follow the footprints visually. Snipers must always remember they are being tracked by a man and a dog team. Tracker dogs track on the tail of the sniper while search dogs track downwind of the trail.

4-122. Wind strength and direction are important factors in tracking. Basically, strong wind inhibits tracking a scent trail but makes it easier for a dog to find a point scent source—like a hide. A general rule is that a dog can smell a man-size source downwind out to 50 meters and a group-size source—a hide—out to 200 meters under ideal conditions. Upwind, a source 1 meter away could be missed.

Wind Direction →

Wind Speed:	Still	Windy
	D------X----------D----------------------D	
Distance:	1 Meter 30 to 50 Meters	Maximum 150 to 200 Meters

D = Dog Team
X = Sniper/Sniper Team

4-123. A strong wind disperses microbes that arise from the ground, hindering a dog's ability to follow a trail. However, a strong wind increases the size of a point scent, helping a dog to find the target in an area search.

4-124. An inflexible rule for the life of a scent trail cannot be provided. In Germany, trackers rate their chance of following a trail that is more than 3 days old as negligible. Terrain, weather, and the sensitivity of the tracking dog are some of the many variables that affect the scent trail. A point smell will last as long as the target emits odors.

4-125. While dogs are mainly scent hunters, they also have good short-range vision. Dogs are colorblind and do not have good distance vision (camouflage works extremely well against dogs). However, they can detect slight movements. Dogs also have a phenomenal sense of hearing, extending far beyond human norms in both the frequency range and in sensitivity. Dogs use smell to approximate a target, and then rely on sound and movement to pinpoint that target.

4-126. Although dogs have tremendous detection abilities, they also have limitations. Following a scent trail is the most difficult task a tracking dog can perform. The level of effort is so intense that most dogs cannot work longer than 20 to 30 minutes at a time, followed by a 10- to 20-minute rest. Dogs can perform this cycle no more than five or six times in a 24-hour period before reaching complete exhaustion. The efficiency of the search also decreases as the dog tires. In wartime, the situation will force the maximum from men and equipment, but times should remain constant for dogs because they always give 100 percent. If the snipers keep moving and stay out of the detection range of the human handlers, then they could outlast the dog-scent trackers.

4-127. When looking for sniper teams, trackers mainly use wood line sweeps and area searches. A wood line sweep consists of walking the dog **upwind** of a suspected wood line or brush line—the key is upwind. If the wind is blowing through the woods and out of the wood line, trackers move 50 to 100 meters inside a wooded area to sweep the woods' edge. Since wood line sweeps tend to be less specific, trackers perform them faster. Trackers perform an area search when a team's location is specific, such as a small wooded area or block of houses. The search area is cordoned off, if possible and the dog-tracker teams are brought on-line about 25 to 150 meters apart, depending on terrain and visibility. The handlers then advance, each moving their dogs through a specific corridor. The handler controls the dog entirely with voice commands and gestures. He remains undercover, directing the dog in a search pattern or to a likely target area. The search line moves forward with each dog dashing back and forth in assigned sectors.

TECHNIQUES TO DEFEAT DOG-TRACKER TEAMS

4-128. Although dog and handler tracking teams are a potent threat, there are counters available to the sniper team. As always, the best defenses are basic infantry techniques: good camouflage and light, noise, and trash discipline. Dogs find a team either by detecting a trail or by a point source such as human waste odors at the hide site. It is critical to try to obscure or limit trails around the hide, especially along the wood line or area closest to the team's target area. Surveillance targets are usually major axes of advance. "Trolling the wood lines" along likely-looking roads or intersections is a favorite tactic of dog-tracker teams. When moving into a target area, the sniper team should take the following countermeasures:

- Remain as far away from the target area as the situation allows.
- Never establish a position at the edge of cover and concealment nearest the target area.
- Minimize the track. Try to approach the position area on hard, dry ground or along a stream or river.
- Urinate in a hole and cover it up. Never urinate more than once in exactly the same spot.
- Deeply bury fecal matter. If the duration of the mission permits, use meals, ready to eat (MRE) bags sealed with tape and take it with you.
- Never smoke.
- Carry all trash until it can be buried elsewhere.

4-129. When dogs are being used against a sniper team, they use other odors left behind or around the team to find it. Sweat from exertion or fear is one of these. Wet clothing or material from damp environments holds in the scent. Soap or deodorant used before infiltration helps the dogs to find the team. Foreign odors such as oils, preservatives, polish, and petroleum products also aid the dogs. Time permitting, the sniper should try to change his diet to that of the local inhabitants before infiltration.

4-130. When the sniper team first arrives in its AO, it is best to move initially in a direction that is from 90 to 170 degrees away from the objective. Objects or items of clothing not belonging to any of the team members should be carried into the AO in a plastic bag. When the team first starts moving, it should drop an item of clothing or piece of cloth out of the bag and leave it on a back trail. This step can confuse a dog long enough to give the team more of a head start. Also, if dogs are brought in late, the team's scent will be very faint while this scent will still be strong.

4-131. While traveling, the team should try to avoid heavily foliaged areas, as these areas hold the scent longer. Periodically, when the situation permits, move across an open area that the sun shines on during the day and that has the potential of being windswept. The wind moves the scent and will eventually blow it away; the sun destroys scent very rapidly.

4-132. When the situation permits, make changes in direction at the open points of terrain to force the dog to cast for a scent. If dogs are very close behind, moving through water does not confuse them, as scent will be hanging in the air above the water. Moving through water will only slow the team down. Throwing CS gas to the rear or using blood, spice mixtures, or any other concoctions will prevent a dog from smelling the team's scent, but it will not be effective on a trained tracker dog.

4-133. While a dog will not be confused by water if he is close, running water, such as a rapidly moving stream, will confuse a dog if he is several hours behind. However, areas with foliage, stagnant air, and little sunlight will hold the scent longer. Therefore, the team should try to avoid any swampy areas.

4-134. The sniper team should move through areas that have been frequently traveled by other people, as this will confuse the scent picture to the dog. Team members should split up from time to time to confuse the dogs. The best place for this is in areas frequently traveled by indigenous personnel.

4-135. If a dog-tracker is on the sniper team's trail, it should not run because the scent will become stronger. The team may attempt to wear out the dog handler and confuse the dog but should always be on the lookout for a good ambush site that it can fishhook into. If it becomes necessary to ambush the tracking party, fishhook into the ambush site and kill or wound the handler, **not** the dog. A tracker dog is trained with his handler and will protect him should he become wounded. This practice will allow the team to move off and away from the area while the rest of the tracking party tries to give assistance to the handler. Also, that dog will not work well with anyone other than his handler.

4-136. If a dog search team moves into the area, the sniper team should first check wind direction and strength. If the team is downwind of the estimated search area, the chances are minimal that the team's point smells will be detected. If upwind of the search area, the team should attempt to move downwind. Terrain and visibility dictate whether the team can move without being detected visually by the handlers. Remember, sweeps are not always conducted just outside of a wood line. Wind direction determines whether the sweep will be parallel to the outside or 50 to 100 meters inside the wood line.

4-137. The team has options if caught inside the search area of a line search. The handlers rely on radio communications and often do not have visual contact with each other. If the team has been generally localized through enemy radio detection-finding equipment, the search net will still be loose during the initial sweep. A sniper team has a small chance of hiding and escaping detection in deep brush or in woodpiles. Larger groups will almost certainly be found. Yet, the team may have the chance to eliminate the handler and to escape the search net.

4-138. The handler hides behind cover with the dog. He searches for movement and then sends the dog out in a straight line toward the front. Usually, when the dog has moved about 50 to 75 meters, the handler calls the dog back. The handler then moves slowly forward and always from covered position to covered position. Commands are by voice and gesture with a backup whistle to signal the dog to return. If a handler is killed or badly injured after he has released the dog, but before he has recalled it, the dog continues to randomly search out and away from the handler. The dog usually returns to another handler or to his former handler's last position within several minutes. This time lapse creates a gap from 25 to 150 meters wide in the search pattern. Response times by the other searchers tend to be fast. Given the high degree of radio "chatter," the injured handler will probably be quickly missed from the radio net. Killing the dog before the handler will probably delay discovery only by moments. Dogs are so reliable that if the dog does not return immediately, the handler knows something is wrong.

4-139. If the sniper does not have a firearm, human versus dog combat is a hazard. One dog can be dealt with relatively easily if a knife or large club is available. The sniper must keep low and strike upward using the wrist, never overhand. Dogs are quick and will try to strike the groin or legs. Most attack dogs are trained to go for the groin or throat. If alone and faced with two or more dogs, the sniper should flee the situation.

4-140. Dog-tracker teams are a potent threat to the sniper team. Although small and lightly armed, they can greatly increase the area that a rear area security unit can search. Due to the dog-tracker team's effectiveness and its lack of firepower, a sniper team may be tempted to destroy such an "easy" target. Whether a team should fight or run depends on the situation and the team leader. Eliminating or injuring the dog-tracker team only confirms to threat security forces that there is a hostile team operating in the area. The techniques for attacking a dog-tracker team should be used only in extreme situations or as a last measure.

COUNTERTRACKING

4-141. There are two types of human trackers—combat trackers and professional trackers. Combat trackers look ahead for signs and do not necessarily look for each individual sign. Professional trackers go from sign to sign. If they cannot find any sign, they will stop and search till they find one. The only way to lose a trained professional tracker is to fishhook into an area and then ambush him.

4-142. If an enemy tracker finds tracks of two men, it tells him that a highly trained specialty team may be operating in his area. However, a knowledge of countertracking enables the sniper team to survive by remaining undetected.

4-143. As with the dogs, to confuse the combat tracker and throw him off the track, the sniper always starts his movement away from his objective. He travels in a straight line for about an hour and then changes direction. Changing will cause the tracker to cast in different directions to find the track.

Evasion

4-144. Evasion of the tracker or pursuit team is a difficult task that requires the use of immediate-action drills mostly designed to counter the threat. A team skilled in tracking techniques can successfully use deception drills to minimize signs that the enemy can use against them. However, it is very difficult for a person, especially a group, to move across any area without leaving signs noticeable to the trained eye.

Camouflage

4-145. The followed party may use two types of routes to cover its movement. It must also remember that travel time reduces when trying to camouflage the trail. Two types of routes include:

- *Most-Used Routes.* Movement on lightly-traveled sandy or soft trails is easily tracked. However, a person may try to confuse the tracker by moving on hard-surfaced, often-traveled roads or by merging with civilians. These routes should be carefully examined. If a well-defined approach leads to the enemy, it will probably be mined, ambushed, or covered by snipers.
- *Least-Used Routes.* These routes avoid all man-made trails or roads and confuse the tracker. They are normally magnetic azimuths between two points. However, the tracker can use the proper concepts to follow the party if he is experienced and persistent.

Reduction of Trail Signs

4-146. A sniper who tries to hide his trail moves at reduced speed; therefore, the experienced tracker gains time. A sniper should use the following methods to reduce trail signs:

- Wrap footgear with rags or wear soft-soled sneakers that make footprints rounded and less distinctive.
- Change into footgear with a different tread immediately following a deceptive maneuver.
- Walk on hard or rocky ground.

Deception Techniques

4-147. Evading a skilled and persistent enemy tracker requires skillfully executed maneuvers to deceive the tracker and cause him to lose the trail. An enemy tracker cannot be outrun by a sniper team that is carrying equipment, because he travels light and is escorted by enemy forces designed for pursuit. The size of the pursuing force dictates the sniper team's chances of success in using ambush-type maneuvers. Sniper teams use some of the following techniques in immediate-action drills and deception drills.

4-148. **Backward Walking.** One of the most basic techniques is walking backward (Figure 4-19) in tracks already made, and then stepping off the trail onto terrain or objects that leave little to no signs. Skillful use of this maneuver causes the tracker to look in the wrong direction once he has lost the trail. This must be used in conjunction with another deception technique. This technique will probably fail if a professional tracker is on your trail.

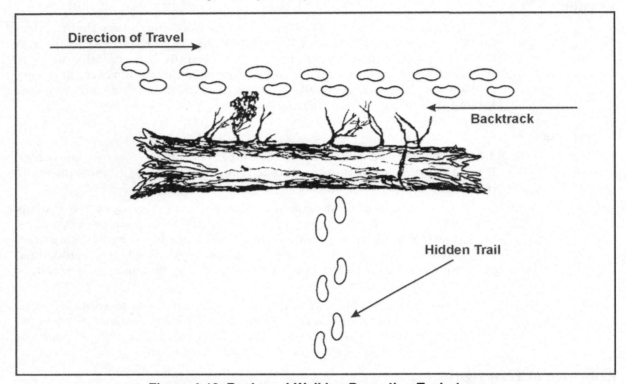

Figure 4-19. Backward-Walking Deception Technique

4-149. **Big Tree.** A good deception tactic is to change directions at large trees (Figure 4-20, page 4-41). To change, the sniper moves in any given direction and walks past a large tree (12 inches wide or larger) from 5 to 10 paces. He carefully walks backward to the forward side of the tree and makes a 90-degree change in the direction of travel, passing the tree on its forward side. This technique uses the tree as a screen to hide the new trail from the pursuing tracker. A variation used near a clear area would be for the sniper to pass by the side of the tree that he wishes to change direction to on his next leg. He walks past the tree into a clear area for 75 to 100 meters and then walks backwards to the tree. At this time he moves 90 degrees and passes on the side away from the tracker. This method could cause the tracker to follow his sign into the open area where, when he loses the track, he might cast in the wrong direction for the track. This technique works only on combat trackers and not professional trackers.

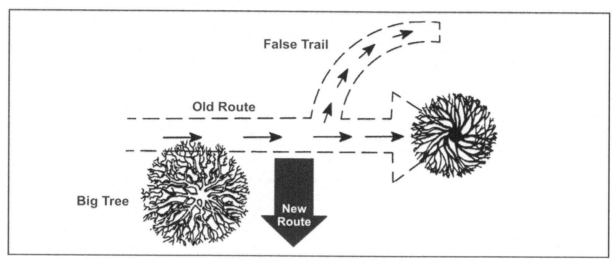

Figure 4-20. Big Tree Deception Technique

NOTE: By studying signs, an observant tracker can determine if an attempt is being made to confuse him. If the sniper team tries to lose the tracker by walking backward, footprints will be deepened at the toe and soil will be scuffed or dragged in the direction of movement. By following carefully, the tracker can normally find a turnaround point.

4-150. **Cut the Corner.** The sniper team uses this deception method when approaching a known road or trail. About 100 meters from the road, the team changes its direction of movement, either 45 degrees left or right. Once the road is reached, the team leaves a visible trail in the same direction of the deception for a short distance down the road. The tracker should believe that the team "cut the corner" to save time. The team backtracks on the trail to the point where it entered the road and then carefully moves down the road without leaving a good trail. Once the desired distance is achieved, the team changes direction and continues movement (Figure 4-21). A combination using the big tree method here would improve the effectiveness of this deception.

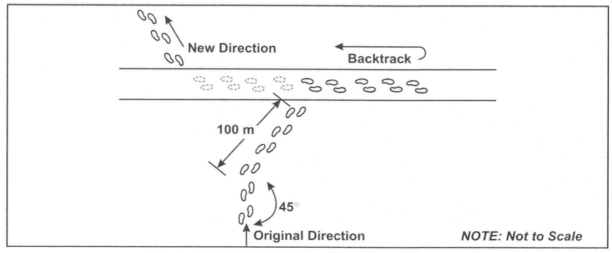

Figure 4-21. Cut-the-Corner Deception Technique

4-151. **Slip the Stream.** The sniper team uses this deception when approaching a known stream. It executes this method the same as the cut-the-corner maneuver. The team establishes the 45-degree deception maneuver upstream, then enters the stream. The team moves upstream and establishes false trails if time permits. By moving upstream, floating debris and silt will flow downstream and cover the true direction and exit point. The team then moves downstream to escape since creeks and streams gain tributaries that offer more escape alternatives (Figure 4-22). False exit points can also be used to further confuse. However, the sniper must be careful not to cause a false exit to give away his intended travel direction.

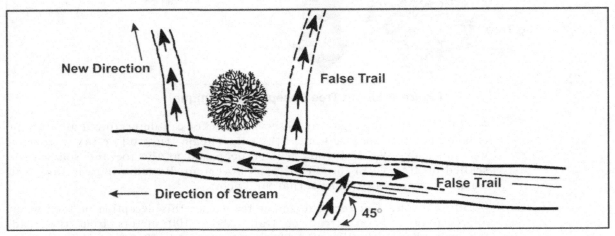

Figure 4-22. Slip-the-Stream Deception Technique

4-152. **Arctic Circle.** The team uses this deception in snow-covered terrain to escape pursuers or to hide a patrol base. It establishes a trail in a circle as large as possible (Figure 4-23). The trail that starts on a road and returns to the same start point is effective. At some point along the circular trail, the team removes snowshoes (if used) and carefully steps off the trail, leaving one set of tracks. The large tree maneuver can be used to screen the trail. From the hide position, the team returns over the same steps and carefully fills them with snow one at a time. This technique is especially effective if it is snowing.

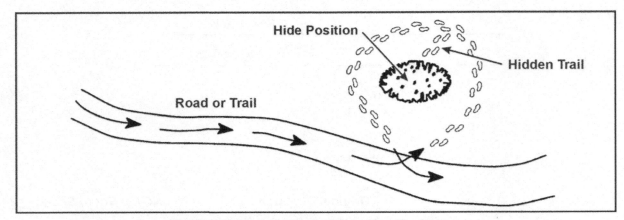

Figure 4-23. Arctic Circle Deception Technique

4-153. **Fishhook.** The team uses this technique to double back on its own trail in an overwatch position (Figure 4-24). It can observe the back trail for trackers or ambush pursuers. If the pursuing force is too large to be destroyed, the team strives to eliminate the tracker. It uses hit-and-run tactics, then moves to another ambush position. The terrain must be used to advantage.

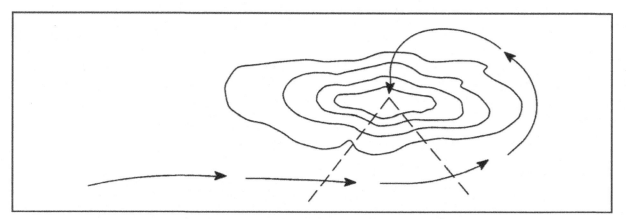

Figure 4-24. The Fishhook Deception Technique

4-154. Dog and visual trackers are not infallible; they can be confused with simple techniques and clear thinking. The sniper should not panic and try to outrun a dog or visual tracker. It only makes it easier for the tracking party. The successful sniper keeps his head and always plans two steps ahead. Even if trackers are not in the area, it is best to always use countertracking techniques.

NOTE: Snipers must always **remember** that there is no way to hide a trail from a professional tracker!

OBSERVATION AND TARGET DETECTION

4-155. The sniper's mission requires that he deliver precision fire to selected targets. He cannot meet this requirement without first observing and detecting the target. During this process, the sniper team is concerned with the significance of the target rather than the number of targets. The sniper team will record the location identification of all targets observed and then fire at them in a descending order of importance.

USE OF TARGET INDICATORS

4-156. As discussed in the camouflage and concealment section, the sniper team must protect itself from target indicators that could reveal its presence to the enemy. It can also use these target indicators to locate the enemy by using the planned and systematic process of observation. The first consideration is toward the discovery of any immediate danger to the sniper team. The team begins with a **hasty search** of the entire area and follows up with a slow, deliberate observation called a **detailed search**. As long as the sniper team remains in position, it will maintain constant observation of the area using the hasty and detailed search methods as the situation requires.

Hasty Search

4-157. This process is the first phase of observing a target area. The observer conducts a hasty search (about 10 seconds) for any enemy activity immediately after the team occupies the firing position. The search is carried out by making quick glances at specific points, terrain features, or other areas that could conceal the enemy. The sniper should not sweep his eyes across the terrain in one continuous movement; it will prevent him from detecting motion. The observer views the area closest to the team's position first since it could pose the most immediate threat. The observer then searches farther out until the entire target area has been searched. The hasty search is effective because the eyes are sensitive to the slightest movement occurring within a wide arc of the object. This spot is called "side vision" or "seeing out of the corner of the eye." The eye must be focused on a specific point to have this sensitivity. When the observer sees or suspects a target, he uses the binoculars or the observation telescope for a detailed view of the suspected target area.

Detailed Search

4-158. After the hasty search, the designated observer starts a detailed search using the overlapping strip method (Figure 4-25). Normally, the area nearest the team offers the greatest danger, therefore, the search should begin there. The detailed search begins at either flank. The observer systematically searches the terrain to his front in a 180-degree arc, 50 meters in depth. After reaching the opposite flank, the observer searches the next area nearest his post. The search should be in overlapping strips of at least 10 meters to ensure total coverage of the area. It should cover as far out as the observer can see, always including areas of interest that attracted the observer during the hasty search.

Figure 4-25. Overlapping Strip Method

4-159. The observer must memorize as much of the area as possible. He should make mental notes of prominent terrain features and other areas that may offer cover and concealment for the enemy. This way, he becomes familiar with the terrain as he searches. These become his key points of interest for his hasty searches.

4-160. This cycle of a hasty search followed by a detailed search should be repeated every 15 to 20 minutes depending upon the terrain and area of responsibility. Repetition allows the sniper team to become accustomed to the area and to look closer at various points with each consecutive pass over the area. After the initial searches, the observer should view the area using a combination of both hasty and detailed searches. While the observer conducts the initial searches of the area, the sniper should record prominent features, reference points, and distances on a range card.

MAINTAINING OBSERVATION

4-161. The team members should alternate the task of observing the area about every 30 minutes. When maintaining observation, the observer keeps movement of his head and body to a minimum. He should not expose his head any higher than is necessary to see the area being observed. After completing his detailed search, the observer maintains observation of the area by using a method similar to the hasty search. He glances quickly at various points throughout the entire area and focuses his eyes on specific features that he had designated during his detailed search.

4-162. While maintaining observation, the observer should devise a set sequence for searching to ensure coverage of all terrain. Since it is entirely possible that his hasty search may fail to detect the enemy, he should periodically repeat a detailed search.

WHY OBJECTS ARE SEEN

4-163. The relative ease or difficulty in seeing objects depends upon several factors. The observer may determine objects by—

- *Shape.* Some objects can be recognized instantly by their shape, particularly if it contrasts with the background. Experience teaches people to associate an object with its shape or outline. At a distance, the outline of objects can be seen well before the details can be determined. The human body and the equipment that a soldier carries are easily identified unless the outline has been altered. Areas of importance when considering shape during observation are—
  - The clear-cut outline of a soldier or his equipment, either partially or fully exposed.
  - Man-made objects, which have geometric shapes.
  - Geometric shapes, which do not occur in nature on a large scale.
- *Shadow.* In sunlight, an object or a man will cast a shadow that can give away his presence. Shadows may be more revealing than the object itself. Care must be taken to detect alterations of the natural shape of a shadow. Where light is excessively bright, shadows will look especially black. Contrast will be extreme, and in this exaggerated contrast the observer's eye cannot adjust to both areas simultaneously.

This requires the observer to "isolate" the shadowed area from the bright sunlight so that his eye can adapt to the shadow.

- *Silhouette.* Any object silhouetted against a contrasting background is conspicuous. Any smooth, flat background, such as water, a field, or best of all, the sky, will cause an object to become well delineated. However, special care must be taken when searching areas with an uneven background, as it is more difficult to detect the silhouette of an object.
- *Surface.* If an object has a surface that contrasts with its surroundings, it becomes conspicuous. An object with a smooth surface reflects light and becomes more obvious than an object with a rough surface that casts shadows on itself. An extremely smooth object becomes shiny. The reflections from a belt buckle, watch, or optical device can be seen over a mile away from the source. Any shine will attract the observer's attention.
- *Spacing.* Nature never places objects in a regular, equally spaced pattern. Only man uses rows and equal spacing.
- *Siting.* Anything that does not belong in the immediate surroundings are obvious and become readily detectable. This evidence should arouse the observer's curiosity and cause him to investigate the area more thoroughly.
- *Color.* The greater the contrasting color, the more visible the object becomes. This point is especially true when the color is not natural for that area. Color alone will usually not identify the object but is often an aid in locating it.
- *Movement.* This final reason why things are seen will seldom reveal the identity of an object, but it is the most common reason an enemy's position is revealed. Even when all other indicators are absent, movement will give a position away. A stationary object may be impossible to see and a slow-moving object difficult to detect, but a quick or jerky movement will be seen.

ELEMENTS OF OBSERVATION

4-164. Four elements in the process of observation include awareness, understanding, recording, and response. Each of these elements may be construed as a separate process or as occurring at the same time.

Awareness

4-165. Awareness is being consciously attuned to a specific fact. A sniper team must always be aware of the surroundings and take nothing for granted. The team should consider the following points because they may influence and distort awareness:

- An object's size and shape can be misinterpreted if viewed incompletely or inaccurately.
- Distractions can occur during observation.
- Active participation or degree of interest can diminish toward the event.
- Physical abilities (five senses) can be limited.

- Environmental changes can affect or occur at the time of observation.
- Imagination or perception can cause possible exaggerations or inaccuracies when reporting or recalling facts.

Understanding

4-166. Understanding is derived from education, training, practice, and experience. It enhances the sniper team's knowledge about what should be observed, broadens its ability to view and consider all factors, and aids in its evaluation of the information.

Recording

4-167. Recording is the ability to save and recall what was observed. Usually, the sniper team has mechanical aids such as writing utensils, logbooks, sketch kits, tape recordings, and cameras to support the recording of events. However, the most accessible method is memory. The ability to record, retain, and recall depends on the team's mental capacity (and alertness) and ability to recognize what is essential to record. Added factors that affect recording include:

- The amount of training and practice in observation.
- Skill through experience.
- Similarity of previous incidents.
- Time interval between observing and recording.
- The ability to understand or convey messages through oral or other communication.
- Preconceived perception of the event as to what or it occurred and who was involved.

Response

4-168. Response is the sniper team's action toward information. It may be as simple as recording events in a logbook, making a communications call, or firing a well-aimed shot.

TARGET INDICATION AT UNKNOWN DISTANCES

4-169. Snipers usually deploy in pairs and can recognize and direct each other to targets quickly and efficiently. To recognize targets quickly, the sniper uses standard methods of indication, with slight variations to meet his individual needs.

4-170. The three methods of indicating targets are the direct method, the reference-point method, and the clock-ray method. It is easier to recognize a target if the area of ground in which it is likely to appear is known. Such an area of ground is called an "arc of fire." An arc of fire is indicated in the following sequence:

- The axis (the middle of the arc).
- The left and right limits of the arc.
- Reference points (prominent objects). These should be as permanent as possible (woods, mounds), a reasonable distance apart, and easy to identify. A specific point of the object is nominated and given a name

and range (mound–bottom left corner; to be known as mound–range 400) the same as on your range card.

Direct Method

4-171. The sniper uses this method to indicate obvious targets. The range, where to look, and a description of the target are given. Terms used for where to look include the following:

- Axis of arc—for targets on or very near the axis.
- Left or right—for targets 90 degrees from the axis.
- Slightly, quarter, half, or three-quarters and left or right—for targets between the axis and the left or right limits.

Reference-Point Method

4-172. To indicate less obvious targets, the sniper may use a reference point together with the direct method, and perhaps the words above and below as well. For example:

- 300-mound (reference point—slightly right—small bush [target]).
- 200-mound (reference point—slightly right and below—gate [target]).

Clock-Ray Method

4-173. To indicate less-obvious targets, a reference-point target with a clock ray may be used. To use this method, it is imagined that there is a clock face standing up on the landscape with its center on the reference point. To indicate a target, the range, the reference point and whether the target is to the left or to the right of it, and the approximate hour on the clock face are given. For example: 300-mound—right—4 o'clock—small bush.

4-174. When indicating targets, the following points must be considered:

- *Range*. Its main purpose is to give an indication of how far to look but it should also be as accurate as possible. The sniper sets the range given to him by his observer as indicated by his shooter's data book.
- *Detailed Indication*. This value may require more detail than a normal indication; nevertheless, it should still be as brief and as clear as possible.

4-175. The sniper can use mil measurements along with the methods of indication to specify the distance between an object and the reference point used (for example, mound—reference point; go left 50 mils, lone tree, base of tree—target). The mil scale in binoculars can assist in accurate indication, although occasionally the use of hand angles will have to suffice. It is important that each sniper is conversant with the angles subtended by the various parts of his hand when the arm is outstretched.

4-176. Sniper teams must always be aware of the difficulties that can be caused when the observer and the sniper are observing through instruments with different magnifications and fields of view (telescope, binoculars). If time and concealment allow it, the observer and the sniper should use the same viewing instrument, particularly if the mil scale in the binoculars is being used to give accurate measurements from a reference point.

4-177. It is necessary that both the observer and the sniper know exactly what the other is doing and what he is saying when locating the target. Any method that is understandable to both and is fast to use is acceptable. They must use short and concise words to locate the target. Each must always be aware of what the other is doing so that the sniper does not shoot before the observer is ready. An example of this dialogue would be:

- Observer: "60—HALF RIGHT, BARN, RIGHT 50 MILS, 2 O'CLOCK, LARGE ROCK, BOTTOM LEFT CORNER, TARGET."
- Sniper: "TARGET IDENTIFIED, READY" (or describe back to observer the target).
- Sniper: "TARGET IS 2 MILS TALL; 1 MIL WIDE."
- Observer: "SET ELEVATION AT 5+1, WINDAGE 0, PARALLAX 2D BALL."
- Sniper: (repeats directions upon setting scope) "READY."
- Observer: "HOLD OF RIGHT" (wind correction). The sniper should have a round downrange within 1 to 2 seconds after the wind call.

4-178. It is extremely important that the sniper fires as soon as possible after the wind call to preclude any wind change that could affect the impact of his bullet. If the wind does change, then the observer stops the firing sequence and gives new wind readings to the sniper. The sniper and the observer must not be afraid to talk to each other, but they should keep everything said as short and concise as possible.

INDEXING TARGETS

4-179. The sniper must have some system for remembering or indexing target locations. He may want to fire at the highest priority target first. He must be selective, patient, and not fire at a target just to have a kill. Indiscriminate firing may alert more valuable and closer targets. Engagement of a distant target may result in disclosure of the sniper post to a closer enemy.

4-180. Since several targets may be sighted at the same time, the observer needs some system to remember all of the locations. To remember, he uses aiming points and reference points and records this information on the sector sketch or range card and observer's log.

4-181. To index targets, the sniper team uses the prepared range card for a reference since it can greatly reduce the engagement time. When indexing a target to the sniper, the observer locates a prominent terrain feature near the target. He indicates this feature and any other information to the sniper to assist in finding the target. Information between team members varies with the situation. The observer may sound like a forward observer (FO) giving a call for fire to a fire direction center (FDC), depending on the condition of the battlefield and the total number of possible targets from which to choose.

4-182. The sniper team must also consider the following factors:

- *Exposure Time.* Moving targets may expose themselves for only a short time. The sniper team must be alert to note the points of disappearance of as many targets as possible before engaging any one of them. By doing so, the sniper team may be able to take several targets under fire in rapid succession.

- *Number of Targets.* When the sniper team is unable to remember and plot all target locations, it should concentrate only on the most important target. By concentrating only on the most important targets, the team will effectively locate and engage high-priority targets or those targets that represent the greatest threat.
- *Spacing.* The greater the space interval between targets, the more difficult it is to note their movements. In such cases, the sniper team should accurately locate and engage the nearest target.
- *Aiming Points.* Targets that disappear behind good aiming points are easily recorded and remembered. Targets with poor aiming points are easily lost. If two such targets are of equal value and a threat to the team, the poor aiming point target should be engaged first, until the target with a good aiming point becomes a greater threat.

TARGET SELECTION

4-183. Snipers select targets according to their value. Certain enemy personnel and equipment can be listed as key targets, but their real worth must be decided by the sniper team in relation to the circumstances in which they are located.

4-184. As stated in the discussion of recording targets, the sniper team may have no choice of targets. It may lose a rapidly moving target if it waits to identify target details. It must also consider any enemy threatening its position as an "extremely high-value" target. When forced to choose a target, the sniper team will consider the following factors:

- *Certainty of Target's Identity.* The sniper team must be reasonably certain that the target it is considering is the key target.
- *Target Effect on the Enemy.* The sniper team must consider what effect the elimination of the target will have on the enemy's fighting ability. It must determine that the target is the one available target that will cause the greatest harm to the enemy.
- *Enemy Reaction to Sniper Fire.* The sniper team must consider what the enemy will do once the shot has been fired. The team must be prepared for such actions as immediate suppression by indirect fires and enemy sweeps of the area.
- *Effect on the Overall Mission.* The sniper team must consider how the engagement will affect the overall mission. The mission may be one of intelligence-gathering for a certain period. Firing will not only alert the enemy to a team's presence, but it may also terminate the mission if the team has to move from its position as a result of the engagement.
- *Probability of First-Round Hit.* The sniper team must determine the chances of hitting the target with the first shot by considering the following:
  - Distance to the target.
  - Direction and velocity of the wind.

- Visibility of the target area.
- Amount of the target that is exposed.
- Length of time the target is exposed.
- Speed and direction of target movement.
- Nature of the terrain and vegetation surrounding the target.

- *Distance.* Although the sniper may be capable of hitting a human target at a range of 800 meters, he should not risk such a distant shot without a special reason. The sniper has been trained to stalk to within 200 meters of a trained observer and plan his retrograde. He must make use of this ability and ensure his first shot hits the target. A clean, one-shot kill is far more demoralizing to the enemy than a near-miss from 600 meters.
- *Multiple Targets.* The sniper should carefully weigh the possible consequences of firing at one of a number of targets, especially when the target cannot be identified in detail. The sniper may trade his life for an unimportant target by putting himself in a position where he must fire repeatedly in self-defense.
- *Equipment as Targets.* A well-placed shot can disable crew-served weapons, radios, vehicles, or other equipment. Such equipment may serve as "bait" and allow the sniper to make repeated engagements of crew members or radio operators while keeping the equipment idle, to be disabled at the sniper's convenience. Retaliation by indirect fire must be considered in these circumstances.
- *Intelligence Collection.* Intelligence is an important collateral function of the sniper team. When in a location near to the enemy, the sniper team must be very judicious in its decision to fire. The sniper may interrupt a pattern of activity that, if observed longer, would allow the pair to report facts that would far outweigh the value of a kill. The well-trained sniper team will carefully evaluate such situations.
- *Key Target Selection.* A sniper selects targets according to their value. A target's real worth is determined by the sniper and the nature of his mission. Key personnel targets can be identified by actions, mannerisms, positions within formations, rank or insignias, and equipment being worn or carried. Key personnel targets are as follows:
    - *Snipers.* Snipers are the number one target of a sniper team. The fleeting nature of a sniper is reason enough to engage him because he may never be seen again.
    - *Dog-Tracking Teams.* Dog-tracking teams pose a great threat to sniper teams and other special teams that may be working in the area. It is hard to fool a trained dog; therefore, the dog-tracking team must be stopped. When engaging a dog-tracking team, the sniper should engage the dog's handler first, unless it is known that the dogs are trained to attack on gunshot.
    - *Scouts.* Scouts are keen observers, provide valuable information about friendly units, and control indirect fires, which make them dangerous on the battlefield.

- *Officers (Military and Political).* These individuals are also targets because in some forces losing key officers is a major disruption and causes coordination loss for hours.
- *Noncommissioned Officers (NCOs).* Losing NCOs not only affects the operation of a unit but also affects the morale of lower-ranking personnel.
- *Vehicle Commanders and Drivers.* Many vehicles are rendered useless or the capabilities are greatly degraded without a commander or driver.
- *Communications Personnel.* Eliminating these personnel can seriously cripple the enemy's communication network, because in some forces only highly trained personnel can operate various radios.
- *Weapons Crews.* Eliminating these personnel reduces the amount and accuracy of enemy fire on friendly troops.
- *Optics on Vehicles.* Personnel who are in closed vehicles are limited to viewing through optics. The sniper can blind a vehicle by damaging these optic systems.
- *Communications and Radar Equipment.* The right shot in the right place can completely ruin a tactically valuable radar or communications system. Also, only highly trained personnel may attempt to repair these systems in place. Eliminating these personnel may impair the enemy's ability to perform field repair.
- *Weapons Systems.* Many high-technology weapons, especially computer-guided systems, can be rendered useless by one well-placed round in the guidance controller of the system.

PRINCIPLES OF VISION

4-185. To fully understand and accomplish the principles of training the eye, the sniper must know its capabilities and limitations. The parts of the eye correspond to the parts of the camera and react in much the same way (Figure 4-26, page 4-53). The eye has a lens like a camera; however, the lens of the eye focuses automatically and more rapidly than the camera lens. The eye also has a diaphragm, called the iris, that regulates the amount of light into the eye. It permits the individual to see in bright light or in dark shadows. Just as with the camera, the eye cannot accomplish both at the same time. The eye's film is the photoreceptor cells located on the back wall, or retina, of the eye. There are two types of cells:

- *Cone Cells.* They are located in the central portion of the retina, used for day vision, and enable one to distinguish color, shape, and sharp contrast. The eye needs a lot of light to activate the cone cells, so these cells are blind during periods of low light.
- *Rod Cells.* They are located peripheral to the cone cells and are used for night vision. They see mostly in black and white and are excellent at seeing movement. These are the cells that give the observer peripheral and night vision.

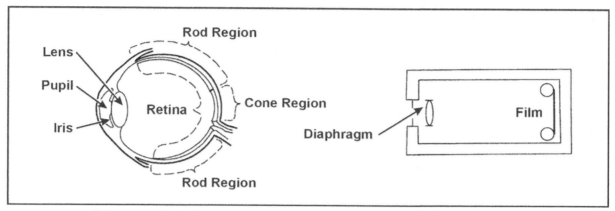

Figure 4-26. Functional Similarities Between the Eye and a Camera

OBSERVATION TECHNIQUES

4-186. Training the eye requires training the mind as well. The sniper's proficiency as an observer will come from a good mental attitude and a trained eye. As an observer, just as with a hunter, the eye must be trained to notice little things, such as the bending of grass when there is no wind, the unnatural shape of a shadow, or the wisp of vapor in cold air. Even when the enemy cannot be seen, his location can be given away by little things, such as a window that is now open when it was closed before, a puff of smoke, signs of fresh soil, or disturbed undergrowth.

4-187. Observers should learn the habits of the animals in the area or watch the domestic animals. A chicken suddenly darting from behind a building; sheep, goats, or cows suddenly moving or just becoming more alert in a field; wild birds flying or becoming quiet; insects becoming quiet at night; or animals startled from their positions should alert the observer of possible enemy activity in his area.

4-188. The observer should study and memorize the AO. Any change will alert the prepared mind to the possibility of the enemy. The observer should inspect all changes to determine the cause. He should also remember some key rules while observing. He must learn to—

- Look for the reasons why things are seen.
- Look for objects that seem out of place. Almost every object in the wild is vertical; only man-made objects such as a gun barrel are horizontal.
- See things in the proper perspective at distances. Learn to see movement, color, shape, and contrast in miniature.
- Look through vegetation, not at it. The observer should not be satisfied until he has seen as far as possible into the vegetation.

4-189. Due to constantly changing clouds and the sun's positions, light is a changing factor in observation. The sniper should always be ready to watch the changing contrast and shadows. An area that the sniper first thought held no enemy may prove different when the light changes. When the sun is to the sniper's back, light will reflect from the enemy's optical devices. But when the light changes and is to the front, the enemy will be able to see the light reflected from the sniper's optical devices.

4-190. It is also more tiring for the sniper to observe when the light shines in his eyes. He should arrange for a relief observer more frequently at this time if possible. If not, the use of some type of shading will help to cut down on the amount of light coming into the eyes.

LIMITED VISIBILITY TECHNIQUES

4-191. Twilight is another time of light changes. The eye begins to produce visual purple and the cone cells begin shutting down. Also, the iris opens more to let more light in. This reaction causes the eye to constantly change focus, and consequently, tires the eye quicker. However, during twilight the enemy will usually become more careless, allowing an alert observer to spot that last change in position or that last cigarette before dark. The sniper should also remember this is not a time for him to become relaxed.

4-192. Limited visibility runs the gamut from bright moonlight to utter darkness. But no matter how bright the night is, the eye cannot function with daylight precision. For maximum effectiveness, an observer must apply the following principles of night vision when training the eye:

- *Night Adaptation*. Allow approximately 30 minutes for the eye to adjust.
- *Off-Center Vision*. Never look directly at an object at night. This look will cause the object to disappear. When it reappears, it could appear to change shape or move.
- *Scanning*. It is important that the eye stops movement for a few seconds during the scan to be able to see an object. When scanning around an object, the temptation to look directly at the object "just to make sure" should be resisted.

4-193. The sniper should remember that the following factors can affect night vision:

- Lack of Vitamin A.
- Colds, headaches, fatigue, narcotics, alcohol, and heavy smoking.
- Exposure to bright light. It will destroy night vision for about 10 to 30 minutes, depending on the brightness and duration of the light.

4-194. Darkness blots out detail, so the eye must be trained to recognize objects by outline alone. While some people can see better than others at night, everyone can use the following techniques to improve their vision at night:

- Train the eye to actually see all the detail possible at nighttime. When the sniper sees a tree, he actually sees the tree, not a faint outline that he thinks may be a tree.
- Open the iris. While the iris of the eye is basically automatic, the eye can be trained to open up the iris even more to gather more light, which allows more detail to be seen.
- Practice roofing, which is silhouetting objects against a light background.

- Maneuver to catch the light. At night, noticeable light will only be in patches where it filters through the trees. The sniper must maneuver to place an object between his eyes and that patch of light.
- Lower the body. By lowering the body or even lying down, the sniper will be able to pick up more light and therefore see things that might otherwise go unnoticed.

OBSERVATION BY SOUND

4-195. Many times sound will warn the sniper long before the enemy is actually seen. Also, the sounds or lack of sounds from birds or animals may alert one to the possible presence of the enemy. It is therefore important to train the ears along with the eyes.

4-196. The ear nearest the origin of the sound will pick up the sound first and will hear it slightly louder than the other ear. The difference is what enables the sniper to detect the direction of the sound. When the sound hits both ears equally then the sound is to his front or rear. The brain will determine front or rear. However, if the sound reaches both ears at the same time and with the same intensity, as in fog or extremely humid weather, then the direction that the sound came from will not be discernible or will be confusing.

4-197. Sound also loses its intensity with distance traveled. The ears must be trained to become familiar with the different sounds at different distances so that the distance to the sound can be estimated. This estimate would then give the sniper a general location of the sound.

4-198. The sniper must learn to actually hear all sounds. Most people rely on sight for most of their information. A trained sniper must learn to use his ears as well as his eyes. The observer must make a conscious effort to hear all of the sounds, so that when a sound changes or a new one occurs, he will be alerted to it. He should close his eyes and listen to the sounds around him. He must categorize the sounds and remember them. Detailed observation includes a recheck of the surrounding sounds.

4-199. By cupping his hand behind one ear, the sniper can increase his ability to hear and pinpoint the direction of a sound.

TARGET LOCATION BY THE "CRACK-THUMP" METHOD

4-200. A trained ear enables the sniper to determine the approximate location of a shot being fired by using the "crack-thump" method. When the sniper is being fired at, he will hear two distinct sounds. One sound is the crack of the bullet as it breaks the sound barrier as it passes by his position. The other sound is the thump created by the muzzle blast of the weapon being fired. The crack-thump relationship is the time that passes between the two sounds. This time interval can be used to estimate the distance to the weapon being fired.

4-201. When the sniper hears the crack, he does not look into the direction of the crack. The sound will give him a false location because the sonic waves of the bullet strike objects perpendicular to the bullet's path (Figure 2-27, page 2-32). The sniper would mistakenly look 90 degrees from the enemy's true position. The crack should instead alert the sniper to start counting seconds.

4-202. The second sound heard is the thump of the weapon being fired. This point is the enemy's location. The time passed in seconds is the distance to the enemy. Sound travels at 340 meters per second at 30 degrees F. The speed of the bullet is twice that, which means it arrives before the sound of the muzzle blast. Therefore, half a second is approximately 300 meters, and a full second 600 meters. It becomes easier to distinguish between the two sounds as the distance increases. By listening for the thump and then looking in the direction of the thump, it is possible to determine the approximate location of the weapon being fired.

4-203. Flash-bang may be used to determine the distance to a weapon fired or and explosion seen. Since the light is instantaneous, the count will equal approximately 350 meters every second or 1,000 meters every 3 seconds or 1 mile every 5 seconds.

4-204. The speed of light is far greater than the speed of sound or of bullets. Remember that the crack-thump and flash-bang relationships are a double-edged sword that may be used against the sniper.

4-205. The speed, size, and shape of the bullet will produce different sounds. Initially, they will sound alike, but with practice the sniper will be able to distinguish between different types of weapons. A 7.62 x 39-mm bullet is just going subsonic at 600 meters. Since the crack-thump sounds differ from weapon to weapon, with practice the experienced sniper will be able to distinguish enemy fire from friendly fire.

4-206. The crack-thump method has the following limitations:
- Isolating the crack and thump is difficult when many shots are being fired.
- Mountainous areas and tall buildings cause echoes and make this method ineffective.

4-207. To overcome these limitations, the innovative sniper team can use—
- *Dummy Targets.* The sniper team can use polystyrene plastic heads or mannequins dressed to resemble a soldier to lure enemy snipers into firing. The head is placed on a stick and slowly raised into the enemy's view while another team observes the area for muzzle blast or flash.
- *The Shot-Hole Analysis.* Locating two or more shot holes in trees, walls, or dummy heads may make it possible to determine the direction of the shots. The team can use the dummy-head method and triangulate on the enemy sniper's position. However, this method only works if all shots come from the same position.

OBSERVATION DEVICE USE AND SELECTION

4-208. The sniper team's success depends upon its powers of observation. In addition to the rifle telescope, which is not used for observation, the team has an observation telescope, binoculars, night vision sight, and night vision goggles to enhance its ability to observe and engage targets. Team members must relieve each other often when using this equipment since prolonged use can cause eye fatigue, which greatly reduces the effectiveness of observation. Periods of observation during daylight should be limited to 30 minutes followed by at least 15 minutes of rest. When using NVDs, the observer should limit his initial period of viewing to 10 minutes followed by a 15-minute rest period. After several periods of viewing, he can extend the viewing period to 15 and then 20 minutes.

4-209. The M19 or M22 binoculars are the fastest and easiest aid to use when greater magnification is not needed. The binoculars also have a mil scale that can aid the sniper in judging sizes and distances. The M19 and M22 binoculars can also be used to observe at twilight by gathering more light than the naked eye. Using this reticle pattern aids the sniper in determining range and adjusting indirect fires. The sniper uses the binoculars to—

- Observe target areas.
- Observe enemy movement and positions.
- Identify aircraft.
- Improve low-light-level viewing.
- Estimate range.
- Call for and adjust indirect fires.

4-210. The M22 binoculars are the latest in the inventory but have several flaws. The M22's flaws are directly attributable to its antilaser protective coating. This coating reflects light like a mirror and is an excellent target indicator. Also, this coating reduces the amount of light that is transmitted through the lens system and greatly reduces the observation capability of the sniper during dawn and dusk.

4-211. The M49 is a fixed 20x observation-spotting telescope and can be used to discern much more detail at a greater distance than the binoculars or the sniper telescope. With good moonlight, the observer can see a target up to 800 meters away. However, the high magnification of the observation scope decreases its field of view. Moreover, the terrain will not be in focus unless it is near the object being inspected. The sniper should use the observation scope for the inspection and identification of a specific point only, not for observation of an area. The M144 is a variable power (15x to 45x) observation scope and a replacement for the M49. More modern and higher-quality spotting scopes are available in limited quantities. The sniper team should research the availability of these improved observation devices.

RANGE ESTIMATION

4-212. Range estimation is the process of determining the distance between two points (Appendix J). The ability to accurately determine range is **the** key skill needed by the sniper to accomplish his mission.

FACTORS AFFECTING ESTIMATION

4-213. Range can be determined by measuring or by estimating. Below are three main factors that affect the appearance of objects when determining range by eye.

Nature of the Target

4-214. Objects of regular outline, such as a house, will appear closer than one of irregular outline, such as a clump of trees. A target that contrasts with its background will appear to be closer than it actually is. A partially exposed target will appear more distant than it actually is.

Nature of the Terrain

4-215. Observing over smooth terrain, such as sand, water, or snow, causes the observer to underestimate distance targets. Objects will appear nearer than they really are when the viewer is looking across a depression, most of which is hidden from view. They will also appear nearer when the viewer is looking downward from high ground or when the viewer is looking down on a straight, open road or along railroad tracks.

4-216. As the observer's eye follows the contour of the terrain, he tends to overestimate the distance to targets. Objects will appear more distant than they really are when the viewer is looking across a depression, all of which is visible. They also appear more distant than they really are when the viewer is looking from low ground toward high ground and when the field of vision is narrowly confined, such as in twisted streets or on forest trails.

Light Conditions

4-217. The more clearly a target can be seen, the closer it will appear. A target viewed in full sunlight appears to be closer than the same target viewed at dusk or dawn or through smoke, fog, or rain. The position of the sun in relation to the target also affects the apparent range. When the sun is behind the viewer, the target appears closer. When the sun appears behind the target, the target is more difficult to see and appears farther away.

MILING THE TARGET FOR RANGE

4-218. When ranging on a human target, the sniper may use two different methods. The first method is to range on the target using the vertical crosshairs and mil dots. The second method is to use the horizontal crosshairs and mil dots.

Vertical Method

4-219. The sniper most often uses this method of range finding when using the M3A. He must become very good at estimating the height of the target in either meters or feet and inches. The sniper has the option of using a 1-meter (head to crotch) target frame or using the entire target (head to toe) as the target frame. To use the vertical method, the sniper places the crosshairs at either the feet or crotch, and measures to the top of the head of the target. The mil value is then read for that target. The sniper must determine the height of the target if he is not using the 1-meter target frame. Since the telescope is graduated in meters, the height of the target must be converted into meters. The sniper then calculates the range using the mil-relation formula. The estimation of the height of the target may be the most important factor in this formula. An error of 3 inches on a 5-foot 9-inch target that is actually 5 feet 6 inches results in a 19-meter error at a reading of 4 mils.

Normal height of the human = 69 inches

$$\frac{69 \text{ inches} \times 25.4}{\text{Size of target in mils (4)}} = \text{Range to target in meters (438.5 or 440 meters)}$$

NOTE: This example may prove to be of specific use when facing an enemy entrenched in bunkers or in dense vegetation.

Horizontal Method

4-220. The horizontal method is based upon a target width of 19 inches at the shoulders. This technique can be very accurate out to ranges of 350 meters, and is very effective in an urban environment. Beyond this range it is no longer effective. The sniper should use this method to double-check ranges derived from groin to head. For example, a range estimate derived from a groin to head (1 meter) measurement of 2 mils would be equal to a 1 mil shoulder to shoulder measurement (horizontal = 1/2 vertical). A good rule of thumb is that if the target is smaller than 1 1/2 mils (322 meters), it is more accurate to use the vertical method in combination with the horizontal method.

4-221. The mil dots in the M3A are 3/4 MOA in diameter. Therefore, it is important to note where on the dots the bottom or the top of the target falls within the mil dot. **The mil dots are spaced 1 mil from center to center, or cross to center of first dot.**

4-222. Objects viewed from an oblique angle may cause the sniper to overestimate the range to that object. Snipers should be aware of this effect and compensate accordingly.

DETERMINING RANGE TECHNIQUES

4-223. A sniper team must accurately determine distance, properly adjust elevation on the SWS, and prepare topographic sketches or range cards. To meet these needs, team members have to be skilled in various range estimation techniques. The team can use any of the following methods to determine distance between its position and the target.

Sniper Telescope

4-224. The M3A has a mil dot reticle and the mil-relation formula is used for range determination. Using the telescope for range estimation is especially helpful when establishing known ranges for a range card or a reference mark. The sniper rifle's inherent stability helps to improve the accuracy of the measurements. The sniper can determine range by using the range feature of the sniping telescope and the following:

- *Personnel.* The distance from the individual's head to his waist is normally 30 inches; from the top of his head to his groin is 1 meter (39.4 inches). The head to groin is the most common measuring point for the human body. The 1-meter measurement will not vary but an inch or two from the 6-foot-6-inch man to the 5-foot-6-inch man.

- *Tanks.* The distance from the ground line to the deck or from the deck to the turret top of a Soviet-style tank is approximately 30 inches.
- *Vehicles.* The distance from the ground line to the fender above the wheel is approximately 30 inches. The distance to the roofline is 3 1/2 to 4 feet.
- *Trees.* The width of the trees in the vicinity of the sniper will be a good indication of the width of the trees in the target area.
- *Window Frames.* The vertical length of a standard frame is approximately 60 inches. This distance is 1.5 meters by 2.0 meters in Europe.

NOTE: Through the process of interpolation, the sniper can range on any object of known size. For example, the head of any individual will measure approximately 12 inches. The M3A has a mil-dot reticle. On this telescope a mil dot equals 3/4 of an MOA, and the space between mil dots equals 1 mil or 3.44 MOA (round to 3.5 in the field). The figure 3.44 is the true number of MOA in a mil as one radian is equal to 57.295 degrees. This makes 6.283 radians in a circle or 6283 mils in a circle. With 21,600 MOA in a circle, the result is 3.44 MOA in a mil.

Mil-Relation Formula

4-225. The sniper can also use the mil-relation formula to determine ranges. The M3A rifle telescope has 10 mils vertical and horizontal measurement between the heavy duplex reticle lines; the space between each dot represents 1 mil. Military binoculars also have a mil scale in the left ocular eyepiece. By using the known measured sizes of objects, the sniper can use the mil-relation formula to determine the range.

NOTE: The size of objects in meters yields ranges in meters; the size of objects in yards yields ranges in yards. Other relationships must also be understood: 1 mil equals 3.44 MOA or 3.6 inches at 100 yards; 1 meter at 1,000 meters or approximately 1 yard at 1,000 yards.

4-226. The sniper uses the following formula to determine the range to the target:

$$\text{Range to target} = \frac{\text{Size of object in meters and yards} \times 1{,}000}{\text{Size of object in mils}}$$

Example 1) Object = 2 meters, Mils = 4 mils (as measured in the M3A scope)

$$\frac{2 \times 1{,}000}{4} = \frac{2{,}000}{4} = 500 \text{ meters} = \text{Range to target}$$

Example 2) Object = 2 yards, Mils = 5 mils (as measured in the M3A scope)

$$\frac{2 \times 1{,}000}{5} = \frac{2{,}000}{5} = 400 \text{ yards} = \text{Range to target}$$

Example 3) Object = 69 inches, Mils = 4 mils (To convert inches to meters, multiply by 25.4.)

$$\frac{69 \times .25.4}{4} = \frac{1752.6}{4} = 438 \text{ meters}$$

NOTE: The distance to the target in yards must be converted to meters to correctly set the M3A's ballistic cam.

4-227. Once the sniper understands the formula, he must become proficient at estimating the actual height of the target in his scope. At longer ranges the measurements must be accurate to within 1/10 mil. Otherwise, the data will be more than the allowable ballistic error. The ability of the sniper team to accurately estimate the height of the target is the single most important factor in using this formula.

Mil Relation (Worm Formula)

Sample Problems:

No. 1: As a member of a sniper team, you and your partner are in your hide site and are preparing a range card. To your front you see a Soviet-type truck that you determine to be 4 meters long. Your team is equipped with an M24 system. Through your binoculars the truck is 5 mils in length. Determine the range to this reference for your system.

Solution: STEP 1. No conversion needed.
STEP 2. Determine the range.

$$\text{Width} = \frac{4 \text{ meters} \times 1,000}{5 \text{ mils}} = 800 \text{ meters}$$

No. 2: You are a member of a sniper team assigned to cover a certain area of ground. You are making a range card and determining ranges to reference points in that area. You see a tank located to your front. Through your binoculars you find the width of the tank to be 8 mils. You determine the length of the tank to be 5 meters. Determine the correct range for your system.

Solution: STEP 1. No conversion needed.
STEP 2. Determine the range.

$$\text{Width} = \frac{5 \text{ meters} \times 1,000}{8 \text{ mils}} = 625 \text{ meters}$$

Military Binoculars

4-228. The sniper can calculate the range to a target by using the M3, M19, and M22 binoculars, or any other optical device that has vertical and horizontal mil reticles.

4-229. **M3 Binoculars.** The graduations between the numbers on the horizontal reference line are in 10-mil graduations. The height of the vertical lines along the horizontal reference line is 2 1/2 mils. The graduation of the horizontal reference lines on the left of the reticle is 5 mils (vertical) between the reference lines. These lines are also 5 mils long (horizontal). The small horizontal lines located above the horizontal reference line in the center of the reticle are 5 mils apart (vertical) and are also 5 mils long (horizontal). The vertical scale on the reticle is not to be used for range finding purposes.

4-230. **M19 Binoculars.** The graduation between the number lines on the horizontal and the vertical lines on the reticle is 10 mils (Figure 4-27). The total height of the vertical lines on the horizontal reference lines is 5 mils. These lines are further graduated 2 1/2 mils above the horizontal line and 2 1/2 mils below the line. The total width of the horizontal lines on the vertical reference line is 5 mils. These lines are further graduated into 2 1/2 mils on the left side of the line and 2 1/2 mils on the right side of the vertical reference line.

4-231. **M22 Binoculars.** The graduation between the numbered lines on the horizontal and vertical reference lines is 10 mils. There are 5 mils between a numbered graduation and the 2 1/2-mil tall line that falls between the numbered graduations. The value of the longer lines that intersect the horizontal and vertical lines on the reticle is 5 mils. The value of the shorter lines that intersect the horizontal and vertical reference lines on the reticle is 2 1/2 mils. These are the lines that fall between the 5-mil lines.

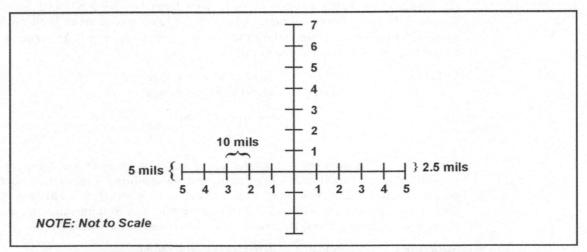

Figure 4-27. The M19 Binocular Reticle Showing the Mil Measurements of the Stadia Lines

Estimation

4-232. There will be times when the sniper must estimate the range to the target. This method requires no equipment and can be accomplished without exposing the observer's position. There are two methods of estimation that meet these requirements: the 100-meter unit-of-measure method and the appearance-of-objects method.

4-233. **The 100-Meter Unit-of-Measure Method.** The sniper must be able to visualize a distance of 100 meters on the ground. For ranges up to 500 meters, he determines the number of 100-meter increments between the two points that he wishes to measure. Beyond 500 meters, the sniper must select a point halfway to the target, determine the number of 100-meter increments of the halfway point, and then double this number to find the range to the target (Figure 4-28).

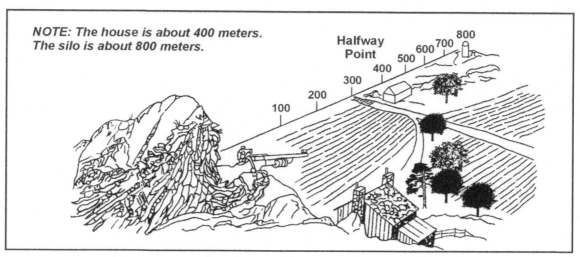

Figure 4-28. The "Halfway Point" Distance Estimation Process

4-234. During training exercises, the sniper must become familiar with the effect that sloping ground has on the appearance of a 100-meter increment. Ground that slopes upward gives the illusion of a shorter distance, and the observer's tendency is to overestimate a 100-meter increment. Conversely, ground that slopes downward gives the illusion of a longer distance. In this case, the sniper's tendency is to underestimate.

4-235. Proficiency in the 100-meter unit-of-measure method requires constant practice. Throughout the training in this technique, comparisons should continuously be made between the range as determined by the sniper and the actual range as determined by pacing or other more accurate means of measurement. The best training technique is to require the sniper to pace the range after he has visually determined it. In this way he discovers the actual range for himself, which makes a greater impression than if he were simply told the correct range.

4-236. The greatest limitation of the 100-meter unit of measure is that its accuracy is directly related to how much of the terrain is visible at the greater ranges. This point is particularly true at a range of 500 meters or more when the sniper can only see a portion of the ground between himself and the target. It becomes very difficult to use the 100-meter unit-of-measure method of range determination with any degree of accuracy.

4-237. **The Appearance-of-Objects Method.** The appearance-of-objects method is the means of determining range by the size and other characteristic details of the object in question. It is a common method of determining distances and used by most people in their everyday living. For

FM 3-05.222

example, a motorist attempting to pass another car must judge the distance of oncoming vehicles based on his knowledge of how vehicles appear at various distances. Of course, in this example, the motorist is not interested in precise distances, but only that he has sufficient road space to safely pass the car in front of him. This same technique can be used by the sniper to determine ranges on the battlefield. If he knows the characteristic size and detail of personnel and equipment at known ranges, then he can compare these characteristics to similar objects at unknown ranges. When the characteristics match, so does the range.

4-238. To use this method with any degree of accuracy, the sniper must be familiar with the characteristic details of objects as they appear at various ranges. For example, the sniper should study the appearance of a man when he is standing at a range of 100 meters. He fixes the man's appearance firmly in his mind, carefully noting details of size and characteristics of uniform and equipment. Next, he studies the same man in a kneeling position and then in a prone position. By comparing the appearance of these positions at known ranges from 100 to 500 meters, the sniper can establish a series of mental images that will help him determine range on unfamiliar terrain. Training should also be conducted in the appearance of other familiar objects such as weapons or vehicles. Because the successful use of this method depends upon visibility, anything that limits the visibility (such as weather, smoke, or darkness) will also limit the effectiveness of this method.

Combination of Methods

4-239. Under proper conditions, either the 100-meter unit-of-measure method or the appearance-of-objects method is an effective way of determining range. However, proper conditions do not always exist on the battlefield and the sniper will need to use a combination of methods. The terrain might limit using the 100-meter unit-of-measure method and the visibility could limit using the appearance-of-objects method. For example, an observer may not be able to see all of the terrain out to the target; however, he may see enough to get a general idea of the distance within 100 meters. A slight haze may obscure many of the target details, but the observer should still be able to judge its size. Thus, by carefully considering the approximate ranges as determined by both methods, an experienced observer should arrive at a figure close to the true range.

Measuring

4-240. The sniper can measure distance on a map or pace the distance between two points. The following paragraphs discuss each method.

4-241. **Map (Paper-Strip Method).** The paper-strip method is useful when determining longer distances (1,000 meters plus). When using this method, the sniper places the edge of a strip of paper on the map and ensures it is long enough to reach between the two points. Then he pencils in a tick mark on the paper at the team position and another at the distant location. He places the paper on the map's bar scale, located at the bottom center of the map, and aligns the left tick mark with the 0 on the scale. Then he reads to the right to the second mark and notes the corresponding distance represented between the two marks.

4-242. **Actual Measurement**. The sniper uses this method by pacing the distance between two points, provided the enemy is not in the vicinity. This method obviously has limited applications and can be very hazardous to the sniper team. It is one of the least desirable methods.

Bracketing Method

4-243. The bracketing method is used when the sniper assumes that the target is no less than "X" meters away, but no more than "Y" meters away. The sniper then uses the average of the two distances as the estimated range. Snipers can increase their accuracy of eye-range estimation by using an average of both team members' estimate.

Halving Method

4-244. The sniper uses this method for distances beyond 500 meters. He selects a point midway to the target, determines the number of 100-meter increments to the halfway point, and then doubles the estimate. Again, it is best to average the results of both team members.

Range Card

4-245. This method is a very accurate means of estimating range. The fact that the sniper has established a range card means he has been in the area long enough to know the target area. He has already determined ranges to indicated reference points. The observer will give his targets to the sniper by giving deflections and distances from known reference points in the target field of view. The sniper can adjust his telescope for a good median distance in the target area and simply adjust fire from that point. **There are multiple key distances that should be calculated and noted with references on the range card.** The first is the point blank zero of the weapon. With a 300-meter zero, the point-blank zero of the M118 ammunition is 375 meters. Targets under this range do not need to be corrected for. The other key distances are merely a point of reference against which further distance determinations can be judged. These are marked as target reference points (TRP) and are also used as reference points for directing the sniper onto a target.

Speed of Sound

4-246. The sniper can estimate the approximate distance from the observer to a sound source (bursting shell, weapon firing) by timing the sound. The speed of sound in still air at 50 degrees F is about 340 meters per second. However, wind and variations in temperature alter this speed somewhat. For practical use, the sniper may assume the speed of sound is 350 meters per second under all conditions. He can time the sound either with a watch or by counting from the time the flash appears until the sound is heard by the observer. The sniper counts "one-1,000, two-1,000," and so on, to determine the approximate time in seconds. He then multiplies the time in seconds by 350 to get the approximate distance in meters to the source of the fire.

Measurement by Bullet Impact

4-247. Another undesirable but potentially useful method is to actually fire a round at the point in question. This practice is possible if you know your target is coming into the area at a later time and you plan to ambush the target. However, this method is not tactically sound and is also very hazardous to the sniper team.

Laser Range Finders

4-248. These can also be used to determine range to a very high degree of accuracy. When aiming the laser at a specific target, the sniper should support it much the same way as his weapon to ensure accuracy. If the target is too small, aiming the laser at a larger object near the target will suffice—that is, a building, vehicle, tree, or terrain feature. The range finder must be used with yellow filters to keep the laser eye-safe for the sniper and observer when observing through optics, as the AN/GVS 5 is not eye safe. This cover limits the range; however, the limitations are well within the range of the sniper. Rain, fog, or smoke will severely limit the use of laser range finders. Laser detectors and NVDs that are set to the correct wavelength may also intercept laser range finders.

> **CAUTION**
> Viewing an "eye-safe" laser through magnifying optics increases the laser's intensity to unsafe levels.

Sniper Cheat Book

4-249. The sniper team should keep a "cheat book" complete with measurements. The team fills in the cheat book during its area analysis, mission planning, isolation, and once in the AO. A tape measure will prove invaluable. Each cheat book should include the following:

- Average height of human targets in AO.
- Vehicles:
    - Height of road wheels.
    - Vehicle dimensions.
    - Length of main gun tubes on tanks.
    - Lengths and sizes of different weapon systems.
- Urban environment:
    - Average size of doorways.
    - Average size of windows.
    - Average width of streets and lanes (average width of a paved road in the United States is 10 feet).

4-250. As the sniper team develops its cheat book, all measurements are converted into constants and computed with different mil readings. These measurements should also be incorporated into the sniper's logbook. The team should use the "worm formula" (paragraph 4-227) when preparing the cheat book.

SELECTION AND PREPARATION OF HIDES

4-251. To effectively accomplish its mission or to support combat operations, the sniper team must select a position called a sniper hide or post. Once constructed, it will provide the sniper team with a well-concealed post from which to observe and fire without fear of enemy detection. Selecting the location of a position is one of the most important tasks a sniper team must accomplish during the mission planning phase of an operation. After selecting the location, the team must also determine how it will move into the area and locate and occupy the final position.

HIDE SELECTION

4-252. Upon receiving a mission, the sniper team locates the target area and then determines the best location for a tentative position by using one or more of the following sources of information:

- Topographic maps.
- Aerial photographs.
- Visual reconnaissance before the mission.
- Information gained from units operating in the area.

4-253. In selecting a sniper hide, maximum consideration is given to the fundamentals and principles of camouflage, cover, and concealment. Once on the ground, the sniper team ensures the position provides an optimum balance between the following considerations:

- Maximum fields of fire and observation of the target area.
- Maximum concealment from enemy observation.
- Covered routes into and out of the position.
- Located no closer than 300 meters from the target area whenever possible.
- A natural or man-made obstacle between the position and the target area.

4-254. A sniper team must remember that if a position appears ideal, it may also appear that way to the enemy. Therefore, the team should avoid choosing locations that are—

- On a point or crest of prominent terrain features.
- Close to isolated objects.
- At bends or ends of roads, trails, or streams.
- In populated areas, unless mission-essential.

4-255. The sniper team must use its imagination and ingenuity in choosing a good location for the given mission. The team must choose a location that not only allows the team to be effective but also must appear to the enemy to be the least likely place for a team position. Examples of such positions are—

- Under logs in a deadfall area.
- Tunnels bored from one side of a knoll to the other.
- Swamps.

- Deep shadows.
- Inside rubble piles.

HIDE SITE LOCATION

4-256. The sniper team should determine the site location by the following factors of area effectiveness:

- Mission.
- Dispersion.
- Terrain patterns.

4-257. Various factors can affect the team's site location. The sniper team should select tentative sites and routes to the objective area by using—

- Aerial photographs.
- Maps.
- Reconnaissance and after-action reports.
- Interrogations of indigenous personnel, prisoners of war, and other sources.
- Weather reports.
- Area studies.

4-258. When the team is selecting a site, it should look for—

- Terrain patterns (urban, rural, wooded, barren).
- Soil type (to determine tools).
- Population density.
- Weather conditions (snow, rain).
- Drainage.
- Types of vegetation.
- Drinking water.

4-259. The sniper team must also consider some additional requirements when selecting the hide site. It should conduct a reconnaissance of the area to determine—

- Fields of fire.
- Cover and concealment.
- Avenues of approach.
- Isolated and conspicuous patterns.
- Terrain features lying between your position and the objectives.

SNIPER HIDE CHECKLIST

4-260. There are many factors to consider in the selection, construction, and use of a sniper hide. The sniper team must remain alert to the danger of compromise and consider its mission as an overriding factor. Figure 4-29, page 4-69, lists the guidelines that the sniper team should use when selecting a site and constructing the sniper hide.

- ❏ Select and construct a sniper hide from which to observe and fire. Because the slightest movement is the only requirement for detection, construction is usually accomplished at night. Caution must still be exercised, as the enemy may employ NVDs, and sound travels greater distances at night.
- ❏ Do not place the sniper hide against a contrasting background or near a prominent terrain feature. These features are usually under observation or used as registration points.
- ❏ Consider those areas that are least likely to be occupied by the enemy.
- ❏ Ensure that the position is located within effective range of the expected targets and that it affords a clear field of fire.
- ❏ Construct or empty alternate hides where necessary to effectively cover an area.
- ❏ Assume that the sniper hide is under enemy observation.
- ❏ Avoid making sounds.
- ❏ Avoid unnecessary movement.
- ❏ Avoid observing over a skyline or the top of cover or concealment that has an even outline or contrasting background.
- ❏ Avoid using the binoculars or telescope where light may reflect from lenses.
- ❏ Observe around a tree from a position near the ground. The snipers should stay in the shadows when observing from a sniper hide.
- ❏ Give careful consideration to the route into or out of the hide. A worn path can easily be detected. The route should be concealed and covered, if possible.
- ❏ Use resourcefulness and ingenuity to determine the type of hide to be constructed.
- ❏ When possible, choose a position that has a terrain obstacle (for example, a river, thick brush) between it and the target and/or known or suspected enemy location.

Figure 4-29. Checklist for Selecting and Constructing a Hide

HIDE SITE OCCUPATION

4-261. During the mission planning phase, the sniper also selects an ORP. From this point, the sniper team reconnoiters the tentative position to determine the exact location of its final position. The location of the ORP should provide cover and concealment from enemy fire and observation, be located as close to the selected area as possible, and have good routes into and out of the selected area.

4-262. From the ORP, the team moves forward to a location that allows the team to view the tentative position area. Once a suitable location has been found, the team member moves to the position. While conducting the reconnaissance or moving to the position, the team—

- Moves slowly and deliberately, using the sniper low crawl.
- Avoids unnecessary movement of trees, bushes, and grass.
- Avoids making any noises.
- Stays in the shadows, if there are any.
- Stops, looks, and listens every few feet.
- Looks for locations to hide spoil if a hide is to be dug into the terrain.

4-263. When the sniper team arrives at the firing position, it—

- Conducts a hasty and detailed search of the target area.
- Starts construction of the firing position, if required.

- Organizes equipment so that it is easily accessible.
- Establishes a system of observing, eating, resting, and using the latrine.

HASTY SNIPER HIDE OR FINAL FIRING POSITION

4-264. The sniper team uses a hasty position when it will be in position for a short time, cannot construct a position due to the proximity of the enemy, or must immediately assume a position. Due to the limited nature of sniper missions and the requirement to stalk, the sniper team will most often use a hasty position.

4-265. This position (fast find) provides protection from enemy fire or observation. Natural cover (ravines, hollows, reverse slopes) and artificial cover (foxholes, trenches, walls) protect the sniper from flat trajectory fires and enemy observation. Snipers must form the habit of looking for and taking advantage of every bit of cover and concealment the terrain offers. They must combine this habit with proper use of movement techniques to provide adequate protection from enemy fire and observation.

4-266. Cover and concealment in a hasty position provide protection from enemy fire and observation. The cover and concealment may be artificial or natural. Concealment may not provide protection from enemy fire. A sniper team should not make the mistake of believing they are protected from enemy fire merely because they are concealed from enemy eyes.

4-267. There should be no limitation on ingenuity of the sniper team in selecting a hasty sniper hide. Under certain circumstances it may be necessary to fire from trees, rooftops, steeples, logs, tunnels, deep shadows, buildings, swamps, woods, and an unlimited variety of open areas. The sniper team's success depends to a large degree on its knowledge, understanding, and application of the various field techniques or skills that allow them to move, hide, observe, and detect the enemy (Table 4-1).

Table 4-1. Hasty Sniper Hide Advantages and Disadvantages

Advantages	Disadvantages
• Requires no construction. The sniper team uses what is available for cover and concealment. • Can be occupied in a short time. As soon as a suitable position is found, the team need only prepare loopholes by moving small amounts of vegetation or by simply backing several meters away from the vegetation that is already there to conceal the weapon's muzzle blast. **Note:** Loopholes may be various objects or constructed by the team, but must provide an adequate view for firing.	• Affords no freedom of movement. Any movement that is not slow and deliberate may result in the team being compromised. • Restricts observation of large areas. This type of position is normally used to observe a specific target area (intersection, passage, or crossing). • Offers no protection from direct or indirect fires. The team has only available cover for protection from direct fires. • Relies heavily on personal camouflage. The team's only protection against detection is personal camouflage and the ability to use the available terrain.
Occupation Time: The team should not remain in this type of position longer than 8 hours; it will only result in loss of effectiveness. This is due to muscle strain or cramps, which is a result of lack of freedom of movement combined with eye fatigue.	

EXPEDIENT SNIPER HIDE

4-268. When a sniper team has to remain in position for a longer time than the hasty position can provide, it should construct an expedient position (Figure 4-30). The expedient position lowers the sniper's silhouette as low to the ground as possible, but it still allows him to fire and observe effectively. Table 4-2 lists characteristics of an expedient sniper hide.

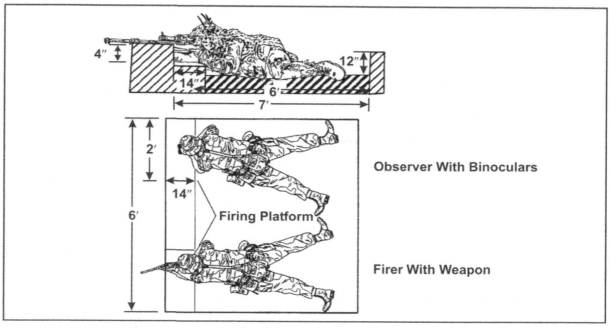

Figure 4-30. Overhead and Side View of the Expedient Sniper Hide Site

Table 4-2. Expedient Sniper Hide Advantages and Disadvantages

Advantages	Disadvantages
• Requires little construction. This position is constructed by digging a hole in the ground just large enough for the team and its equipment. Soil dug from this position can be placed in sandbags and used for building firing platforms. • Conceals most of the body and equipment. The optics, rifles, and heads of the sniper team are the only items that are above ground level in this position. • Provides some protection from direct fires due to its lower silhouette.	• Affords little freedom of movement. The team has more freedom of movement in this position than in the hasty position. However, teams must remember that stretching or reaching for a canteen causes the exposed head to move unless controlled. Team members can lower the head below ground level, but this movement should be done slowly to ensure a target indicator is not produced. • Allows little protection from indirect fires. This position does not protect the team from shrapnel and debris falling into the position. • Exposes the head, weapons, and optics. The team must rely heavily on the camouflaging of these exposed items.
Construction Time: 1 to 3 hours (depending on the situation). **Occupation Time:** 6 to 12 hours.	

BELLY HIDE

4-269. The belly hide (Figure 4-31) is similar to the expedient position, but it has overhead cover that not only protects the team from the effects of indirect fires but also allows more freedom of movement. A belly hide is most useful in mobile situations or when the sniper does not intend to be in the position for extended periods of time. This position can be dug out under a tree, a rock, or any available object that will provide overhead protection and a concealed entrance and exit. Table 4-3, pages 4-72 and 4-73, lists the belly hide characteristics.

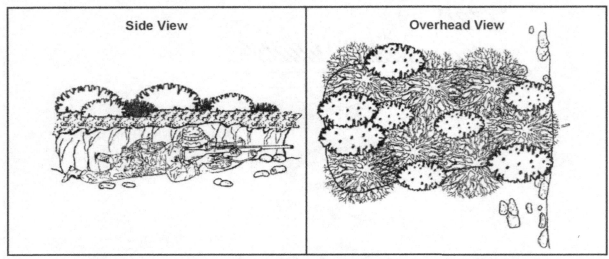

Figure 4-31. Overhead and Side View of the Expedient Belly Hide Site

Table 4-3. Belly Hide Advantages, Disadvantages, and Construction

Advantages	Disadvantages
• Allows some freedom of movement. The darkened area inside this position allows the team to move freely. The team should cover the entrance/exit hole with a poncho or piece of canvas so outside light does not silhouette the team inside the position. • Conceals all but the rifle barrel. All equipment is inside the position except the rifle barrels, but the barrels could be inside, depending on the room available to construct the position. • Provides protection from direct and indirect fires. The team should try to choose a position that has an object that will provide good overhead protection (rock, tracked vehicle, rubble pile, and so forth), or prepare it in the same manner as overhead cover for other infantry positions. • Is simple and can be quickly built. This hide can be used when the sniper is mobile, because many can be built.	• Is uncomfortable. • Cannot be occupied for long periods of time. • The sniper is exposed while firing. • Provides limited protection from the weather or fire. • Requires extra construction time. • Requires extra materials and tools. Construction of overhead cover will require saws or axes, waterproof material, and so forth. • Has limited space. The sniper team will have to lie in the belly hide without a lot of variation in body position due to limited space and design of the position.

Table 4-3. Belly Hide Advantages, Disadvantages, and Construction (Continued)

Construction

• Dig a pit (shallow) for the prone position.
• Build an overhead cover using:
  ▪ Dirt and sod.
  ▪ A drop cloth.
  ▪ Woven saplings.
  ▪ Corrugated metal, shell boxes, scrap metal, doors, chicken wire, or scrap lumber.

Construction Time: 4 to 6 hours. **Occupation Time:** 12 to 48 hours.

SEMIPERMANENT SNIPER HIDE

4-270. The sniper uses the semipermanent hide mostly in a defensive or outpost situation (Figure 4-32). Construction of this position requires additional equipment and personnel. However, it will allow sniper teams to remain there for extended periods or be relieved in place by other sniper teams. Like the belly hide, the sniper can construct this position by tunneling through a knoll or under natural objects already in place. This prepared sniper hide should provide sufficient room for movement without fear of detection, some protection from weather and overhead or direct fire, and a covered route to and from the hide.

Figure 4-32. The Semipermanent Sniper Hide

4-271. A semipermanent hide can be an enlargement of the standard one- or two-man fighting position with overhead cover. The sniper constructs this type of hide when in a defensive posture, since construction requires considerable time. It would be suitable when integrated into the perimeter defense of a base camp, during static warfare, or during a stay-behind infiltration. It can be constructed as a standing or lying type of hide.

4-272. The construction of loopholes requires care and practice to ensure that they afford an adequate view of the required fields of fire. The sniper should construct the loopholes so that they are wide at the back where he is and narrow in the front, but not so narrow that observation is restricted. Loopholes may be made of old coffee cans, old boots, or any other rubbish, provided that it is natural to the surroundings or that it can be properly and cleverly concealed.

4-273. Loopholes may be holes in windows, shutters, roofs, walls, or fences, or they may be constructed by the sniper team. Loopholes must blend in with the surrounding area. Table 4-4 lists the semipermanent sniper hide characteristics.

Table 4-4. Semipermanent Sniper Hide Advantages and Disadvantages

Advantages	Disadvantages
• Offers total freedom of movement inside the position. The team members can move about freely. They can stand, sit, or even lie down. • Protects against direct and indirect fires. The sniper team should look for the same items as mentioned in the belly hide. • Is completely concealed. Loopholes are the only part of the position that can be detected. They allow for the smallest exposure possible; yet, they still allow the sniper and observer to view the target area. The entrance and exit to the position must be covered to prevent light from entering and highlighting the loopholes. Loopholes that are not in use should be covered from the inside with a piece of canvas or suitable material. • Is easily maintained for extended periods. This position allows the team to operate effectively for a longer period.	• Requires extra personnel and tools to construct. This position requires extensive work and more tools. Very seldom can a position like this be constructed near the enemy, but it should be constructed during darkness and be completed before dawn. • Increases risk of detection. Using a position for several days or having teams relieve each other in a position always increases the risk of the position being detected. Snipers should never continue to fire from the same position.
Construction Time: 4 to 6 hours (4 personnel). **Occupation Time:** 48 hours plus (relieved by other teams).	

TREE OR STUMP HIDES

4-274. Nature can provide these types of hides but they also require the sniper to do some heavy construction (Figure 4-33). Table 4-5 lists the tree or stump hide characteristics.

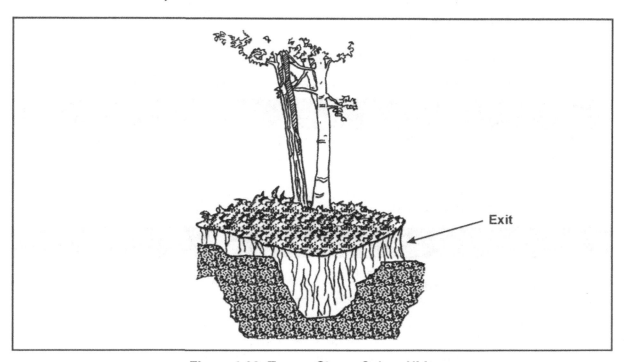

Figure 4-33. Tree or Stump Sniper Hide

Table 4-5. Tree or Stump Hide Advantages, Disadvantages, and Construction

Advantages	Disadvantages
• Can be rapidly occupied. • The sniper team is protected from fire and shrapnel. • The sniper team has freedom of movement. • Provides comfort.	• Takes time to construct. • The sniper team requires pioneer equipment for construction of the hide (picks, shovels, axes).
Construction	
• Use trees that have a good, deep root such as oak, chestnut, or hickory. During heavy winds these trees tend to remain steady better than a pine tree, which has surface roots and sways a bit in a breeze. • Use a large tree that is set back from the woodline. This location may limit the view but will provide better cover and concealment.	

TYPES OF HASTY SNIPER HIDES

4-275. The sniper can also use different types of deliberate hides to increase his chances for mission success and maintain the sniper training objectives. The various positions are explained below.

Enlarged Fire Trench Hides

4-276. This hide is actually an enlarged fighting position. Table 4-6 lists the enlarged fire trench hide characteristics.

Table 4-6. Enlarged Fire Trench Hide Advantages, Disadvantages, and Construction

Advantages	Disadvantages
• The sniper team is able to maintain a low silhouette. • Simple to construct. • Can be occupied for a moderate period of time with some degree of comfort.	• Is not easily entered into or exited from. • The sniper team has no overhead cover when in firing position. • The sniper team is exposed while firing or observing.
Construction	
• Enlarge and repair the sides and the parapet. • Camouflage the hide with a drop cloth.	

Shell-Hole Hides

4-277. This sniper hide is a crater improved for kneeling, sitting, or prone firing positions (Figure 4-34). Table 4-7, page 4-77, lists the shell-hole hide characteristics.

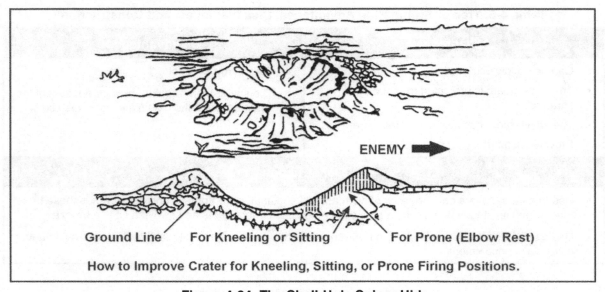

Figure 4-34. The Shell-Hole Sniper Hide

Table 4-7. Shell-Hole Hide Advantages, Disadvantages, and Construction

Advantages	Disadvantages
• Requires little digging.	• Requires material to secure the sides. • Affords no drainage.
Construction	
• Dig platforms for either the prone, the kneeling, or the sitting positions. • Reinforce the sides of the craters.	

HIDE SITE CONSTRUCTION CONSIDERATIONS

4-278. A sniper mission always requires the team to occupy some type of position. These positions can range from a hasty position to a more permanent position. When choosing and constructing positions, the sniper team must use its imagination and ingenuity to reduce the time and difficulty of position construction. The team should always plan to build its position during limited visibility.

4-279. Whether a sniper team will be in a position for a few minutes or a few days, the basic considerations in choosing a type of position remain the same (Table 4-8).

Table 4-8. Hide Site Construction Considerations

Location	Time
• Type of terrain and soil. Digging and boring of tunnels can be very difficult in hard soil or in fine, loose sand. The team needs to take advantage of what the terrain offers (gullies, holes, hollow tree stumps, and so forth). • Enemy location and capabilities. Enemy patrols in the area may be close enough to the position to hear any noises that may accidentally be made during any construction. The team also needs to consider the enemy's night vision and detection capabilities.	• Amount of time to be occupied. If the sniper team's mission requires it to be in position for a long time, the team must consider construction of a position that provides more survivability. This allows the team to operate more effectively for a longer time. • Time needed for construction. The time needed to build a position must be a consideration, especially during the mission planning phase.
Personnel and Equipment	
• Equipment needed for construction. The team must plan the use of any extra equipment needed for construction (bow saws, picks, axes). • Personnel needed for construction. Coordination must take place if the position requires more personnel to build it or a security element to secure the area during construction.	

STEPS USED IN THE CONSTRUCTION OF A SNIPER HIDE

4-280. When the sniper team is en route to the objective area, it should mark all material that can be used for constructing a hide. The team should establish an ORP, reconnoiter the objective area, select a site, and mark the fields of fire and observation. After collecting additional material, the team returns to the sniper hide site under the cover of darkness and begins constructing the hide. Team personnel—

- Post security.
- Remove the topsoil (observe construction discipline).
- Dig a pit. Dispose of soil properly and reinforce the sides. Ensure the pit has—
  - Loopholes.
  - A bench rest.
  - A bed.
  - A drainage sump (if appropriate).
- Construct an overhead cover.
- Construct an entrance and exit by escape routes selected.
- Camouflage the hide.
- Inspect the hide for improper concealment (continuous).

HIDE SITE CONSTRUCTION TECHNIQUES

4-281. The sniper can construct belly and semipermanent hide sites of stone, brick, wood, or turf. Regardless of material, he should ensure the following measures are taken to prevent enemy observation, provide adequate protection, and allow for sufficient fields of fire:

- *Frontal Protection*. Regardless of material, every effort is made to bulletproof the front of the hide position. The most readily available material for frontal protection is the soil taken from the hide site excavation. It can be packed or bagged. While many exotic materials can be used, including Kevlar vests and armor plate, weight is always a consideration. Several dozen empty sandbags can be carried for the same weight as a Kevlar vest or a small piece of armor plate.
- *Pit*. Hide construction begins with the pit since it protects the sniper team. All excavated dirt is removed (placed in sandbags, taken away on a poncho, and so forth) and hidden (plowed fields, under a log, or away from the hide site).
- *Overhead Cover*. In a semipermanent hide position, logs should be used as the base of the roof. The sniper team places a dust cover over the base (such as a poncho, layers of empty sandbags, or canvas), a layer of dirt. The team spreads another layer of dirt, and then adds camouflage. Due to the various materials, the roof is difficult to conceal if not countersunk.

- *Entrance.* To prevent detection, the sniper team should construct an entrance door sturdy enough to bear a man's weight. The entrance must be closed while the loopholes are open.
- *Loopholes.* The construction of loopholes requires care and practice to ensure that they afford adequate fields of fire. These loopholes should have a large diameter (10 to 14 inches) in the interior of the position and taper down to a smaller diameter (4 to 8 inches) on the outside of the position. A position may have more than two sets of loopholes if needed to cover large areas. Foliage or other material that blends with or is natural to the surroundings must camouflage loopholes. The loopholes must be capable of being closed when the door is open.
- *Approaches.* It is vital that the natural appearance of the ground remains unaltered and camouflage blends with the surroundings. Remember, construction time is wasted if the enemy observes a team entering the hide; therefore, approaches must be concealed whenever possible. Teams should try to enter the hide during darkness, keeping movement around it to a minimum and adhering to trail discipline. In built-up areas, a secure and quiet approach is needed. Teams should avoid drawing attention to the mission and carefully plan movement. A possible ploy is to use a house search with sniper gear hidden among other gear. Sewers may be used for movement, also.

> **WARNING**
> When moving through sewers, teams must be alert for booby traps and poisonous gases.

TOOLS AND MATERIALS NEEDED TO CONSTRUCT A SNIPER HIDE

4-282. The tools and materials needed to build a sniper hide depend on the soil, the terrain, and the type of hide to be built. Figure 4-35 lists various items that the sniper team should consider during construction.

- Entrenching tools.
- Bayonets.
- GP nets.
- Ponchos.
- Waterproof bags.
- Rucksacks.
- Shovels.
- Axes and hatchets.
- Hammers.
- Machetes.
- Chisels.
- Saws (Hacksaws).
- Screwdrivers, pliers, garden tools.
- Garbage bags.
- Wood glue.
- Nails.
- Chicken wire, newspapers, flour, water.

Figure 4-35. Items Used to Construct the Sniper Hide

HIDE SITE ROUTINES

4-283. Although the construction of positions may differ, the routines while in position are the same. The sniper should have a stable firing platform for his weapon and the observer needs a steady platform for the optics. When rotating observation duties, the sniper weapon should remain in place, and the optics are handed from one member to the other. Data books, observation logs, range cards, and the radio should be placed within the site where both members have easy access to them. The team must arrange a system of resting, eating, and making latrine calls. All latrine calls should be done during darkness, if possible. A hole should be dug to conceal any traces of latrine calls.

SNIPER RANGE CARD, OBSERVATION LOG, AND MILITARY SKETCH

4-284. The sniper team uses range cards, observation logs, and military sketches to enable it to rapidly engage targets. These items also enable the sniper to maintain a record of his employment during an operation.

RANGE CARD

4-285. The range card (Figure 4-36, page 4-81) represents the target area as seen from above with annotations indicating distances throughout the target area. It provides the sniper team with a quick-range reference and a means to record target locations since it has preprinted range rings on it. These cards can be divided into sectors by using dashed lines (Figure 4-37, page 4-82). This break provides the team members with a quick reference when locating targets. A field-expedient range card can be prepared on any paper the team has available. The sniper team position and distances to prominent objects and terrain features are drawn on the card. There is not a set maximum range on either range card, because the team may also label any indirect fire targets on its range card. Information contained on both range cards includes the—

- Sniper's name and method of obtaining range.
- Left and right limits of engageable area.
- Major terrain features, roads, and structures.
- Ranges, elevation, and windage needed at various distances.
- Distances throughout the area.
- Temperature and wind. (Cross out previous entry whenever temperature, wind direction, or wind velocity changes.)
- Target reference points (azimuth, distance, and description).

4-286. Relative locations of dominant objects and terrain features should be included. Examples include—

- Houses.
- Bridges.

- Groves.
- Hills.
- Crossroads.

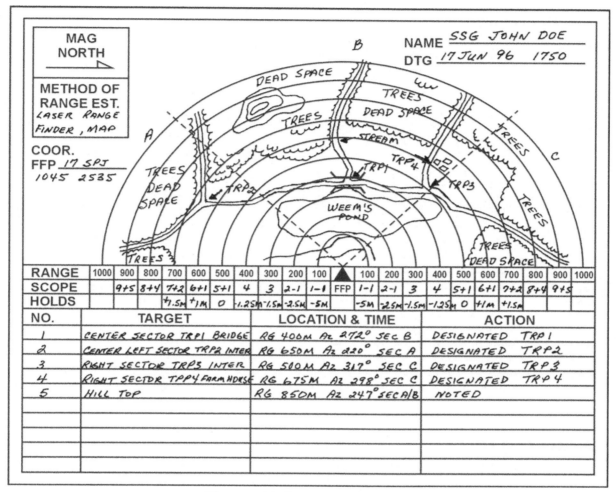

Figure 4-36. Sample Range Card

4-287. The sniper team will indicate the range to each object by estimating or measuring. All drawings on the range card are from the perspective of the sniper looking straight down on the observation area.

4-288. The range card is a record of the sniper's area of responsibility. Its proper preparation and use provides a quick reference to key terrain features and targets. It also allows the sniper team to quickly acquire new targets that come into their area of observation. The sniper always uses the range card and the observation log in conjunction with each other.

FM 3-05.222

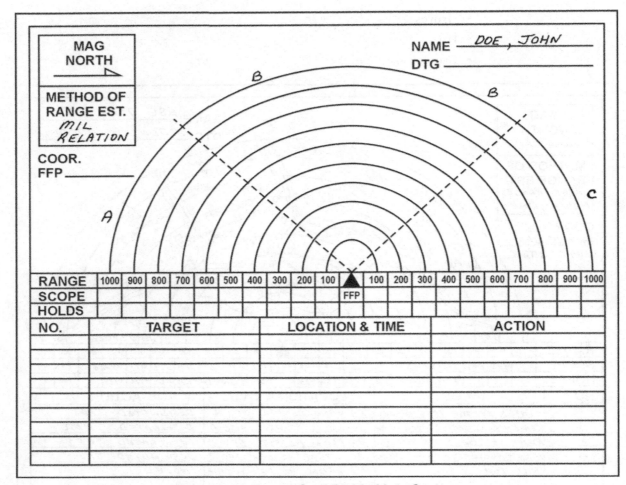

Figure 4-37. Range Card Divided Into Sectors

OBSERVATION LOG

4-289. The observation log (Figure 4-38, page 4-83) is a written, chronological record of all activities and events that take place in a sniper team's area (Appendix K). The log starts immediately upon infiltration. It is used with military sketches and range cards; this combination not only gives commanders and intelligence personnel information about the appearance of the area, but it also provides an accurate record of the activity in the area. Information in the observation log includes the—

- Grid coordinates of the sniper team's position.
- Observer's name.
- Date and time of observation and visibility.
- Sheet number and number of total sheets.
- Series number, time, and grid coordinates of each event.
- Events that have taken place.
- Action taken.

SNIPER'S OBSERVATION LOG			SHEET ____ OF ____ SHEETS	
ORIGINATOR: DOE, JOHN R.	DATE: 17 JUN 96	TOUR OF DUTY: 17 JUN 96 - 18 JUN 96	LOCATION: 17SPJ 10452535	
SERIAL	TIME	GRID COORDINATE	EVENT	ACTIONS OR REMARKS
1	0300	17SPJ 10452535	OCCUPIED POSITION	OBSERVATION
2	0340	SAME	TRUCK DROVE BY HEADING NORTH	EMPTY
3	0420	SAME	PFC JUDSON ASSUMED OBSERVATION	I RESTED
4	0530	SAME	TRUCK DROVE BY HEADING SOUTH	WITH 4 SOLDIERS
5	0630	SAME	PREPARED RANGE CARD AND TOPOGRAPHIC SKETCH	LIGHT ENOUGH TO SEE
6	0655	17SPJ 10452535	BRM CROSSED BRIDGE GONE SOUTH	OBSERVED
7	0700	17SPJ 10452535	PREPARED SKETCH OF BRIDGE GL03117631	COMPLETE
8	0900	17SPJ 10452535	MISSION COMPLETED - RETURN TO CP	END OF MISSION

Figure 4-38. Sample Observation Log

4-290. The sniper log will always be used in conjunction with a military sketch. The sketch helps to serve as a pictorial reference to the written log. If the sniper team is relieved in place, a new sniper team can easily locate earlier sightings using these two documents as references. The observer's log is a ready means of recording enemy activity, and if properly maintained, it enables the sniper team to report all information required.

4-291. Sniper observation logs will be filled out using the key word SALUTE for enemy activity and OAKOC for terrain. When using these key words to fill out the logs, the sniper should not use generalities; he should be very specific (for example, give the exact number of troops, the exact location, the dispersion location).

- The key word SALUTE:
    - **S** - Size.
    - **A** - Activity.
    - **L** - Location.
    - **U** - Unit/Uniform.
    - **T** - Time.
    - **E** - Equipment.

FM 3-05.222

- The key word OAKOC:
  - **O** - Observation and fields of fire.
  - **A** - Avenues of approach.
  - **K** - Key terrain.
  - **O** - Obstacles.
  - **C** - Cover and concealment.

MILITARY SKETCH

4-292. The sniper uses a military sketch (Figure 4-39) to record information about a general area, terrain features, or man-made structures that are not shown on a map. These sketches provide the intelligence sections a detailed, on-the-ground view of an area or object that is otherwise unobtainable. These sketches not only let the viewer see the area in different perspectives but also provide detail such as type of fences, number of telephone wires, present depth of streams, and other pertinent data. There are two types of military sketches: road or area sketches and field sketches. The sniper should not include people in either of these sketches.

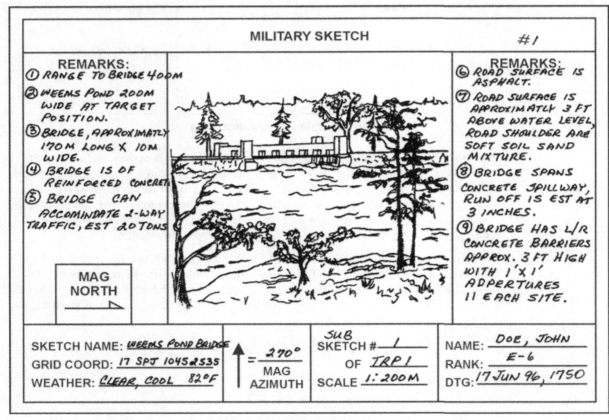

Figure 4-39. Sample Military Sketch

4-84

Road or Area Sketch

4-293. This sketch is a panoramic representation of an area or object drawn to scale as seen from the sniper team's perspective. It shows details about a specific area or a man-made structure (Figure 4-40). Information considered in a road or area sketch includes—

- Grid coordinates of sniper team's position.
- Magnetic azimuth through the center of sketch.
- Sketch name and number.
- Scale of sketch.
- Remarks section.
- Name and rank.
- Date and time.
- Weather.

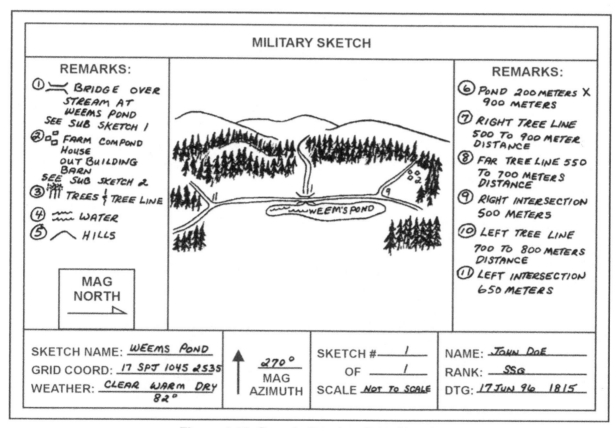

Figure 4-40. Sample Road or Area Sketch

Field Sketches

4-294. A field sketch is a topographic representation of an area drawn to scale as seen from above. It provides the sniper team with a method for describing large areas while showing reliable distance and azimuths between major features. This type of sketch is useful in describing road systems, flow of streams and rivers, or locations of natural and man-made obstacles. The field sketch can also be used as an overlay on the range card. Information contained in a field sketch includes—

- Grid coordinates of the sniper team's position.
- Left and right limits with azimuths.
- Rear reference with azimuth and distance.
- Target reference points.
- Sketch name and number.
- Name and rank.
- Date and time.
- Weather and visibility.

4-295. The field sketch serves to reinforce the observation log. A military sketch is either panoramic or topographic.

PANORAMIC SKETCH

4-296. The panoramic sketch is a picture of the terrain in elevation and perspective as seen from one point of observation (Figure 4-41).

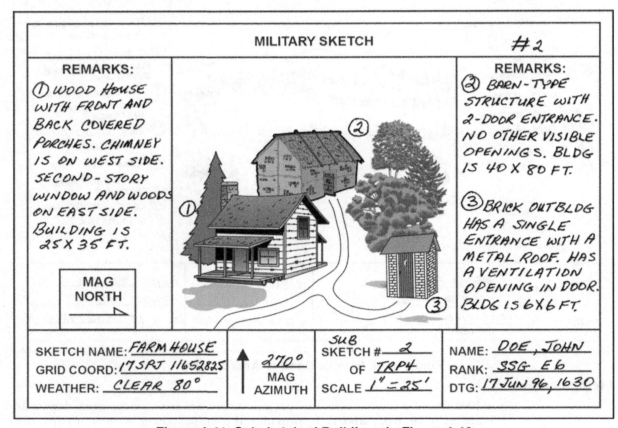

Figure 4-41. Subsketch of Buildings in Figure 4-40

TOPOGRAPHIC SKETCH

4-297. The topographic sketch is similar to a map or pictorial representation from an overhead perspective. It is generally less desirable than the panoramic sketch because it is difficult to relate this type of sketch to the observer's log. It is drawn in a fashion similar to the range card. Figure 4-42 represents a topographic sketch or an improvised range card.

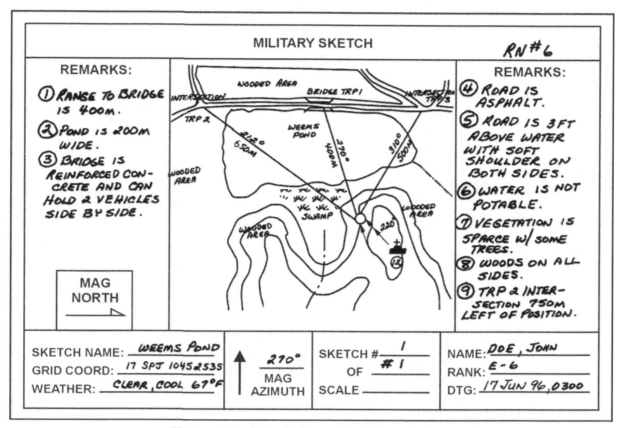

Figure 4-42. Sample Improvised Range Card

GUIDELINES FOR DRAWING SKETCHES

4-298. As with all drawings, artistic skill is an asset, but satisfactory sketches can be drawn by anyone with practice. The sniper should use the following guidelines when drawing sketches:

- *Work from the whole to the part.* First determine the boundaries of the sketch. Then sketch the larger objects such as hills, mountains, or outlines of large buildings. After drawing the large objects in the sketch, start drawing the smaller details.

- *Use common shapes to show common objects.* Do not sketch each individual tree, hedgerow, or wood line exactly. Use common shapes to show these types of objects. Do not concentrate on the fine details unless they are of tactical importance.

- *Draw in perspective; use vanishing points.* Try to draw sketches in perspective. To do this, recognize the vanishing points of the area to be sketched. Parallel lines on the ground that are horizontal vanish at a point on the horizon. Parallel lines on the ground that slope downward away from the observer vanish at a point below the horizon. Parallel lines on the ground that slope upward, away from the observer, vanish at a point above the horizon. Parallel lines that recede to the right vanish on the right and those that recede to the left vanish on the left (Figure 4-43).

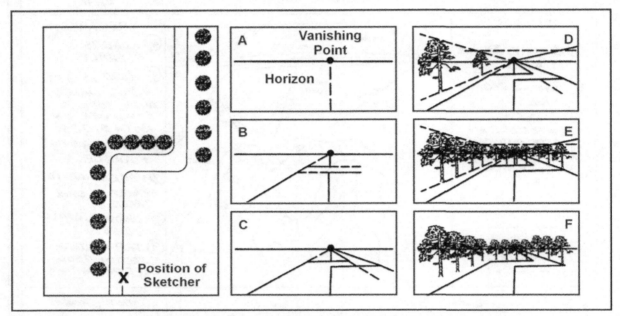

Figure 4-43. Sample Sketch Using Vanishing Points

4-299. For the sniper team to thoroughly and effectively observe its area of responsibility, it must be aware of the slightest change in the area. These otherwise insignificant changes could be an indicator of targets or enemy activity that needs to be reported. By constructing a panoramic sketch, the team has a basis for comparing small changes in the surrounding terrain. Updating data permits the team to better report intelligence and complete its mission.

GENERAL PRINCIPLES OF SKETCHING

4-300. The sniper initiates the panoramic sketch only after the observer's log and range card have been initiated and after the sniper team has settled into the AO.

4-301. The sniper studies the terrain with the naked eye first to get an overall impression of the area. After he obtains his overall impression, he uses binoculars to further study those areas that attracted his attention before the first mark is made on a sketch pad.

4-302. Too much detail is not desirable unless it is of tactical importance. If additional detail is required on a specific area, the sniper can make subdrawings to supplement the main drawing.

Principles of Perspective and Proportionality

4-303. Sketches are drawn to perspective whenever possible. To be in perspective, the sketcher must remember that the farther away an object is, the smaller it will appear in the drawing. Vertical lines will remain vertical throughout the drawing; however, a series of vertical lines (such as telephone poles or a picket fence) will diminish in height as they approach the horizon. Proportionality is representing a larger object as larger than a smaller object. This gives depth along with perspective.

Using Delineation to Portray Objects or Features of the Landscape

4-304. The sketcher forms a horizontal line with the horizontal plane at the height of his eye. This is known as the eye-level line and the initial control line. The skyline or the horizon and crests, roads, and rivers form other "control lines" of the sketch. These areas are drawn first to form the framework within which the details can be placed. The sketcher should represent features with a few, rather than many, lines. He should create the effect of distance by making lines in the foreground heavy and making distance lines lighter as the distance increases. He can use a light "hatching" to distinguish wooded areas, but the hatching should follow the natural lines of the object (Figure 4-44).

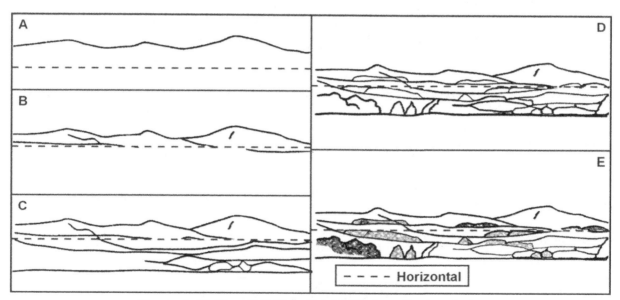

Figure 4-44. Use of Delineation to Portray the Landscape

Using Conventional Methods to Portray Objects

4-305. If possible, the sketcher should show the actual shape of all prominent features that may be readily selected as reference points. These features may be marked with an arrow and with a line to a description; for example, a prominent tree with a withered branch. The sketcher should also show—

- Rivers and roads as two lines that diminish in width to the vanishing point as they recede.
- Railroads in the foreground as a double line with small crosslines (which represent ties). The crosslines will distinguish them from roads. To portray railroads in the distance, a single line with vertical ticks to represent the telegraph poles is drawn. When rivers, roads, and railroads are all present in the same sketch, they may have to be labeled to show what they are.
- Trees in outline only, unless a particular tree is to be used as a reference point. If a particular tree is to be used as a reference point, the tree must be drawn in more detail to show why it was picked.
- Woods in the distance by outline only. If the woods are in the foreground, the tops of individual trees can be drawn.
- Churches in outline only, but it should be noted whether they have a tower or a spire.
- Towns and villages as definite rectangular shapes to denote houses. he also shows the locations of towers, factory chimneys, and prominent buildings in the sketch. Again, detail can be added in subdrawings or hatchings (Figure 4-45).
- Cuts, fills, depressions, swamps, and marshes are shown by using the usual topographic symbols.

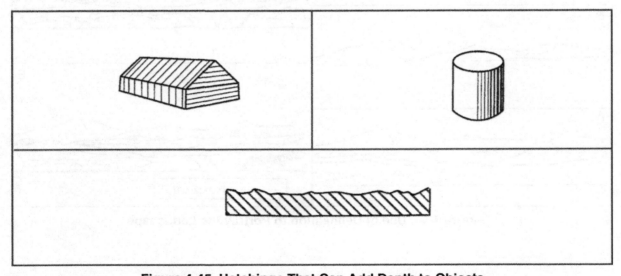

Figure 4-45. Hatchings That Can Add Depth to Objects

Using a Legend to Label the Sketch

4-306. The legend includes the title of the sketch, the date-time group, and the sketcher's signature. It also includes an explanation of the topographic symbols used in the sketch.

KIM GAMES

4-307. The name of the game comes from the book *Kim* by Rudyard Kipling. The story is about a young Indian boy who was trained to remember intelligence information during the British occupation of India. To assist some in remembering the name of the game, it has been misnamed—Keep in Mind (KIM). Sniper operations encompass a much larger scope than hiding in the woods, spotting targets of opportunity, and engaging them. The sniper must observe vast areas and accurately record any and all information. Because many situations occur suddenly and do not offer prolonged observation, snipers must learn to observe for short periods of time and extract the maximum amount of information from any situation.

4-308. KIM games are a series of exercises that can help increase the sniper's abilities to both perceive reality and retain information. They can be conducted anywhere, in very little time, with a large return for the trainer's investment of effort and imagination. Although the various time limits of viewing, waiting, and recording the objects are often not reflected in tactical reality, KIM games are designed to exercise the mind through overload (much the same as weight training overloads the muscles).

4-309. Advancement in KIM games is measured by shortening the viewing and recording times and lengthening the waiting time. Greater results can be realized by **gradually** adding additional elements to increase confusion and uncertainty. In the sniper's trade, the perception of reality often means penetrating the enemy's deception measures. These measures may include, but are not limited to—

- Misdirection.
- Disguise.
- Exchange.

4-310. There is a marked similarity between the above list and the principles of stage magic. Just as knowing how a magician performs a trick takes the "magic" from it, knowing how one is being deceived negates the deception.

THE BASIC GAME

4-311. The instructor will require a table, a cover, and an assortment of objects. He selects ten objects and randomly places them on the table. He should not place the objects in orderly rows, since studies have shown that objects placed in rows make memorization easier. He then covers the objects. The instructor briefs the students on the following rules before each iteration:

- No talking is allowed.
- Objects may not be touched.
- Students will not write until told to do so.

4-312. The students gather around the table. The instructor removes the cover, and the time for viewing begins. When the time is up, he replaces the cover, and students return to their seats. After a designated interval, the students begin to write their observations within a designated time limit. To aid in retaining and recording their observations, the following standardized categories are used throughout:

- Size.
- Shape
- Color.
- Condition.
- What the object appears to be.

4-313. It must be stressed that the above categories are not intended for use in a tactical setting.

THE SAVELLI SHUFFLE

4-314. A variation of the KIM game that trains the eyes to "look faster" and coordinates hand-to-eye movement is the Savelli Shuffle. Two individuals face each other approximately 5 meters apart. The first man has a bag containing a number of yellow rubber balls and a smaller number of red rubber balls. The second man has an empty bag. The first man reaches into his bag and picks out a ball, concealing it from the second man. The first man tosses the ball to the second man. Speed will depend on level of experience.

4-315. The second man has a quick decision to make—catch yellow balls with the left hand and red balls with the right hand. The second man then places the caught balls into his bag.

4-316. This process is repeated until the first man's bag is emptied. Positions of the first man and the second man are exchanged. Advancement in this exercise is measured by the speed at which the balls are thrown and the distance between men.

INTEREST AND ATTENTION

4-317. When learning to observe, team members must make a distinction between interest and attention. Interest is a sense of being involved in some process, actual or potential. Attention is a simple response to a stimulus, such as a loud noise. Attention without interest cannot be maintained for very long. During long periods of uneventful observation, attention must be maintained through interest. Deception at the individual level can be thought of as manipulation of interest.

Chapter 5

Employment

SF sniper employment is complex. When employed intelligently, skillfully, and with originality, the SF sniper will provide a payoff far greater than would be expected from the assets used. For this to happen, the planner must have more than a basic knowledge of the SWS. He must understand the capabilities and limitations of the sniper. However, sniping is an individual talent and skill that varies with each sniper. This trait compounds the planner's challenge, but he can minimize these variables with careful planning. The SF sniper, when properly trained and employed, can be one of the SO forces' most versatile weapons systems. See Appendix L for some specific tricks of the trade that a sniper must master to maintain employment proficiency.

METHODS

5-1. The sniper planner must apply methods of interdiction in relation to the necessary target and the desired effects against it. The employment of SF snipers generally falls into the following four categories.

SURVEILLANCE AND RECONNAISSANCE

5-2. Sniping, by nature of its execution (stealthful movement, infiltration, use of long-range optics, and limited-visibility operations), is closely related to reconnaissance and surveillance. The techniques a sniper uses to hunt a target are similar to those the scout uses to conduct surveillance—only the end results are different. Also, human intelligence (HUMINT) collection is a secondary function to sniping. Operational planners should refrain from employing snipers in solely HUMINT roles but should take advantage of the HUMINT function when possible. Combining both functions would be analogous to using a long-range guard—the sniper provides needed information and can intercede if necessary.

POINT INTERDICTION

5-3. The sniper's goal is to interdict targets for the purpose of impeding, destroying, or preventing enemy influence in a particular area. A point interdiction is essentially hunting a specific target. The SF sniper can interdict both personnel and material point targets in support of SO missions. Such missions tend to be complex and may require difficult infiltration, precise navigation to the target, evasion of enemy forces, the broaching of sophisticated security systems, and external mission support systems (safe houses, special intelligence). Normally, the more complex the target or the more protected it is, the greater the degree of sophistication required to defeat it. For instance, a protected personnel target may require detailed intelligence and a highly skilled sniper for successful interdiction. Point interdiction also includes firing situations like those encountered in counterterrorist situations.

LONG-RANGE HARASSMENT

5-4. Long-range harassment is not intended to be decisive; creating psychological fear in the enemy and restricting his freedom of action are the sniper's primary goals. The sniper has the greatest latitude of employment in harassment missions. He can often engage opportunity targets at his discretion but always within the constraints of the mission. This method may include harassing specific kinds of targets to disrupt key functions such as command and control (C2) procedures. In some situations, the sniper can afford to engage targets at extreme ranges and risk nonfatal or missed shots, which maximizes harassment by interdicting more targets. Snipers normally conduct harassment at extended ranges to take advantage of their ability to engage targets at distances beyond the enemy's small-arms fire. This practice normally means they will not engage targets closer than 400 meters—100 meters beyond the common effective-fire range of conventional small arms. The average range for harassing fire is 600 meters.

SECURITY OPERATIONS

5-5. Snipers can provide long-range security to deny an enemy freedom of action in a particular area. The sniper security mission can take the form of a series of mutually supporting sniper outposts or cordons. An example of security operations where snipers proved invaluable was during the USMC operations in Beirut, Lebanon. The Marine snipers were interwoven with traditional defenses and proved to be effective in long-range protection of local U.S. facilities and interests. Security and cordon missions normally entail static, defenselike operations. However, with the austere firepower of sniper teams and their inability to maneuver in defensive warfare, they are vulnerable to becoming decisively engaged. Therefore, security operations are best integrated into conventional security and reaction forces to help snipers increase their defensive capability. Without such support, snipers can easily be suppressed and maneuvered upon with fire and maneuver tactics.

PLANNING

5-6. When employing snipers, the operational planner must consider many factors. Tactical planning considerations of the sniper include hide selection, deception plans, and movement techniques. However, the planner must consider sniper employment from an even higher level of operational perspective. He must realize that snipers are a unique weapon system and possess entirely different attributes from conventional forces, among which (the one most frequently misunderstood) is the sniper's firepower. Unlike conventional small-arms fire that emphasizes volume, the sniper's firepower emphasizes precision. Sniper fire is most effective when combined with a mind that can exploit long-range precision. A two-man sniper team can deliver only limited volumes of fire, and no matter how accurate, the volume seldom equals that of even the most austere military units. If employed incorrectly, the sniper easily becomes another soldier on the battlefield—except that he is handicapped with a slow-firing weapon. The sniper's unique employment considerations should be guided by the following factors.

STANDOFF

5-7. The planner should base employment around the sniper's ability to engage targets at extended ranges. The maximum effective range will vary with each sniper. However, planners can establish nominal engagement ranges based on the sniper's ability to group his shots into a specified area or shot group. This measurement can, in turn, be applied to specific targets. The sniper should be able to keep his fire within 2 MOA shot groups under simulated combat conditions.

5-8. The application of group size is important for determining maximum standoff in relation to target size. For planning, SF snipers should be expected to provide instant incapacitation (nonreflexive impact) first-round shots on personnel to 200 meters; personnel interdiction with 90-percent probability to 600 meters on stationary targets; and 50-percent probability to 800 meters. Engagement of more complicated targets, such as those moving or in adverse environmental conditions, depends on the individual sniper's skill and his weapon's capability. The sniper should be able to hit 100 percent of assigned moving targets at 200 meters and 90 percent at 300 meters. The sniper should also be able to hit 100 percent of short, 3-second, exposure targets at 200 meters and 90 percent at 300 meters. Sniper employment planning should also consider the probability of error against the risks incurred if the shot misses. Such analysis will help determine the minimum standoff range for a reliable chance of a hit on the target.

DECEPTION

5-9. The sniper's most critical tools are his deceptive talents. To planners, deception is also important for operational needs. The SF sniper may use a weapon from another country to duplicate using that weapon's characteristic signature (ballistic characteristics, cartridge case, bullets) for cover. He must consider both operational and tactical deception methods when conducting each mission.

Operational

5-10. Planners may center operational deception on infiltrating the target area using a clandestine (concealed) sniper weapon. Operational deception may also require plausible deniability of the operation and lead the enemy to believe the target damage was the result of normal failure, accident, or some other form of sabotage. With such interdiction requirements, the sniper can use special weapons and munitions and aim for vulnerable points to purposely obtain such results. (Such targets include those that tend to burn, detonate, or self-destruct when shot.) However, this kind of deception is not possible with many targets and is especially difficult to conceal in personnel interdiction. Deception also means a sniper can seldom fire more than twice from any location as the sound of shots (even suppressed) is increasingly easier to locate with repetition. (This concept differs greatly from many media and war stories, where the sniper engages his enemy on a protracted basis from the same location—firing shot after shot with apparent impunity.) In reality, snipers locked in a decisive dual with enemy forces and firing defensively will normally lose as sniping seldom succeeds in such situations. Planners should refrain from employing snipers in missions that will not allow deception or concealment after firing.

Tactical

5-11. The sniper's use of tactical deception is often his only real security. Employment planners must consider security from operational aspects of the mission when using snipers. These include infiltration means, communications procedures, and methods of C2. These procedures are important because the sniper must remain covert before interdiction to ensure success. Normally, once the sniper fires, he is no longer covert and must rely on other plans to facilitate escape. Many environments may permit sniper employment, but few allow plausible denial for the sponsor or operation after interdiction. In other words, such covert operations may be easy to perform, but the risk of compromise, no matter how small, may overshadow the mission. Missions to collect information concerning another country's hostile intentions may themselves provoke serious repercussions if discovered. Moreover, using snipers will assuredly indicate an alternative motive to actually interdicting a target—which could compromise the mission even more.

TIME

5-12. The sniper's mission normally requires more time than conventional operations. Because the sniper normally moves on foot with stealth, his only defense is that of remaining unseen. If the sniper does not have enough time to execute the mission, he may hurry and unnecessarily compromise the mission or fail to reach the target.

TEAM EMPLOYMENT

5-13. Teams provide limited security for self-protection and allow near-continuous operations, yet are small enough to allow concealment for execution. In practice, one sniper fires while the other observes. The sniper-observer identifies and selects targets, adjusts the sniper-firer to environmental factors, provides security, and helps correct missed shots. However, the greatest advantage is the sniper-observer's detachment from the firing process, leaving the sniper-firer to concentrate on the act of firing. In other words, firing does not complicate the sniper-observer's decision process—a task requiring total concentration. Mission needs may also require snipers to be part of a larger force or in multiple sniper teams to engage the same target. Both techniques of employment can enhance the sniper's effect; however, the basic sniper team should always be retained. Snipers will never be employed in elements smaller than a two-man team, larger elements of three or four men may be required depending upon mission, duration, visibility expected in the target area, and size of target area.

TERRAIN

5-14. Terrain features are extremely important to the sniper's mission. Some areas, such as those that are densely wooded, tightly compartmented, or heavily vegetated, are not suitable for sniper employment because they reduce the sniper's ability to employ the full standoff capability of his weapons system. The threat can quickly suppress snipers that engage targets inside their minimum standoff envelope (usually 400 meters). Moreover, restrictive terrain offers the threat cover and concealment that can mask his attack against the sniper. The sniper must always consider both maximum and minimum engagement ranges; he must never get so close to the target as to compromise the mission.

INNOVATION

5-15. A sniper's most important attribute is his ability to improvise. The operational planner must also be innovative in the planning process. The sniper is a weapon of opportunity, not one to be used as a matter of course. Planners must actively seek missions and opportunities to apply the sniper's unique attributes of long-range precision rifle fire and concealment. Often the sniper's greatest handicap is the planner's inability to fully use his potential because of the planner's lack of familiarity with the sniper's true role and capability. Planners should include the sniper team in the planning process. Multiple sniper teams can often suggest a better solution when planning from the bottom up. Staff officers with little practical sniper experience or lacking innovative thought will never be able to fully take advantage of the sniper's capabilities.

ORGANIZATION

5-16. Organizational grouping of snipers above the sniper-team level normally occurs through expedient pooling of sniper pairs into larger organizations. Such centralized grouping of sniper assets can prove beneficial to their employment for specific missions. In all cases, the sniper specialist within the unit should manage control of the snipers. Regardless of any provisional or temporary sniper grouping, sniper teams should not be split. They function best in the pairs in which they have trained, with all members being fully qualified snipers.

5-17. The level at which sniping is organized and managed directly influences the ability of sniping to provide direct or indirect support to friendly operations. Centralized organization and management of sniping provides a great degree of flexibility regarding deployment. This flexibility permits snipers to be deployed to areas or locations where they will have the greatest influence on the enemy and provide the maximum support to friendly operations.

5-18. The organization of sniper teams will magnify their effectiveness against the enemy. Sniping, like any other supporting arm, is an individual specialty that requires independent action to achieve its greatest potential effect on the enemy. Requiring special organization, snipers may be organized into teams, squads, sections, and platoons.

SNIPER TEAM

5-19. The base element of any sniper unit is that the team consists of two equally trained snipers and is assigned to the company. The company is the lowest level at which sniping can be centralized and still maintain operational effectiveness. Sniper teams should not be attached to the tactical subunits of the company. However, a subunit, squad or platoon, may be attached to the sniper team as security or cover for a stay-behind type mission. When organized into a team, snipers are able to—

- Provide mutual security.
- Diminish stress.
- Lengthen their duration of employment.
- Aid in the engagement of targets more rapidly.

SNIPER SQUAD

5-20. A sniper squad is composed of three to four sniper teams and is located at battalion level. The organization of the sniper squad is as follows:

- Squad leader.
- Assistant squad leader.
- Three senior snipers.
- Three junior snipers.
- The sniper pairs include a senior sniper and a junior sniper.

5-21. The mission of the sniper squad is to support the operations of the battalion. The squad may be broken and the separate teams attached to any company in the battalion.

SNIPER SECTION

5-22. The sniper section's mission is to directly or indirectly support the combat operations of brigade or regiment subordinate units. In direct support, sniper teams are attached to company or battalion headquarters elements as needed, and employment considerations are identical to those of company sniper teams. Indirect support gives the sniper teams assigned sectors of responsibility as part of the battalion fire plan. The sniper section is attached to the brigade regiment headquarters S-2 or S-3, and the section commander acts as the brigade sniper coordinator. The sniper section consists of a **command element** (section commander, assistant section leader), a **support element** (armorer, radio operator), and **8 to 10 operational two-man sniper teams** (per team—senior sniper, junior sniper).

SNIPER PLATOON

5-23. The mission of a sniper platoon is to support division combat and intelligence operations independently or by attachment to division subunits. When attached, sniper squads should remain intact and should be attached no lower than battalion level. The sniper platoon is composed of a platoon leader, a platoon sergeant, a radiotelephone operator or driver, an armorer, and three sniper squads consisting of a squad leader and five two-man sniper teams. The sniper platoon falls under direct operational control of the division intelligence officer or indirect control through liaison with the sniper platoon leader. Sniper platoon operations may include deep penetration of the enemy rear areas, stay-behind operations, and rear-area protection.

COMMAND AND CONTROL

5-24. C2 of snipers is accomplished using indirect and direct control procedures. These procedures complement the sniper's self-discipline in executing his assigned mission. The sniper team will often operate in situations where direct control methods will not be possible. Therefore, the sniper must execute his mission (within the parameters of the commander's intent) on personal initiative and determination. This reaction is a major reason (in the sniper-selection process) why personnel with motivation and self-determination are required as snipers. Without these personal traits, the sniper's decentralized execution allows total disregard for the mission and its completion. In other words, he can go out to perform a mission and merely stay out of sight until time to return.

INDIRECT CONTROL OF SNIPERS

5-25. Commanders can accomplish indirect control of snipers through a variety of methods, the simplest being rules of engagement (ROE) and fire control measures. Even with strict direct control (voice radio, wire) of sniper teams, commanders should establish ROE and fire control to maximize flexibility and prevent unnecessary engagements. The ROE will normally designate combatant forces and situations that will allow the sniper to engage the enemy.

5-26. One significant problem with contemporary ROE is the restrictive measures used in peacetime operations. Often, such ROE will specify enemy personnel as only those presenting a direct threat to friendly forces or requiring verbal warning before engagement. The paradox is that a sniper's modus operandi is to engage targets that are not a direct threat to him (outside small-arms effective fire range) at the moment, but which later may be. It is extremely difficult for the sniper to stay within ROE because once the enemy gets within his minimum standoff, the conflict can become one of close-quarters battle and a 12-pound, scope-sighted sniper rifle is no match for an AK47 or M16 at close quarters—despite the fact that it may be a semiautomatic rifle. Therefore, ROE for the sniper must provide for his safety by adding security forces or by removing him from the operation.

5-27. Fire control measures are just as important for the sniper as they are for indirect-fire weapons and aircraft. As with any long-range weapons system, positive target identification is difficult at extended ranges—even with the advanced optics the sniper will carry. Establishment of no-fire zones or times, fire coordination lines, and free-fire zones or times will help in sniper C2 by establishing guidelines for when and where he can fire. If positive target identification is required, then appropriate security measures are required to prevent decisive engagement to the sniper.

DIRECT CONTROL OF SNIPERS

5-28. Commanders can maintain direct control of SF snipers by using technical and nontechnical systems, including radio and wire communications. In some circumstances, direct control means may include commercial telephones or other nontraditional tactical forms of communications. The mission and the operational environment will determine the exact methods of control.

5-29. Nontechnical control of snipers involves using prearranged methods including rendezvous, message pickups and drops, and other clandestine methods of secure communications. In denied areas, or those with electronic interception capabilities, these methods may be the only secure techniques for communicating with the sniper teams. These systems, although often quite secure, tend to be slow, and execution is complex.

5-30. Snipers can also use many forms of technical communications systems such as radio and wire. Both radio and wire offer near-instant message traffic and facilitate C2 with two-way communications. Snipers most often use radios as their method of communications because they are responsive and provide real-time control and reporting capabilities. Also, radio (voice, data burst, or satellites) provides the mobility that snipers require for their mobile-employment methodology. The major advantage to radio is its ability to transmit mission changes, updates, and intelligence in a timely manner. However, when properly arrayed, enemy direction-finding assets can determine the location of even the most focused and directional transmissions. To avoid detection, SO must use specialized communications techniques and procedures. Even then, the deployed teams will still have the problem of transmitting from their location to confirm messages or send data.

5-31. The major drawback to radio communications is the transmitter's electronic signature. In the sniper's operational area, enemy detection of any electronic signature can be just as damaging as reception of a message. Once the enemy is aware of the sniper's presence (through spurious transmissions), it becomes an academic problem to hunt him down. Even with successful evasion from threats (for example, scent- and visual-tracking dogs), the sniper team will be preoccupied with evasion and escape (E&E) instead of the target. Of course, this act can also be an objective—to divert enemy internal security forces to a rear-area sniper threat.

5-32. Under all conditions, the sniper team must have a method for immediate recall. This method allows for immediate reaction forces to be available for the sniper. This factor is imperative in an unstable battlefield.

5-33. Wire communications can provide protection from enemy deception, jamming, and interception. Static security operations, defensive positions, and extended surveillance posts are suitable for the use of wire communications. However, the sniper team must also calculate the disadvantages of wire, such as time to emplace, lack of mobility, and relative ease of compromise if found by the enemy. When possible, the team should back up wire communications by more flexible forms of control, such as radio.

5-34. Certain environments (FID or CBT missions) may allow for more flexible communications techniques. For example, the use of commercial telephones may be more appropriate than traditional military communications. Also, many environments possess a low threat from actual message interception or direction-finding assets, which allows the sniper team more liberal use of the radio. However, planners would be wise to remember the time-tested proverb: Never underestimate your enemy.

COORDINATION

5-35. There must be meticulous coordination with both supported and unsupported units that fall within the sniper team's AO. This coordination is the sniper coordinator's main focus; however, the sniper team must ensure that the coordination has been accomplished. Coordination with supported and unsupported units includes the following:

- Nature, duration, and extent of local and extended patrols.
- Friendly, direct, and indirect fire plans.
- Local security measures.
- Location and extent of obstacles and barrier plan.
- Rendezvous and linkup points.
- Passage and reentry of friendly lines.
- Unit mission and area of responsibility.
- Routes and limits of advance.
- Location and description of friendly units.
- Communication plan.

5-36. Although it is important that the sniper team receives as much information as possible for mission success, the sniper coordinator must not tell the team so much that, if captured, the entire sector would be compromised. This objective demands that everyone involved—the sniper teams, the sniper coordinator, supported and unsupported units in the area of operations—communicate and remain "coordinated." The sniper coordinator will establish control measures to assist in the deployment of the teams. This will keep the teams and units from committing fratricide, while not compromising the units if the teams are captured.

5-37. Once coordination begins, the team must establish control measures to protect the sniper and the supported and unsupported friendly units. Also, if the situation changes, there will be a recall capability to prevent the sniper team from unnecessary danger. The sniper team must also receive warning that friendly operations are in the area and it could be subjected to friendly fire. The team must have enough latitude to avoid engagement with the enemy by remaining mobile, elusive, and unpredictable. However, the team must understand that operational areas with lines of advance, exclusion areas, and no-fire zones are designed to protect him and friendlies and are not to be violated.

SUPPORT RELATIONSHIPS

5-38. Sniping is a combat support activity. Snipers should augment only those units that have a specific need for it. Sniping provides either indirect or direct support. Deployed as human intelligence (HUMINT) assets, snipers indirectly support friendly units and operations. There are two types of direct support.

Operational Control

5-39. Snipers are under the operational control (OPCON) of the supported unit only for the duration of a specific operation. At the end of the mission, they return to the control of the parent unit. This practice is the optimal method of supporting operations, as it is flexible and efficient toward the unit to which the snipers are attached.

Attachment

5-40. For extended operations or distances, snipers may support a specific unit. This unit responds to all the sniper team's requirements for the duration of attachment. The sniping specialist or coordinator should also be attached to advise the unit on assignment of proper employment methods. If it is not possible to attach the sniper coordinator, then the senior and most experienced sniper on the attachment orders must assume the job as sniper coordinator for the period of attachment. The receiving unit must also understand the status of the sniper coordinator and the importance of his position. Normally, attachment for extended periods will include supply and logistics support to the sniper element from the unit of attachment.

5-41. The support given to the unit and support received from the unit can also determine the planning, coordination, and control requirements. The four types of support given to a unit are as follows:

- Offensive operational.
- Defensive operational.
- Retrograde operational.
- Special operations.

TARGET ANALYSIS

5-42. There are two general classes of sniper targets—personnel and material. The sniper coordinator can further categorize these targets as having either tactical or strategic value. Tactical targets have local, short-term value to the current battle or situation. Tactical personnel targets are normally of enough significance to warrant the risk of detection when firing. Such targets include enemy snipers, key leaders, scouts, and crew-served weapons personnel. Tactical material targets are of particular importance to the current operation.

5-43. Strategic personnel targets are not as well-defined as tactical personnel targets because of problems with the concept and definition of assassination. The definition of assassination versus the elimination of a military target is based on the end result. If the end result is military, then the target is classified as military ambush. However, if the end result is political in nature, then the target is classified as assassination and possibly illegal. This is a simplified definition of a complex issue and further discussion is beyond the scope of this manual.

5-44. Strategic material targets consist of all types of objects of a military nature, including components or systems within a target (such as a turbine in an aircraft). The sniper must always consider criticality, accessibility, recuperability, vulnerability, effect, recognizability (CARVER) in evaluating the target.

TARGET SYSTEMS AND CRITICAL NODES

5-45. SF snipers should be directed at the enemy's C2 facilities and the critical nodes supporting them. Snipers can frequently regard targets as being in an interrelated system; that is, any one component may be essential to the target's entire operation. These interrelated and essential components are known as critical nodes. Critical C2 nodes are components, functions, or systems that support a military force's C2. These will differ for each target, but they will generally consist of the following:

- *Procedures.* Snipers can easily interdict the procedures, routines, and habits the enemy uses to conduct operations. Of most significance, snipers can create fear in the enemy that will cause him to take extreme measures in security or to modify procedures to keep from being shot. The enemy may curtail certain functions, divert assets for security, or restrict movement in his own rear areas to prevent interdiction.

- *Personnel.* Personnel targets are critical, depending on their importance or function. The target does not necessarily need to be a high-ranking officer but may be a lower-ranking person or a select group of people, such as a skill or occupational group, who are vital to the enemy's warfighting apparatus.

- *Equipment.* Equipment is critical when the loss of it will impact the enemy's conduct of operations. Seldom will singular equipment targets be so critical as to impact the enemy in any significant fashion. However, targets or components that are not singularly critical may, collectively, be vital to the enemy. Common targets include objects common to all other similar targets or systems vulnerable to interdiction, such as a particular component (a radar antenna) which is common to many other radars. Interdicting only one antenna would have limited effect; it would merely be replaced. However, interdicting other radar components would significantly impair the enemy's logistics.

- *Facilities.* These activities and complexes support the enemy's operations or C2 functions. In the larger context, snipers are not suited for such interdiction. However, where possible, the sniper can focus on critical elements, such as C2 nodes or logistics capabilities of the larger facility (power generation systems or transportation equipment).

- *Communications.* Communications nodes often are the most fragile components of C2 systems. Snipers can usually interdict these nodes because they are easy to recognize and frequently quite vulnerable. Attacking other targets, not critical in some fashion, serves no purpose in using SF snipers and only wastes resources without a definable objective. Target analysis helps determine which critical nodes to interdict and predict how effective the sniper will be.

CARVER PROCESS

5-46. Target analysis includes selecting the appropriate method to use against a target, such as an aircraft, a strike force, or snipers. In doing so, the planner can match the sniper's capabilities to the potential target. Sniper capabilities include using special weapons and performing covert operations.

5-47. Attacking targets by sniper fire requires detailed planning and coordination; the sniper should not attack these targets indiscriminately. Interdiction must occur within the parameters of the assigned mission from higher headquarters to the stated results, maximizing the target's vulnerabilities and the priority of interdiction (on multiple targets or components).

5-48. The target analysis system that snipers use is based on the acronym CARVER. The CARVER analysis process is a generic model for SO interdiction missions. It is also suitable for sniper interdiction, particularly during material interdiction planning, which is similar to interdiction with special munitions or demolitions. The sniper can apply sniper fire within the framework of the CARVER model to better determine if sniping would be the appropriate interdiction method and precisely how and where to apply it. Mission planners apply the CARVER analysis to sniper interdiction based on the following criteria:

- *Criticality*. A target is critical in relation to the impact its destruction would have on the enemy. The mission order will largely determine critical targets. However, within a target system there may be components that may be critical for the operation of the entire target. For example, a turbine is a critical component of a jet aircraft. The concept of attacking a critical component (using accurate fire) allows the sniper to engage a much greater variety of targets than commonly accepted.

- *Accessibility*. This factor is based on how readily the target can be attacked. For the sniper, target accessibility includes getting through the target's security systems (security police or intrusion detectors) and knowing what the reaction will be to the sniper's stand-off interdiction. Accessibility for sniper interdiction is unique, because the sniper can frequently engage targets without violating security systems, and in turn, reduce the enemy's ability to detect the sniper before the interdiction. Again, the sniper must base accessibility on both maximum and minimum ranges. He must also be able to get off the target after he shoots.

- *Recuperability*. The sniper measures recuperability of a target based on the time it takes the target to be repaired, replaced, bypassed, or substituted. Frequently, planners think only in terms of total destruction as opposed to a lesser degree of destruction. However, the same effect can be achieved by simply shutting down the target or destroying one vulnerable component. The advantage of interdiction short of total destruction is in the application of force; complete destruction normally requires a more elaborate force and more units. Also, the ability to control target destruction with precision fire can prevent unnecessary damage. It can limit adverse effects to systems that the local populace may depend on for electrical power, food, or water. However, the planners must take into consideration repair time. If the target can be repaired in less time than the target window allows, the another destruction technique must be considered.

- *Vulnerability.* A target (or component) is vulnerable to the sniper if he has the weapons and skill required to interdict the critical points that the target analysis has identified. The key to target vulnerability is identifying the weakest critical link in the target system and destroying it. The sniper must match the weapon to the target.

- *Effect.* A wide range of interdiction effects are possible. Target effect is the desired result of attacking the target, including all possible implications—political, economic, and social effects of the interdiction. Occasionally, the planner must decide what is the desired effect. It may be the removal of key personnel, the psychological impact of the interdiction, or the threat of interdiction. Planners must always consider the balance of effect on the overall mission and the effect on the populace. When an adverse effect on the populace outweighs the effect on the mission, the sniper must reconsider the mission.

- *Recognizability.* A target is recognizable if it can be effectively acquired by the sniper. A target may be well within the sniper's standoff range but cannot be effectively engaged because the target is masked or concealed. For example, the sniper's recognition of targets using night-vision equipment might be restricted because of the technological limitations of the device. Positive identification of targets, as well as small target components, is difficult given the characteristics of the phosphor screen in NVDs. Other factors complicating recognizability include the time of day, light conditions, terrain masking, environmental factors, and similar nontargets in the area.

5-49. The fear of interdiction is evident in the German attempts to kill Winston Churchill in World War II, which forced him to remain hidden for some time. Conversely, John M. Collins' book, *Green Berets, SEALs, and SPETSNAZ*, details the political implications of a DA mission to kill a figure such as Emperor Hirohito of Japan, the "emperor-god," during World War II. Such action, Collins states, "would have had an adverse impact by rallying the Japanese people." A similar reaction was seen when the U.S. bombed Libya in 1986. During the raid, U.S. bombs seriously injured one of Colonel Mu`ammar Gadhafi's children and resulted in negative media and international backlash. (Despite Gadhafi's unscrupulous acts, endangering his family was unacceptable to the international public.)

5-50. Material target interdiction by sniper fire is much more limited than it is with personnel targets. The SF sniper's abilities could increase by his choice of special weapons to interdict material targets, but he might still be limited by the relative vulnerability of the target. The greatest obstacle for successful interdiction of material targets rests primarily with the identification of the vulnerable nodes. The goal of the sniper's fire on these nodes is to be as effective as more powerful weapons—using precision fire at key points instead of brute force in the general area.

MISSION PLANNING

5-51. Successful accomplishment of a sniper mission relates directly to the planning and preparation that takes place. Each mission requires the expertise of different people at each planning level.

LEVELS OF PLANNING

5-52. The two levels of mission planning are above-team level and team level. At above-team level the sniper employment officer (SEO) or sniper leader plans and coordinates the actions of more than one sniper team. The intent of this directive is to have several teams carry out coordinated or independent missions toward the same objective. At team level, the members of the sniper team will carry out the planning, preparation, and coordination for the mission. Therefore, warning orders are not necessary at this level, and the following sniper operation order (Figure 5-1, pages 5-14 through 5-16) itself is a mission planning tool.

---

1. **SITUATION**
    a. Enemy forces.
        (1) Weather. Light data, precipitation, temperature, effect on the enemy and the sniper team.
        (2) Terrain. Terrain pattern, profile, soil type, vegetation, and fauna, effect on the enemy and the sniper team.
        (3) Enemy. Type unit(s), identification, training, and presence of countersnipers, significant activities, and effect on the sniper team.
    b. Friendly forces. Adjacent units, left, right, front, and rear. Since sniper teams are vulnerable to capture, they should not receive this information. Rather, they should receive information such as the location of free-fire and no-fire zones.

2. **MISSION**
   Who, what, where, when.

3. **EXECUTION**
    a. Commander's intent. This paragraph relates specifically what is to be accomplished, in a short, precise statement. It should include the commander's measure of success.
    b. Concept of the operation. This paragraph relates step-by-step how the mission will proceed. Breaking the mission down into phases works best. Specific tasks will be carried out in each phase, usually starting from preinfiltration to exfiltration and debriefing.
    c. Fire support. Normally, in a deep operation, fire support will not be available. However, in other situations, the assets may exist.
    d. Follow-on missions. This paragraph will outline any follow-on missions that may be needed. Once the primary mission is accomplished, the sniper team may be tasked to carry out another mission in the AO before exfiltration. This duty may consist of another sniper mission or a linkup with another team, unit, or indigenous persons as a means of exfiltration.

Figure 5-1. Sniper Operation Order

e. Coordinating instructions. Consist of the following:
  (1) Actions at the objective. This paragraph contains specifically the duties of each member of the team and their rotation to include—
    (a) Security.
    (b) Selection and construction of the hide.
    (c) Removal of spoils.
    (d) Camouflage and fields of fire.
    (e) Observer's log, range card, and military sketch.
    (f) Placement of equipment in the hide.
    (g) Maintenance of weapons and equipment.
    (h) Observation rotation.
  (2) Movement techniques. This paragraph will cover the movement techniques, security at halts, and responsibilities during movement to and from the ORP and the hide and during the return trip.
  (3) Route. This paragraph covers the primary and alternate routes to and from the objective area. It may also include the fire support plan if it is not included in the fire support annex.
  (4) Departure and reentry of friendly positions. This information is normally used in the support of conventional forces, but it could be used when dealing with indigenous persons, for example, during a linkup.
  (5) Rally points and actions at rally points. In some instances, these points can be used, but for a two-man element a rendezvous is much more advisable. For example, several rendezvous points en route should be preplanned with a specific time or period for linkup. These designations are set so that movement is constantly toward the objective, preventing the lead man from backtracking and wasting time. It is not advisable for a two-man team to attempt to use en route rally points.
  (6) Actions on enemy contact. This paragraph stresses minimal contact with the enemy. The team should avoid contact and not engage in a firefight. It is best to avoid contact, even in an ambush; evade as best as possible. The team should not attempt to throw smoke or lay down a base of fire. This action calls attention to the team's position and will cause the enemy to pursue with a much larger element. If an air attack occurs, hide. It is not possible for a two-man team to successfully engage an enemy aircraft.
  (7) Actions at danger areas. Avoid danger areas by moving around them, unless this is not possible or time is critical. When moving across large open areas, stalk across; do not move in an upright posture. Linear danger areas are best crossed by having both team members move across the area at the same time after an extended listening halt. This will avoid splitting the team in case of enemy contact and lower the risk of compromise while traversing the danger area.
  (8) Actions at halts. Security is critical even when taking a break and nobody is expected in the area. Stay alert.

Figure 5-1. Sniper Operation Order (Continued)

(9) Rehearsals. If time is not available, at the minimum, always practice actions at the objective. During rehearsals, practice immediate action drills (IADs) and discuss actions at rally and rendezvous points. The team must know these points and the routes on the map. It is also important that the team rehearses any previously untrained actions.

(10) Inspections. Team members should inspect their equipment. Use a checklist for equipment and ensure that everything works. The team should have the proper equipment and camouflage for the terrain and the environment that it will encounter. Inspections should be conducted prior to infiltration, after infiltration, and finally in the ORP before occupying the FFP.

(11) Debriefing. This paragraph covers who will attend the debriefing, where it will occur, and when it will take place. The observer's log and military sketches become useful information-gathering tools during the debriefing (Appendix M).

(12) Priority intelligence requirements (PIR)/information requirements (IRs). These requirements are given to the sniper team as information that should be gathered when the team is employed.

(13) Annexes. This section contains specific maps and sketches showing items such as routes, the fire support plan, the tentative ORPs, and the hide sites. It will also include the evasion and recovery (E&R) plan, sunrise sunset overlay, and terrain profiles.

4. **SERVICE SUPPORT**

This paragraph covers, but is not limited to, administrative items such as—

a. Rations.

b. Arms and ammunition that each team member will carry.

c. Uniform and equipment that each team member will carry.

d. Method of handling the dead and wounded.

e. Prisoners and captured equipment. This paragraph is not likely to be used, unless the equipment can be carried, photographed, or sketched.

f. Caches and mission support sites (MSSs).

5. **COMMAND AND SIGNAL**

a. Frequencies and call signs. It is not necessary to list all the frequencies and call signs. You need only to refer to the current signal operating instructions (SOI).

b. Pyrotechnics and signals, to include hand and arm signals. It is best to have a team SOP to which you can refer. Otherwise, you must list all the pyrotechnics and hand and arm signals.

c. Challenge and password. The challenge and password will be necessary when linking up at rendezvous points and passing through friendly lines.

d. Code words and reports. This refers to any contact made with higher headquarters or possibly a linkup with indigenous persons.

e. Chain of command.

Figure 5-1. Sniper Operation Order (Continued)

TERRAIN PROFILE

5-53. A terrain profile is an exaggerated side view of a portion of the earth's surface between two points. The profile will determine if LOS is available. The sniper leader can use line of sight to determine—

- Defilade positions.
- Dead space.
- Potential direct-fire weapon positions.

5-54. The sniper leader can construct a profile from any contoured map. Its construction requires the following steps:

- Draw a line from where the profile begins to where it ends.
- Find the highest and lowest value of the contour lines that cross or touch the profile line. Add one contour value above the highest and one below the lowest to take care of hills and valleys.
- Select a piece of notebook paper with as many lines as contours on the profile line. The standard Army green pocket notebook or any paper with quarter-inch lines is ideal. If lined paper is not available, draw equally spaced lines on a blank sheet.
- Number the top line with the highest value and the rest of the lines in sequence with the contour interval down to the lowest value.
- Place the paper on the map with the lines parallel to the profile line.
- From every point on the profile line where a contour line, a stream, an intermittent stream, or a body of water crosses or touches, drop a perpendicular line to the line having the same value. Where trees are present, add the height of the trees to the contour.
- After all perpendicular lines are drawn and tick marks placed on the corresponding elevation line, draw a smooth line connecting the marks to form a horizontal view or profile of the terrain. (The profile drawn may be exaggerated. The space of lines on the notebook paper will determine the amount of exaggeration.)
- Draw a straight line from the start point to the finish point on the profile. If the straight line intersects the curved profile, line of sight is not available.

SUNRISE/SUNSET OVERLAY

5-55. A sunrise/sunset overlay (SSO) is a graphic representation of the angle to the rising and setting sun and the objective. An SSO enables a team to plan a line of advance or tentative hide sites to take best advantage of the light. An SSO requires a table showing the true azimuth of the rising sun and the relative bearing of the setting sun for all months of the year. An SSO is constructed in the following manner:

- Using the projected date of the mission and the latitude of the target, determine the true azimuth of the sunrise from Table 5-1, pages 5-19 and 5-20.
- Using a protractor and a straightedge, draw a line from the objective along the true azimuth.

- Subtract the true azimuth from 360 to find the sunset azimuth.
- Using a protractor and a straightedge, draw another line from the objective along the sunset azimuth.
- Convert each azimuth to a back azimuth and write it on the appropriate line.
- Label the appropriate lines sunrise and sunset.
- Write down the latitude and the date that was used to construct the overlay.

SNIPER SUPPORT IN SPECIAL OPERATIONS MISSIONS AND COLLATERAL ACTIVITIES

5-56. Special operations (SO) forces plan, conduct, and support activities in all operational environments. The following paragraphs explain how the SF sniper supports each mission and activity.

CIVIL AFFAIRS AND PSYCHOLOGICAL OPERATIONS

5-57. The misuse of sniper interdiction can adversely affect Civil Affairs (CA) and civic action programs sponsored by friendly organizations. The sniper is a very efficient killer and given a target will go to extreme efforts to interdict it. Therefore, planners must temper the use of force with common sense and the future goals of the operation. It may be easier to eliminate threats than to negotiate, but in the long run, negotiations may open the door for settlement where sniping may close it or may set the stage for undesirable reactions.

5-58. Planners must also consider the psychological operations (PSYOP) aspects of the mission, including both positive and negative impacts. The sniper can project not only accurate weapons fire but also tremendous psychological destruction. Such impact was given as rationale for the Vietnam My Lai massacre. There, in defense of their actions, some soldiers claimed that enemy sniper fire (and friendly casualties) over prolonged time drove them to commit the war crimes. On the other end of the spectrum, U.S. use of snipers can also cause adverse reaction on enemy forces. As at My Lai, the enemy may focus on innocent noncombatants and commit inappropriate reprisals in response to intense sniper pressure. This practice is especially true in UW and FID environments where U.S. SO forces may use local populations as guerrillas and security forces.

5-59. The psychological impact of sniping has received little attention in the overall scheme of war. Historians often focus on the large weapons systems and overlook the stress and fear that sniping adds to the battlefield. Yet, this psychological impact can ruin the fiber and morale of an entire army; for example, in World War I, the sniper's bullet was often feared far more than many other ways of dying.

5-60. The U.S. military has only recently recognized the psychological impact of sustained combat, although the sniper has always contributed as much to fear as he has to fighting. Operational planners may consider this PSYOP capability when planning sniper missions, especially when using PSYOP in UW where it plays a vital role.

Table 5-1. Finding Direction From the Rising or Setting Sun

	Date	Angle to North From the Rising or Setting Sun (level terrain) Latitude												
		0°	5°	10°	15°	20°	25°	30°	35°	40°	45°	50°	55°	60°
January	1	113	113	113	114	115	116	117	118	121	124	127	155	141
	6	112	113	113	113	114	115	116	118	120	123	127	132	140
	11	112	112	112	113	113	114	115	117	119	122	125	130	138
	16	111	111	111	112	112	113	114	116	118	120	124	129	136
	21	110	110	110	111	111	112	113	115	117	119	122	127	133
	26	109	109	109	109	110	111	112	113	115	117	120	124	130
February	1	107	107	108	108	108	109	110	111	113	115	117	121	126
	6	106	106	106	106	107	107	108	109	111	113	115	118	123
	11	104	104	105	105	105	106	107	108	100	110	112	116	120
	16	103	103	103	103	103	104	105	106	107	108	110	112	116
	21	101	101	101	101	101	102	102	103	104	105	107	109	112
	26	99	99	99	99	100	100	100	101	102	103	104	106	108
March	1	98	98	98	98	99	99	99	100	100	101	102	104	106
	6	96	96	96	96	96	97	97	97	98	98	99	100	102
	11	94	94	94	94	94	94	95	95	95	96	96	97	98
	16	92	92	92	92	92	92	92	92	93	93	93	93	94
	21	90	90	90	90	90	90	90	90	90	90	90	90	90
	26	88	88	88	88	88	88	88	88	87	87	87	87	96
April	1	86	86	86	86	85	85	85	85	84	84	83	82	81
	6	84	84	84	83	83	83	83	82	82	81	80	79	77
	11	82	82	82	82	81	81	81	80	80	79	77	76	74
	16	80	80	80	80	79	70	78	78	77	76	74	72	70
	21	78	78	78	78	78	77	76	76	75	73	72	69	66
	26	77	77	76	76	76	75	75	74	72	71	69	66	63
May	1	75	75	75	74	74	73	73	72	70	69	66	63	59
	6	74	74	73	73	73	72	71	70	68	67	64	61	56
	11	72	72	72	72	71	70	69	68	67	64	62	58	52
	16	71	71	71	70	70	69	68	67	65	63	60	55	49
	21	70	70	70	69	69	68	67	65	63	61	58	53	47
	26	69	69	69	68	68	67	66	64	62	60	56	51	44
June	1	68	68	68	67	66	66	64	63	61	58	54	49	40
	6	67	67	67	67	66	65	64	62	60	59	53	48	40
	11	67	67	67	66	66	64	63	62	59	56	53	47	39
	16	67	67	67	66	65	64	63	62	59	56	53	47	39
	21	67	67	67	66	66	64	63	62	59	56	53	47	39
	26	67	67	67	66	65	64	63	62	59	56	53	47	39

NOTES 1: When the sun is rising, the angle is reckoned from east to north. When the sun is setting, the angle is reckoned from the west to north.
2: This chart is for the Northern Hemisphere.

Table 5-1. Finding Direction From the Rising or Setting Sun (Continued)

		Angle to North From the Rising or Setting Sun (level terrain) Latitude												
	Date	0°	5°	10°	15°	20°	25°	30°	35°	40°	45°	50°	55°	60°
July	1	67	67	67	66	65	64	63	62	59	56	53	47	39
	6	67	67	67	66	66	65	64	62	60	57	53	48	40
	11	68	68	68	67	66	65	64	63	61	58	54	49	41
	16	69	68	68	68	67	66	65	64	62	59	55	50	43
	21	69	69	69	69	68	67	66	65	63	60	57	52	45
	26	70	70	70	70	69	68	67	66	64	62	59	54	48
August	1	72	72	72	71	71	70	69	68	66	64	61	57	51
	6	73	73	73	73	72	71	71	69	68	68	63	60	55
	11	75	75	74	74	74	73	72	71	70	68	66	63	58
	16	76	76	76	76	75	75	74	73	72	70	68	65	61
	21	78	78	77	77	77	76	76	75	74	72	71	68	65
	26	79	79	79	79	79	78	78	77	76	75	73	71	68
September	1	82	82	82	81	81	81	80	80	79	78	77	75	73
	6	83	83	83	83	83	83	82	82	81	81	80	78	77
	11	85	85	85	85	85	85	85	84	84	83	83	82	81
	16	87	87	87	87	87	87	87	86	86	86	85	85	84
	21	89	89	89	89	89	89	89	89	89	89	88	88	88
	26	91	91	91	91	91	91	91	91	91	91	92	92	92
October	1	93	93	93	93	93	93	93	94	94	94	95	95	96
	6	95	95	95	95	95	96	96	96	97	97	98	98	100
	11	97	97	97	97	97	98	98	99	99	100	101	102	104
	16	99	99	99	99	99	100	100	101	101	102	104	105	108
	21	101	101	101	101	101	102	102	103	104	105	107	109	112
	26	102	102	193	103	103	104	104	105	106	108	109	112	115
November	1	104	104	105	105	105	106	107	108	109	110	113	116	120
	6	106	106	106	107	107	108	109	110	111	113	115	119	123
	11	107	107	108	108	108	109	110	111	113	115	117	121	126
	16	109	109	109	109	110	111	112	113	115	117	120	124	130
	21	110	110	110	111	111	112	113	114	116	119	122	126	133
	26	111	111	111	112	112	113	114	116	118	120	124	128	135
December	1	112	112	112	113	113	114	115	117	119	122	125	130	138
	6	112	112	113	113	114	115	116	118	120	123	126	132	140
	11	113	113	113	114	115	116	117	118	121	124	127	133	141
	16	113	113	113	114	115	116	117	118	121	124	127	133	141
	21	113	113	113	114	115	116	117	118	121	124	127	133	141
	26	113	113	113	114	115	116	117	118	121	124	127	133	141

NOTES 1: When the sun is rising, the angle is reckoned from east to north. When the sun is setting, the angle is reckoned from the west to north.
2: This chart is for the Northern Hemisphere.

UNCONVENTIONAL WARFARE OPERATIONS

5-61. In a UW environment, the SF sniper provides an additional capability to the resistance force. The primary mission of the resistance force is to support conventional forces during times of war. Therefore, the SF sniper must know conventional sniper tactics as well as unconventional techniques to effectively train a U.S.-sponsored resistance force. During peacetime, mobile training teams (MTTs) can train foreign military or paramilitary forces. In times of war, the training takes place during the organization and training phase of the resistance force after linkup.

5-62. The importance of a sniper in UW cannot be measured only by the number of casualties he inflicts upon the enemy. Realization of the sniper's presence instills fear in enemy troops and influences their decisions and actions. Selective and discriminate target interdiction not only instills fear in the enemy, but can lead to general confusion and relocation of significant enemy strengths to counter such activity.

5-63. In UW, the SF sniper can perform as a fighter and a trainer. Not only can he teach sniper skills to the force he is training; he can act as a direct action asset when needed. The sniper's ancillary skills in camouflage, stalking, surveillance, and deception are also useful in the UW environments. The impact of these talents is magnified when the sniper acts as a trainer. By training others he is, in effect, performing interdiction much more efficiently than he could alone.

5-64. UW or guerrilla warfare (GW) consists of three major phases: buildup, consolidation, and linkup. Snipers will play an important role in all three phases.

Buildup

5-65. During initial contact and buildup, SF snipers will mainly train the indigenous force snipers and then act as sniper coordinators.

5-66. During the buildup, snipers are extremely effective when used in the harassing and sniper ambush role. By using the snipers' ability to deliver long-range precision rifle fire, the UW force can accomplish the following objectives all at once:

- Be able to strike at the enemy forces while minimizing their own exposure.
- Deny the comfort of a secure area to the enemy.
- Build UW force morale with successes while minimizing the amount of UW force exposure.
- Since the fires are discriminatory, maintain a positive effect on the civilian population, as civilian casualties are minimized. This also reinforces in the civilians' minds the inability of government forces to control that part of the countryside.

5-67. However, it is very important that the snipers go after targets with a military objective only. The line between sniper ambush and assassination at this point can be unclear. The sniper must remember that an ambush is for military gain, while an assassination is for political gain. Assassination, under any guise, is illegal due to Executive Order 12333, Part II, paragraph 2-11, dated 4 December 1981.

5-68. During the end of the buildup and before the consolidation phase, the UW snipers will be used the same as strike operations snipers; that is, in support of small raids and ambushes. As the size of the UW force grows, so will the size of the missions that are similar to strike missions.

Consolidation

5-69. During consolidation, as the UW force becomes larger, the role of the sniper reverts to that of the conventional sniper. The same missions, tactics, and employment principles apply.

Linkup

5-70. During and after linkup, the snipers will mainly act as part of the security force and rear area protection (RAP) force. The UW force snipers will be particularly suited for this role. They have spent their time in that area and should know most, if not all, of the main areas that could support the enemy during infiltration and rear area attacks.

5-71. During the initial contact phase of a resistance movement, sniper employment will normally be limited to supporting small-unit operations and will include such actions as—

- *Harassment of Enemy Personnel*. When performed at ranges greater than 500 meters, harassment serves to lower the enemy's morale and inhibit his freedom of movement.
- *Infiltration*. Before an attack, snipers may infiltrate enemy units' positions and establish themselves in the enemy's rear area. During the attack, the infiltrated snipers engage specific targets of opportunity to divert the enemy's attention from the attacking units and to disrupt his freedom of movement in his rear areas.
- *Interdiction*. The snipers will delay or interdict reinforcing elements to a target and deny the enemy use of an area or routes by any means.
- *Multiple Team, Area Sniper Ambush*. This type of ambush involves multiple sniper teams operating together to engage targets by timed or simultaneous fire. Each sniper will fire a fixed number of rounds, and the ambush will end when either the targets have been successfully engaged or the predetermined number of shots have been fired. Planning considerations must include how the ambush is to be initiated, how the snipers will communicate with each other, and what methods the snipers will use to engage the targets.
- *Security and Surveillance*. Snipers are employed to gather information or to confirm existing intelligence by long-term surveillance of a target site. They may also be used to provide early warning of impending counterattacks. Snipers will normally establish a hide position to conduct their surveillance.
- *Offensive Operations*. During the advanced stages of the combat phase of a resistance movement, snipers may be used to detect and engage long-range targets that could impede the progress of the offensive

element. The teams must be ready to assume the defensive role immediately after the offensive operation.

- *Defensive Operations.* Snipers are best used in defensive operations outside the forward line of troops (FLOT) to provide early warning of enemy approach, disorganize his attack, and cause him to deploy early. Snipers may also be used to delay the enemy's advance by interdicting enemy movements using a series of interlocking delay positions, thus allowing the friendly forces to withdraw.

FOREIGN INTERNAL DEFENSE OPERATIONS

5-72. The primary role of SF snipers in FID is that of a teacher. During the passive FID role, SF snipers will be in-country for training and advising only and will not have an active role. During active FID, the SF snipers could find themselves in both a trainer's role and an active role. In either case, passive or active, the primary tactics will be that of conventional warfare—offense, defense, and withdrawal.

5-73. During active FID, the SF sniper will conduct counterguerrilla operations, sniper cordons and periphery observation posts (OPs), sniper ambushes, urban surveillance, and civil disorders.

5-74. Sniper participation in RAP is the main line of attack in accomplishing counterguerrilla operations. Snipers can enhance the protective measures surrounding sensitive facilities or installations by setting up observation posts along routes of access, acting as part of a reaction force when the rear area has been penetrated or patrolling the area (as members of established security patrols). They can then operate in a stay-behind role once the security patrol has moved on. In RAP operations, the sniper—

- Protects critical installations and sites.
- Covers gaps between units to avoid infiltration.
- Prevents removal of obstacles.
- Tracks enemy patrols known to have penetrated into the rear area.

5-75. The sniper's ancillary skills in camouflage, stalking, surveillance, and deception are also useful in the FID environment. The impact of these talents is magnified when the sniper acts as a trainer. By training others he is, in effect, performing interdiction much more efficiently than he could alone. Appendix N provides a sample sniper range complex (SRC) for the trainer's use.

SNIPER ELEMENT ORGANIZATION IN UW AND FID

5-76. In a UW or FID role, the sniper elements organize above-team-level size with elements under the control of the commander and the S-2. Depending upon the availability of trained personnel, the sniper elements should organize as a squad at battalion level (10 men or 5 teams) and as a section at regimental or brigade level. A sniper coordinator is required at regimental level and desirable at battalion level. He should be assigned to the S-2/G-2 staff for intelligence purposes. However, he must work closely with the S-3/G-3 staff for planning purposes. The sniper coordinator should be a sniper-qualified senior NCO, warrant officer, or officer who is well versed in mission planning. He must also be strong enough to ensure that the sniper teams are not improperly deployed. All other members of the squads, the platoon, and the platoon headquarters element must be sniper-qualified.

DIRECT ACTION OPERATIONS

5-77. DA operations are short-duration strikes and other small-scale offensive actions conducted by SOF to seize, destroy, or inflict damage on a specified target. When employed in DA missions, snipers will perform one or more of the following four functions.

Harassment

5-78. Snipers use deliberate harassment to impede, destroy, or prevent movement of enemy units. The degree of harassment depends on the amount of time and planning put into the operation. Harassment is best suited for protracted or unconventional operations. During such operations sniper casualties will be high, and provisions for their replacement must be included in the harassment plan.

Multiple Team, Area Sniper Ambush

5-79. The "sniper ambush" is when multiple sniper teams operate together to engage targets by timed or simultaneous fire. Each sniper fires a fixed number of rounds; the ambush ends either after target engagement or after all shots are fired. The planners for each ambush should always consider how the ambush will start, how the snipers will communicate with each other, and how they will engage the targets in the kill zone.

Sniper Cordon

5-80. A sniper cordon is a series of outposts surrounding a specific area. A sniper cordon can prevent the enemy from entering or leaving a target location. Snipers may operate in cordon operations by being integrated into the overall fire plan as a supporting force or in cordon areas as independent elements. Snipers should be used during cordon operations to maximize their precision long-range fire capabilities. Due to the snipers' limited volume of fire and reliance on stealth, they possess little capability to become decisively engaged during such operations. Once the snipers have been located, they may be suppressed by fire and maneuver or indirect fire. Therefore, the snipers' ability to hold or cordon an area will be directly commensurate to the enemy force encountered and the support from friendly units.

Interdiction

5-81. Interdiction is preventing or hindering enemy use of an area or route by any means. When deployed for interdiction, the snipers can restrain dismounted avenues of approach. Their ability to interdict vehicular traffic is limited to harassment unless armed with large-caliber SWSs. Snipers can deploy with vehicular interdiction elements to harass the enemy when it is forced to dismount. They can also cause armor vehicles to "button up," making them more vulnerable to antitank weapons.

THE STRIKE FORCE OF DA OPERATIONS

5-82. The size of the strike force depends on the mission, location of the target, and enemy situation. Planners tailor the strike force in size and capability to perform a specific mission. It can be a small team to interdict a personnel target or a larger force to destroy a large facility or plant. Regardless of size, most strike operations consist of command, security, support, and assault elements. Snipers can provide support to any of these elements depending on the objectives and needs of the commander. The requirements for the SF sniper in strike operations may include the elements discussed below.

Command Element

5-83. This element forms the primary command post and normally consists of the strike force commander and, as a minimum, his S-2/S-3 and fire support element controllers. The sniper coordinator also works with the command element. The snipers assigned to the command element are formed by the expedient pooling of strike force snipers. They are under the control of the sniping specialist. Regardless of their origin, pooled snipers stay in their original teams. Under the command element, snipers will be able to conduct reconnaissance and DA missions supporting the entire strike force or multiple missions supporting one or more strike force elements throughout the operation. Examples of these missions may include—

- Reconnoitering the ORPs, routes, or exfiltration sites.
- Reconnoitering and observing the objective (once action is initiated, covert OP snipers may perform a DA function in support of the strike force).
- Establishing a reserve to intervene or reinforce elements with precision rifle fire.
- Screening danger areas and vulnerable flanks or sealing off the enemy rear.

Security Element

5-84. Snipers may operate in conjunction with a larger security force or independently in support of the security mission. When sniper teams work with a larger security force, they should not collocate with crew-served weapons. This step will ensure that sniper fire is not suppressed by enemy fire directed at the crew-served weapons. The element will determine sniper employment by the scope of the operation and personnel constraints. The security element's missions include—

- Securing rallying points.
- Providing early warning of enemy approach.
- Blocking avenues of approach into the objective area.
- Preventing enemy escape.
- Acting as left, right, and rear security elements for the strike force.

5-85. In smaller operations, the security element could consists entirely of snipers. This would reduce personnel requirements. In larger operations, a larger, more flexible (antiarmor, demolitions) security force would be necessary, and snipers would serve to complement this security element's capabilities. For example, armored threats require augmentation by appropriate antiarmor weapons. Snipers can provide accurate long-range suppressive fire to separate infantry from their armored units and to force tanks to button up, which will hinder their ability to detect the launch of wire-guided missiles. The sniper team can employ large-bore sniper weapons to help delay and interdict light material targets.

5-86. Snipers performing security missions in DA operations are well suited to perform successive or simultaneous missions. They also provide early warning of delaying and harassing reaction forces. Reaction forces located some distance from the objective will approach using vehicles or aircraft. The mobility assets of the reaction force can be dedicated to that mission and can subsequently present an actual threat to the strike force. Snipers may operate as part of the security force to interdict or harass reaction force avenues of approach or landing zones (if known or obvious). In addition to the main role of security, the snipers may also—

- Report information before an assault.
- Support the assault force by fire (caution must be used here).
- Assist in sealing the objective during the assault.
- Maintain contact after the assault.
- Act as a rear guard during the withdrawal of the assault force.

Support Element

5-87. This element of the strike force must be capable of placing accurate supporting fire on the objective. It must deliver a sufficient volume of fire to suppress the objective and provide cover to the assault element. It also provides fire support to cover the withdrawal of the assault element from the objective.

5-88. Snipers in the support element provide discriminate fire in support of the assault force. The sniper's optics facilitate positive target identification and acquisition, which allows him to fire in close proximity to friendly forces with reduced risk of fratricide. This practice is opposed to more traditional automatic or indirect supporting fire that must terminate or shift as friendly forces approach the target area (referred to as "lift and shift"). At night, friendly troops can wear distinctive markings such as reflective tapes or infrared devices (visible to the sniper's night-vision equipment) to aid identification.

5-89. When assigned to the support element, snipers should organize into four-man sniper teams (two pairs working together). There are several reasons for this type of organization. First, the sniper team leader can better control the snipers' rate and control of supporting fire. Second, sniper elements centrally located can better redeploy to critical locations to delay pursuing forces. Third, limited vantage points from which to deliver precision rifle fire may exist. Concentrating snipers at these vantage points may be the only effective way to maximize their capabilities of long-range precision rifle fire. Again, as in the support role, snipers should not be collocated with crew-served weapon systems.

5-90. When snipers are assigned to the support element, their mission should be specific. The effectiveness of sniper fire is not in the volume, but the precision with which it is delivered. Sniper missions include—

- Disrupting C2 by engaging officers or NCOs directing the defense.
- Suppressing guards and enemy security forces.
- Providing precision covering force to the assault element.
- Precision reduction of hard points.
- Delaying pursuing forces after withdrawal.
- Maintaining contact with displaced enemy forces after the attack.
- Observing for enemy counterattacks or continued harassment of the enemy to disorganize any counterattack efforts.

5-91. One advantage of snipers in the support element is that they do not have to lift and shift as crew-served weapons do once the assault element is on the objective. The snipers can continue to support through precision rifle fire.

Assault Element

5-92. Snipers seldom operate with the assault element, mainly because of the need for rapid movement combined with suppressive fire. This type of maneuver seldom allows for the snipers' deliberate (sedentary) firing process. In addition, the assault element often participates in close-quarter battle—nullifying the snipers' standoff capability. However, snipers can support the element when C2 would be better effected or in circumstances where they can enhance the element's mission. They may provide cover fire when the assault element must pass through an area that is dead space from other supporting elements. However, the snipers would then support the assault element's movement to the objective and not be an actual part of it. They may also provide support by using aerial platforms (Appendix O).

ENEMY CONSIDERATIONS DURING DIRECT ACTION OPERATIONS

5-93. The type and number of enemy security forces likely to be manning the target or available for reaction must be considered in the plan. These forces may be static, foot-mobile, vehicle-mounted, or airmobile.

Enemy Security Forces

5-94. Mission planners will generally position armored vehicles on the perimeter; light vehicles will normally remain in a vehicle park. Armored vehicles are likely to become centers of resistance, around which defenders will concentrate during the action. This position will present the snipers with a high density of targets, particularly officers and NCOs who will tend to use static-armored vehicles as rally points. The lack of vehicular mobility on the part of the strike force renders them vulnerable to a mobile threat. In such circumstances, snipers should be delegated the task of interdicting routes of access to vehicle parks. Drivers of light vehicles are the primary targets; track or tank commanders are the prime armored-vehicle targets.

On-Site Defensive Positions

5-95. DA targets deep within enemy lines will generally have less protection and a lower defensive posture than those located close to the main battle area. Target site defenses can be either hasty or permanent.

5-96. Hasty defensive positions provide less protection to defending personnel than prepared ones. Strike force snipers are able to engage such positions at a greater distance with more effectiveness due to the limited protection to the targets. Snipers should consider any object or location at the target site that affords protection to the enemy (for example, behind light vehicles or in buildings) as a hasty defensive position.

5-97. Permanent defensive positions consist of bunkers, sandbagged fighting positions, or prepared buildings. Such targets present unique circumstances to the snipers. These well-protected targets, which often have narrow firing ports and are mutually supportive, make engagement difficult and require the snipers to move closer to the targets than normal. As the range to the targets decreases, the probability of detection and engagement from the enemy forces increases.

Enemy Reaction Force

5-98. Strike force snipers functioning in a support capacity, or as part of the strike force security element, will primarily target the enemy reaction force.

SPECIAL RECONNAISSANCE OPERATIONS

5-99. The SF sniper offers some advantages to SR missions. He is well trained in surveillance and his ability to interdict material targets at extended range is often complementary to follow-on SR missions. If interdiction of C2 systems is the goal of the follow-on mission, then snipers can carry significant potential destruction in the form of large-bore sniper rifles.

5-100. Snipers make extensive use of fixed and roving surveillance to acquire targets or assess their vulnerabilities. They will normally establish a hide position to conduct their surveillance. Once hidden, they will continue noting detailed information in their observation log. The log will serve as a record of events and assist in mission debriefing. The snipers will report all PIR and IRs as required.

5-101. Because of their mission-essential equipment, snipers are ideally suited to perform reconnaissance in conjunction with their primary DA mission. They can obtain information about the activity and resources of an enemy or potential enemy and secure data concerning the meteorological, hydrographic, or geographic characteristics of a particular area.

5-102. Snipers may need to reconnoiter enemy positions that are of specific interest to supported units. Information gathered by snipers includes, but is not limited to the locations of—

- FDCs.
- Crew-served weapons.

- Tactical operations centers (TOCs).
- Gaps in enemy wire.
- Listening posts (LPs) and OPs.
- Gaps between enemy units and positions.
- Infiltration routes.

5-103. Snipers may also infiltrate through enemy positions in support of offensive operations or to harass enemy rear areas. Once sniper teams have infiltrated enemy positions, their tasks may be to report information on—

- Troop strength and movements.
- Concentrations and reserve locations.
- OPs and weapons locations.
- Command, control, and communications facilities.

COUNTERTERRORISM OPERATIONS

5-104. The primary mission of SOF in counterterrorism (CT) is to apply specialized capabilities to preclude, preempt, and resolve terrorist incidents abroad. Snipers provide three primary functions in CT operations. They—

- Deliver discriminate fire to interdict hostile targets.
- Cover the entry teams into the objective area with rifle fire.
- Provide the CT force commander with his most accurate target intelligence.

5-105. In the last case, the commander will normally position the sniper in an ideal position to observe the enemy. Most frequently, this position will be the commander's only view of the target.

5-106. Counterterrorism operations require extensive training and coordination. Most important, the sniper teams must know the plans and actions of the entry teams to avoid possible injury to friendly personnel, and they must fire when told to do so. Failure to engage and neutralize a target can have devastating consequences, similar to what occurred in the 1972 Olympic games in Munich, Federal Republic of Germany. Snipers did not neutralize their terrorist targets on command. The result was that the terrorists were free to execute the hostages. To compound the problem, the snipers were so confused that they shot and killed several of their own men. Of course, overzealous snipers can create results similar to what occurred in Los Angeles, California. Police snipers shot and killed a bank president who was indicating a gunman by pointing his finger. The overanxious police sniper thought the man was pointing a gun and shot him. Obviously, the line between shoot and do not shoot is thin and can be stretched thinner by haste or indecisiveness.

5-107. Part of the solution to these problems lies in the selection and training process. During the selection process, an individual's mind is the one variable that a psychologist cannot effectively measure. In fact, oftentimes psychologists cannot agree on what traits to look for in a sniper. How does one pick a man to deliberately kill another man who presents no immediate threat to him personally? Unfortunately, the real test of a sniper comes only when it is time to pull the trigger. Only then will the sniper's reliability definitely be known.

5-108. Another problem that seems to manifest itself in CT scenarios is the Stockholm Syndrome. This type of reaction occurs when the sniper is unable to shoot a person who has become familiar to him. The syndrome manifests itself when the sniper has conducted constant surveillance of his target and becomes so familiar with the target's actions, habits, and mannerisms that the target becomes more human, almost well acquainted—too familiar to shoot. On the other hand, some reports have indicated the opposite to be true; some snipers hope to have the opportunity to shoot someone from some twisted, personal motivation. Perhaps this happened in Los Angeles. Nevertheless, these psychological extremes—eager or reluctant firers—are inappropriate to the sniper's function; the sniper must be somewhere in between.

COMBAT SEARCH AND RESCUE

5-109. In CSAR operations, the sniper's role is extremely limited because the mission is to rescue and not to interdict. However, the sniper can provide traditional long-range security and early warning to rescue forces. His ability to operate in denied areas can greatly assist the rescue forces by providing accurate information regarding the rescue. The sniper can infiltrate before the rescue and conduct surveillance of the rescue area unnoticed. The U.S. Air Force is considering using snipers with their pararescue units (in place of machine guns) to provide long-range security during rescue operations. This method would give them the benefit of selectively interdicting threat targets while not endangering innocent bystanders.

COUNTERSNIPER

5-110. A sniper team is the best asset available to a commander for a countersniper operation. The team plans and coordinates the operation to eliminate the enemy sniper threat. A countersniper operation occurs between two highly trained elements—the sniper team and the enemy sniper—each knowing the capabilities and limits of the other.

5-111. A sniper team's first task is to determine if there is a sniper threat. If so, it then identifies information that may be gained from the unit in the operations area, such as—

- Enemy soldiers in special camouflage uniforms.
- Enemy soldiers with weapons in cases or drag bags, which includes:
  - Rifles of unusual configuration
  - Long-barrel rifles.
  - Mounted telescopes.
  - Bolt-action rifles.
- Single-shot fire at key personnel (commanders, platoon leaders, senior NCOs, or weapons crews).
- Lack or reduction of enemy patrols during single-shot fires.

- Light reflecting from optical lenses.
- Reconnaissance patrols reporting of small groups of enemy (one to three men) by visual sighting or tracking.
- Discovery of single, expended casings (usually of rifle calibers 7.62x54R, 7.62NM, 300WM, 338 Lapua)

5-112. The sniper team next determines the best method to eliminate the enemy sniper. It—

- Gathers information, which includes:
  - Times of day precision fire occurs.
  - Locations where enemy sniper fire was encountered.
  - Locations of enemy sniper sightings.
  - Material evidence of enemy snipers such as empty brass casings or equipment.
- Determines patterns.

5-113. The sniper team evaluates the information to detect the enemy's established patterns or routines. It conducts a map reconnaissance, studies aerial photos, or carries out a ground reconnaissance to determine travel patterns. The sniper must picture himself in the enemy's position and ask, "How would I accomplish this mission?"

5-114. Once a pattern or routine is detected, the sniper team determines the best location and time to engage the enemy sniper. It also requests—

- Coordinating routes and preplanned fires (direct and indirect).
- Additional preplotted targets (fire support).
- Infantry support to canalize or ambush the sniper.
- Additional sniper teams for mutual supporting fire.
- Baiting of likely engagement areas to deceive the enemy sniper into commitment by firing.
- All elements be in place 12 hours before the expected engagement time.

5-115. During a countersniper operation, the team must ignore battle activity and concentrate on one objective—the enemy sniper. When an enemy sniper is operating in a unit's area, the sniper team ensures that the unit uses the following passive countermeasures to defend against enemy sniper fire:

- Do not establish routines—for example, consistent meal times, ammunition resupply, assembly area procedures, or day-to-day activities that have developed into a routine.
- Conduct all meetings, briefings, or gatherings of personnel under cover or during limited visibility.
- Cover or conceal equipment.
- Remove rank from helmets and collars. Do not salute officers. Leaders should not use authoritative mannerisms.

- Increase OPs and use other methods to increase the unit's observation capabilities. All information should be consolidated at the S-2 for analysis and logged-in regardless of insignificance.
- Brief patrols on what to look for, such as single, expended rounds or different camouflage materials.
- Do not display awareness of the enemy's presence at any time.
- Be aware that some of the enemy snipers may be women. Patrols and OPs must not be misled when sighting a woman with a mounted telescope on her rifle. She is a deadly opponent.
- Be aware of resupply operations by women and children into suspected or possible sniper locations. Watch for movement and scheduled patterns.

CONVENTIONAL OFFENSIVE OPERATIONS

5-116. Snipers can add deception to the battlefield and provide economy-of-force to allow the conventional force commander to focus combat power elsewhere. Commanders must also think of sniper operations in unilateral terms. The effect of snipers on a scale of ones and twos is small. However, when employed in coordinated actions on a broad front, their effect can be substantial, not only throughout the battlefield but also before, during, and after the battle. They can provide support to conventional units in the following critical phases of offensive operations:

PREOFFENSIVE MISSIONS

5-117. Any missions before offensive operations will primarily be in the deep battle area to gather information on the enemy's disposition. Snipers can help collect this information and interdict selected targets, if necessary. If the objective is to divert enemy assets from the main effort, then snipers can imitate the actions that the Russian partisans conducted against the Germans in World War II. The result of such actions can impair logistics operations and demoralize enemy soldiers in their own rear areas. The preoffensive missions are generally HUMINT-oriented. However, the sniper can perform the following DA functions as a natural consequence of his proximity to the enemy as a HUMINT asset.

Reconnoitering

5-118. The sniper's tasks can vary with each reconnoitering mission. Some of his functions are to—

- Gather (real-time) information on enemy dispositions, terrain, and weather.
- Penetrate enemy security zones in an effort to determine the extent and nature of enemy deception efforts.
- Confirm or deny existing intelligence as requested by the commander or S-2.
- Locate securable routes or axes of advance.

- Locate enemy reserve forces and the possible routes they could use to reinforce the objective.
- Establish or modify preplanned fires of indirect weapons to more effectively reduce TOCs, FDCs, crew-served weapons, hard points, avenues of approach, and retreat.
- Locate enemy security measures, such as mines, obstacles, or barriers.

Harassment

5-119. This function serves to lower the enemy's morale and inhibit his freedom of movement within his own lines. It takes the feeling of a secure area away from the enemy and inhibits his ability to rest his troops. The sniper generally performs this type of harassment at ranges greater than 500 meters.

Infiltration

5-120. Before an attack, snipers infiltrate the gaps between enemy units and positions and establish themselves in the enemy's rear area. During the attack the infiltrated snipers will engage specific targets and targets of opportunity both on the main line of resistance and in the rear area. This method diverts the enemy's attention from the attacking units and disrupts the freedom of movement in its own rear areas. Specific targets engaged by infiltrating snipers include—

- Enemy snipers.
- Command, control, and communications facilities and personnel.
- Crew-served weapons personnel.
- Artillery and forward air controllers.
- Dismounted reserve forces.
- Military policemen.
- Wire repair and resupply parties.

MISSIONS DURING THE OFFENSE

5-121. Sniping during the offensive is DA-oriented. Snipers are attached to friendly units to provide immediate direct support by means of precision rifle fire. The main function of attached snipers will be the suppression of enemy crew-served weapons, enemy snipers, and C2 personnel. Snipers can also support the offensive by interdicting follow-on or reserve forces (such as second-echelon combat forces or logistics). Conventional snipers, assigned to their parent units, can also interdict key targets in the main battle area. Also, attached snipers can be used to screen the flanks of advancing units, cover dead space from supporting crew-served weapons, and engage specific selected targets of the defending enemy units. Snipers maintain pressure on the retreating forces to prevent assembly and reconsolidation. The sniper will pursue retreating forces until he reaches his limit of advance. Then he will prepare for postoffensive operations.

POSTOFFENSIVE MISSIONS

5-122. Snipers' postoffensive role begins during the consolidation of the objective. Snipers are deployed forward of the consolidating unit's OP or LP line. The snipers will observe for enemy assembly for counterattack and either harass with direct fire or call for indirect fire. Once the enemy begins movement to the line of departure, the sniper will interdict the advance of dismounted counterattacking forces or button up advancing armor. This interdiction will give the antitank weapons a better chance of success and survival. When sufficient numbers of snipers are available, hasty sniper ambushes are established to interdict patrols, probing elements, and enemy sniper teams that normally precede a counterattack. Snipers can also use these ambushes to harass the displaced enemy to prevent him from establishing a base to counterattack. One of the primary jobs of the sniper is to get the enemy to deploy early in the attack formation. This will cost the enemy positive control of his attack formation.

Interdiction

5-123. In the interdiction mission, snipers push out beyond the range of friendly support in an effort to preinfiltrate reestablished first-echelon defenses, infiltrate second-echelon defenses, or engage counterattacking forces from the rear. They will interdict lines of communication in the enemy's rear areas and force him to commit more troops to the rear areas and weaken his forward lines. This can also cause the enemy to reinforce the wrong areas before the next attack.

Security

5-124. Because of their ability to remain undetected in close proximity to the enemy, snipers can maintain contact with displaced enemy forces. During consolidation, snipers range ahead of the main LP or OP line, determine the enemy's whereabouts, and continue to harass until the attack is resumed. Forward deployment also permits snipers to provide early warning of impending counterattacks.

Countersniping

5-125. Displaced enemy forces will often result in individuals or small groups getting cut off from their parent units. Oftentimes snipers will stay behind to disrupt the attacker's consolidation efforts. As these threats are small, snipers can track down and eliminate stay-behinds and isolated pockets of resistance. At the very least, snipers can suppress them until suitable forces can be spared to deal with them.

RESERVE MISSIONS

5-126. In a reserve role, snipers can give support where needed. They can reinforce success or react to enemy incursions. They can also provide stopgap measures until the commander can rally forces that are more appropriate. Snipers can maintain security in their own rear areas by using stealth and unconventional skills to seek out enemy forces. Their main support roles are as follows:

- Reinforcement involves attaching themselves directly to the unit engaged and adding their fires to those of the unit.

- Intervention enables the sniper to outflank the local resistance and suppress it with precision rifle fire.

5-127. Snipers may also conduct a dismounted movement to contact by deploying before the movement. Once deployed, they will move along the route to reconnoiter the route and select sniper hide positions to secure the route for the moving element. Depending on the number of snipers available, it is possible to secure a corridor over 1,500 meters at the widest (depending on the terrain) and as deep as permitted by the number of sniper teams and terrain. During reconnaissance and combat patrols, snipers may function as part of the security or support elements.

CONVENTIONAL DEFENSIVE OPERATIONS

5-128. The SF sniper's support to conventional defensive operations is similar to offensive operations. He can lend support anywhere on the battlefield including deep, rear, and main battle areas. However, conventional snipers normally operate in the main battle area in concert with their parent units—making SF sniper support seldom necessary in this area. The SF sniper's most important role is in the deep battle area. The rear battle area is also an area of employment, providing a rear-area threat exists.

5-129. Sniper operations in the deep battle area can be used to keep enemy efforts off-balance and directed toward rear area protection. The more enemy assets the sniper eliminates from the deep battle area, the fewer forces the enemy will have to execute attacks against the main effort. The sniper can also provide information on enemy strengths, location of reserves, and intentions.

SNIPER INTERDICTION

5-130. Just as in offensive operations, SOF units using snipers should deploy on a broad front to disrupt the enemy's order of battle. The main goal is to disrupt follow-on forces in the deep battle area. Snipers can assist in interdicting the enemy's soft underbelly—his unarmored logistics columns, fragile C2 nodes, and critical military weapons such as missiles and fire control equipment.

5-131. Defensive operations that could involve the sniper are—
- Area defense.
- Perimeter defense.
- Security forces.
- Reverse slope defense.
- Defense of built-up or fortified positions.
- River line defense.
- Mobile defense.
- Economy-of-force.
- Withdrawal operations.

5-132. Threat doctrine calls for simultaneous attacks at critical nodes located in U.S. rear areas. The sniper is ideally suited to locate and interdict the threat of enemy SO units that conduct such operations. The sniper uses the following methods to achieve these objectives:

Harassment

5-133. Snipers operate best in defensive operations beyond the FLOT to provide early warning of the approaching enemy, disorganize his attack, and cause him to deploy early. If armored vehicles are being used, it will cause the vehicle commanders to button up early. Snipers should closely integrate in the security force while performing this mission.

5-134. Snipers can also work directly into the FLOT defensive positions or assume their positions after withdrawal of the security fire. Snipers in defense of the FLOT should operate similarly to the crew-served weapons. Snipers can obtain optimum results by maximizing their standoff range to the targets, positioning on lucrative avenues of approach, and engaging targets of opportunity. Sniper positions should not be emplaced near obvious indirect fire targets. No matter how well concealed a hide is, if it is in the bursting radius of an indirect fire weapon, it can be compromised and destroyed.

5-135. The use of skilled marksmen will enhance the overall combat effectiveness of the defensive positions. Skilled marksmen are not necessarily snipers. They are simply skilled rifle shots who, for whatever reason, have neither the inclination nor the background skill to be successful snipers. However, they do possess the ability to engage targets at long ranges. When equipped with special weapons, such as caliber .50 or high-powered target rifles, they are particularly useful for conducting long-range harassment.

Delay

5-136. When friendly forces need to withdraw from contact with the enemy, snipers can delay and impede the enemy's advance. They deploy throughout the withdrawing unit's sector. By using a series of interlocking delay positions, a handful of snipers can interdict dismounted avenues of approach and severely impede advancing enemy forces. They can use successive delay positions to permit the withdrawing forces to reassemble and establish new defensive positions. Sniper elements must remain mobile to avoid decisive engagement with the attacking enemy. They can operate during the withdrawal to cover obstacles with precision rifle fire and thus increase the effectiveness of the obstacles. They can also be the stay-behind element and attack the enemy forces' rear area and supply columns.

Rear Area Protection

5-137. In this mission, snipers can enhance the protective measures surrounding sensitive facilities or installations. They can strengthen these measures by either establishing OPs along routes of access, acting as a reaction force to rear area penetrations, or by patrolling. Snipers will not normally patrol by themselves but as members of established security patrols.

5-138. The role of sniping in security operations is that of extending the depth and scope of the security effort. Specific roles include—

- Protecting critical installations, sites, or projects from infiltration.

- Dominating the gaps between units to prevent infiltration by enemy combat elements or patrols.
- Preventing the removal or breaching of obstacles.
- Tracking enemy patrols known to have penetrated into the rear area.

SNIPER SUPPORT TO DEFENSIVE HUMINT COLLECTION

5-139. Using snipers in defensive operations provides a variety of means to maintain constant offensive pressure on the enemy. Sniping in the defense is dependent on the collection and use of information. When the snipers collect information for their personal use, it is known as targeting. Information collected for organizational use is but an element of the total HUMINT collection effort of the snipers' unit. OPs are the snipers' primary means of collecting information in the defense. In the role of the observers, the snipers establish a series of OPs that dominate their sector. These OPs are of two types—overt and covert.

Overt Observation Post

5-140. This OP is not overt in that its location or function is known to the enemy, but that the snipers may engage high-priority targets from it. While firing from the OP may not necessarily reveal its exact location, it will certainly reveal the snipers' presence and the fact that such a location exists.

Covert Observation Post

5-141. The sniper uses this OP because it offers a commanding view of enemy positions. These posts should remain unknown to the enemy and should never be fired from, regardless of the temptation to do so. The information that the sniper collects from a well-sited covert OP is far more valuable than any targets that may appear.

CIVIL DISTURBANCE ASSISTANCE

5-142. The U.S. Army provides military assistance to civil authorities in civil disturbances when it is requested or directed IAW prevailing laws. When such assistance is requested, the military forces assist local authorities in the restoration and maintenance of law and order.

5-143. Military assistance is considered as a last resort. When committed, involvement is to the degree justified by the circumstances to restore law and order with a minimum loss of life and property. When using force, the guiding principle should be minimum force consistent with mission accomplishment.

5-144. The sniper team's precision fire and observation abilities give authorities a way to detect and eliminate criminal threats with low risk to innocent personnel. The use of sniper teams in civil disorders must be planned and controlled. They may be an important factor in the control and elimination of weapons fire directed against riot control authorities. Snipers functioning in this role must operate under strict ROE. However the team must never allow itself to be overrun. The team should always plan its multiple covert positions.

CHARACTERISTICS OF URBAN VIOLENCE

5-145. Crowd behavior during a civil disturbance is essentially emotional and without reason. The feelings and the momentum generated have a tendency to make the whole group act like its worst members. Skillful agitators or subversive elements exploit these psychological factors during these disorders. Regardless of the reason for violence, the results may consist of indiscriminate looting and burning or open and violent attacks on officials, buildings, and innocent passersby. Rioters may set fire to buildings and vehicles to—

- Block the advance of troops.
- Create confusion and diversion.
- Achieve goals of property destruction, looting, and sniping.

5-146. In addition, organized rioters or agitators may use sniper fire to cause government forces to overreact.

SNIPER SUPPORT DURING CIVIL DISTURBANCES

5-147. The sniper team uses planning factors to estimate the amount of time, coordination, and effort that it will take to support local authorities, when faced with an enemy sniper threat or any type of civil disturbance such as a riot. For the team's mission to run smoothly and be a success, all participants should consider the following factors.

Briefings

5-148. Sniper teams must receive a detailed briefing on the areas and routes within the riot area. Representatives of local authorities should be assigned to the sniper teams for protection and communications with local indigenous personnel.

Adequate Personnel

5-149. The civil authorities should have sufficient sniper teams to provide maximum versatility to the riot control personnel. Sniper teams should also have at least one reaction team assigned to them. This capability will permit the team to direct a reaction team to a troublemaker for apprehension without the requirement to fire a weapon. These teams should consist of both military and local authority personnel.

Observation Areas and Fields of Fire

5-150. Observation areas and fields of fire are clearly defined by streets and highways. However, surveillance and detection are complicated by the numerous rooftops, windows, and doorways from which hostile fire may be directed. Sniper teams take maximum advantage of dominant buildings or rooftops to maintain continuous observation of a riot scene. Mutually supporting teams cover blind spots or dead space within the area. Sniper teams must place themselves at various heights to give them view into the different multistoried buildings.

Cover and Concealment

5-151. Built-up areas offer excellent cover and concealment for both the rioters and the sniper teams.

Avenues of Approach

5-152. The best avenues of approach to a riot scene, or to points of observation and firing positions, are through building interiors. Movement through streets may be difficult and easily detected by rioters. Sniper teams should also consider underground passages such as cableways.

Operations

5-153. Sniper teams should operate in each established area. The teams remain at a sufficient distance from control troops to keep from getting involved in direct riot actions.

Firing Positions

5-154. The firing position should provide the maximum stability, because precision fire is used to wound and not to kill. A stray shot that wounds or kills a woman, child, or unarmed rioter may only inflame an already riotous situation. When firing from a window, the sniper team should fire, if possible, from a supported position in the back of the room. The distance will muffle the muzzle blast and keep the muzzle flash from being noticed. If the sniper shows his rifle or part of his body, it may invite fire from weapons-equipped rioting personnel. When possible, he should use a silencer on his sniper rifle.

Camouflage

5-155. Sniper teams should dress in drab or blending clothing to prevent identification or observation. However, snipers must wear an identifying mark so as not to be engaged by friendly forces.

Civil Authorities

5-156. Since civil authorities are in charge, snipers maintain a direct line of communications (LOC) with the civilian who permits or directs snipers to engage. Civil authorities also determine the caliber of weapon as well as the type of ammunition. However, usually anything within 300 meters is engaged with 5.56-mm ammunition unless special penetration capability is required.

Sniper Team Control

5-157. A key to effective sniper team use is control. When directed to engage in countersniping activities, the sniper team's actions must be swift and precise. The sniper leader must maintain positive control over the teams at all times.

Rules of Engagement

5-158. When countersniping is required, the sniper team should direct its precision fire to wound rather than to kill, if possible, unless in direct defense of human life.

5-159. Snipers employed to counteract sniper fire from a street disorder require quick and decisive action. When directed to support the control forces during a street disorder, the sniper team—

- Deploys to rooftops or vantage points providing observation and fields of fire into the riot area.
- Institutes communications with the commander.
- Begins observation immediately and continues it.
- Relays information continuously to the commander.
- Conducts countersniping actions as directed.

5-160. During civil disorders, rioters may seize control of buildings for the purpose of using the vantage points of rooftops or windows from which to direct hostile sniper fire on riot control forces. The sniper team may have to provide covering fire to allow the searching or clearing team to approach and clear the building. On the other hand, the sniper may have to use precision fire to engage the hostile sniper if the hostile sniping is directed at control authorities in mob control actions.

5-161. Upon identifying or locating a riotous sniper who is directing fire at fire-fighting personnel, the sniper immediately reacts to reduce the hostile sniper fire. He directs this countersniper fire with accuracy to kill.

5-162. Civil authorities must try to quickly control looting because it may also lead to more serious acts of murder and arson, often against innocent nonparticipants. The sniper team's employment to assist in looting control is mainly for observation, communication, and to act as a covering force should the looters fire upon the control forces. When control forces are fired upon, the sniper team immediately engages the riotous sniper to facilitate apprehension by the control forces.

5-163. The sniper team's role in support of riot control forces is equally important during the hours of darkness. Optical equipment, to include NVDs, allows the sniper team to provide prolonged night observation. Therefore, the team can sufficiently accompany patrol forces, man observation posts and roadblocks, or cover control troops during mob control activities.

5-164. Use of snipers during civil disturbances can become a source of greater agitation among the rioters. Civil authorities should publicly remove compromised snipers while leaving the other snipers in place. In many instances this removal will embolden the agitators and permit rapid identification for quick apprehension by control personnel. In the same vein, firers may become more relaxed and show themselves for easier identification by the posted countersnipers.

Chapter 6

Sniper Operations in Urban Terrain

Snipers are extremely effective in urban terrain. Their long-range precision fire can engage targets at a distance; their advanced optics can discriminate individual point targets to save innocent bystanders or protect property; and their observation skills can offer superior intelligence-collection capabilities. In an urban environment, the sniper is both a casualty producer and an intimidating psychological weapon.

URBAN TERRAIN

6-1. Urban terrain consists mainly of man-made structures. Buildings are the main components of urban terrain. They provide cover and concealment, limit fields of fire and observation, and impair movement. Thick-walled buildings provide excellent protection from hostile fire.

6-2. Urban streets are generally avenues of approach. However, forces moving along streets are often channalized by buildings and terrain that offer minimal off-road maneuver space. Obstacles on streets prove difficult to bypass, due to these restrictive avenues of approach.

6-3. Underground systems found in some urban areas are easily overlooked but can be important to the outcome of operations. They include subways, sewers, cellars, and utility systems.

6-4. Civilians will be present in urban operations, often in great numbers. Concern for the safety of noncombatants may restrict fire and limit maneuver options available to the commander.

CATEGORIES OF URBAN TERRAIN

6-5. The world is largely urban in terms of population concentration. Commanders categorize urban terrain as large cities, towns and small cities, villages, and strip areas.

Large Cities (population greater than 100,000)

6-6. In Europe, other than the former Soviet Union, there are approximately 410 cities with a population of more than 100,000. Large cities frequently form the core of a larger, densely populated urban complex consisting of the city, its suburban areas, and small towns. Such complexes have the appearance of a single large and continuous city containing millions of people and occupying vast areas of land.

Towns and Small Cities (population of 3,000 to 100,000)

6-7. These areas are mostly located along major lines of communications and situated in river valleys. Similar to larger cities, these areas are continuing to expand and will eventually form new concentrations or merge with existing ones.

Villages (population of less than 3,000)

6-8. In most cases, villages are agriculturally oriented and usually exist among the more open cultivated areas.

Strip Areas

6-9. These built-up areas generally form connecting links between villages and towns. These areas also exist among LOCs leading to larger complexes.

DESCRIPTIONS OF URBAN TERRAIN

6-10. Within the city, urban terrain differs based on size, location, and history. The areas within the city are generally categorized as follows:

- *Industrial Areas and Residential Sprawl.* Residential areas consist of some houses or small dwellings with yards, gardens, trees, and fences. Street patterns are normally rectangular or curving. Industrial areas consist of one- to three-story buildings of low, flat-roofed factories or warehouses, generally located on or along major rail and highway routes. In both regions, there are many open areas.
- *Core Periphery.* The core periphery consists of narrow streets (12 to 20 meters wide) with continuous fronts of brick and heavy-walled concrete buildings. The height of the buildings is generally uniform, two to three stories in small towns and five to ten stories in large cities.
- *City Cores and Outlying High-Rise Areas.* Typical city cores of today are made of high-rise buildings that vary greatly in height and allow for more open space between buildings. Outlying high-rise areas are dominated by this open-construction style to a greater degree than city cores. Generally, streets form a rectangular pattern.
- *Commercial Ribbons.* These are rows of stores, shops, or boutiques built along either side of major streets through the built-up areas. Generally, these streets are 25 meters wide or wider. The buildings are uniformly two to three stories tall.

NATURE OF URBAN COMBAT

6-11. Urban combat usually occurs when a city is between two natural obstacles and it cannot be bypassed, the seizure of the city contributes to the attainment of an overall objective, or political or humanitarian concerns require the seizure or retention of the city.

6-12. In the city, the ranges of observation and fields of fire are reduced by the structures as well as the smoke and dust of combat. Targets will generally be seen briefly at ranges of 200 meters or less.

6-13. Units fighting in urban areas often become isolated by an enemy. Therefore, snipers must have the skill, initiative, and courage to operate effectively while isolated from their unit. Combat in more up-to-date nations can no longer avoid urban areas; therefore, snipers must train and be psychologically prepared for the demands of urban combat.

6-14. The defender will generally have the advantage over the attacker in urban combat. The defender occupies strong positions, whereas the attacker must expose himself to advance. Also, the greatly reduced LOS ranges, built-in obstacles, and compartmented terrain require the commitment of more troops for a given frontage. Troop density may be three to five times greater for both attacker and defender in urban combat than in natural environments.

6-15. Density of structures degrades radio communications. This factor, combined with limited observation, makes control of forces difficult. The well-established defender will probably use wire communications to enhance control, thus adding to his advantage.

6-16. Soldiers may encounter a greater degree of stress during urban combat. Continual close combat, intense pressure, high casualties, the fleeting nature of targets, and fire from an unseen enemy may produce increased psychological strain and physical fatigue.

6-17. Commanders may have to restrict their use of weapons and tactics to minimize collateral damage. This restriction may be necessary to preserve a nation's cultural heritage and gain the support of the population. In such cases, snipers are ideally suited to deliver discriminatory fire against selected targets.

6-18. Attacks will generally limit artillery fires to the direct fire mode. Units use this method to avoid reducing the city to rubble. Direct fire causes few casualties and tends to enhance the defender's fortifications and concealment. It also restricts the attacker's avenues of approach.

6-19. Forces engaged in urban fighting use large quantities of munitions. Units committed to urban combat must also have special equipment, such as grappling hooks, ropes, snaplinks, construction materials, axes, sandbags, and ladders.

6-20. Urban combat historically has presented chances for looting. Looting can break down discipline, reduce alertness, increase vulnerability, and delay the progress of the unit. It also alienates the civilian population.

EVALUATING URBAN TERRAIN

6-21. When the sniper evaluates urban terrain, he should consider the following factors:

- *Observation and Fields of Fire.* Buildings on the edge of a city provide better fields of fire than buildings in the interior. In the city, tall buildings with numerous windows often provide the best fields of fire, especially if the buildings have spaces between them. However, the sniper should never choose the outermost buildings as they are usually subjected to the greater amount of fire and preparatory bombardment.

- *Avenues of Approach.* The best way to gain entry into a building is from the top. Therefore, the most important avenue of approach to look for is one that quickly leads to the top (fire escapes, drainpipes, or adjacent buildings). Personnel must protect these when the sniper is in the defense and allow him use when required.
- *Key Control Points.* The key points in a building are entrances, hallways, and stairs; troops that control these areas control the building.
- *Obstacles.* Doors and fire barriers are common in commercial buildings. They become obstacles if they are shut and secured. Furniture and appliances can also be obstacles in a building. Snipers can also use barbed wire effectively inside a building because it further restricts movement.
- *Cover and Concealment.* Buildings with brick walls and few, narrow windows provide the best balance between cover and concealment and fields of fire. Roofs provide little protection; snipers usually have better protection in the lower stories than directly under the roof. (An exception to this rule is the parking garage.) Floor layouts with many small rooms provide more protection than floor layouts with larger rooms.
- *Intra-City Distribution of Building Types.* The sniper can generally determine the layout of a city by the distribution of the buildings within the city. Types and layout are as follows:
    - Mass construction buildings (older apartments and hotels) are the most common structures in old city cores and older built-up areas (two-thirds of the total area). They are usually constructed of bricks or cement block.
    - Frame and heavy clad, steel and concrete-framed, as well light clad, glass, multistory buildings are found in the core area—a city's most valuable land—where, as centers of economic and political power they have potentially high military significance.
    - Open spaces (for example, parks, athletic fields, and golf courses) account for about 15 percent of an average city's area. Most of this area is suitable for airmobile operations.
    - Frame and light clad, wood, and cosmetic brick structures dominate residential sprawl areas.
- *Environmental Considerations.* Environmental factors will influence the effectiveness of the sniper. He should closely evaluate these factors during the selection and preparation of the urban sniper hide site.

6-22. Population density will affect the ease of movement to and from the hide as well as the ability of the team to remain undetected. The sniper must also consider the safety of the local civilian population. Dependent upon the type of operation, eliminating civilian collateral damage may be an overriding factor for measuring success. In urban areas, the sniper team must be prepared to deal with pet animals. If these pets pose a threat to the sniper team (detection or actual attack), it may be necessary to eliminate or silence the pets. Snipers should be aware of the possible consequences if these animals should suddenly disappear.

6-23. The media, in the form of international news television and radio commentators, will probably be present in some strength in all future conflicts. Their presence may compromise or negate the effectiveness of the snipers' mission and must be a consideration.

6-24. Glass or windows can cause problems for the sniper. Depending on the mission, the sniper may be able to remove the glass during hide construction. If not, he must devise a method of emergency glass removal.

6-25. Natural and artificial lighting will impact on the effectiveness of standoff optics and NVDs. All lights in the hide should be off and secured or deactivated to avoid inadvertent activation.

6-26. Ambient noise levels may aid in the occupation and construction of the hide. It could also provide a desirable time window for the snipers to engage targets. In urban areas, most noise levels will go in cycles from high levels during the day to low levels at night.

LINE-OF-SIGHT FACTORS

6-27. Streets serving areas composed mostly of one type of building normally have a common pattern. Street widths are grouped into three major classes:

- Narrow (7 to 15 meters)—such places as medieval sections of European cities.
- Medium (15 to 25 meters)—newer, planned sections of most cities.
- Wide (25 to 50 meters)—areas where buildings are located along broad boulevards or set far apart on large parcels of land.

6-28. When a street is narrow, observing or firing into windows of a building across the street can be difficult because an observer must look along the building rather than into the windows. When the street is wide, the observer has a better chance to look and fire into and out of the window openings.

6-29. The same limitation on LOS occurs when looking up or down tall buildings.

SOURCES OF INFORMATION IN URBAN TERRAIN

6-30. Operations in urban terrain require detailed intelligence. Snipers should have the following materials for planning operations:

- *Maps and Aerial Photos.* Although tactical maps do not show man-made objects in enough detail for tactical operations in urban terrain, they do show the details of terrain adjacent to urban areas. The sniper should supplement tactical maps with both vertical and oblique aerial photos. From the aerial photos, the sniper should construct plan view sketches to locate the best LOS positions.
- *Civil Government and Local Military Information.* The sniper can obtain considerable current information on practically all details of a city from civil governments and local military forces. Items include:
  - Large-scale city maps.
  - Diagrams of underground sewer, utility, transport, and miscellaneous systems.

FM 3-05.222

- Key public buildings and rosters of key personnel.
- Size and density of the population.
- Police and security capabilities.
- Civil defense, air raid shelters, and firefighting capabilities.
- Utility systems, medical facilities, and mass communications facilities.

CAMOUFLAGE TECHNIQUES FOR URBAN TERRAIN

6-31. To survive in urban combat, the sniper must supplement cover and concealment with camouflage. He must study the surroundings in the area to properly camouflage himself. He must make the firing positions look like the surrounding terrain. For instance, if there is no damage to buildings, he will not make loopholes for firing and will use only the materials needed. Any excess material can reveal his position. For example, if defending the city park, the sniper will use the entire park for resources; he will not denude a small area near the position for camouflage material.

6-32. Buildings provide numerous concealed positions. Thick masonry, stone, or brick walls offer excellent protection from direct fire and provide concealed routes. If the tactical situation permits, the sniper will inspect positions from the enemy's viewpoint. He will conduct routine checks to see if the camouflage remains material-looking and actually conceals the position. He should not remove his shirt because exposed skin reflects light and could attract the enemy's attention.

6-33. When using urban camouflage techniques, the sniper must consider the following:

- *Use of Shadows.* Buildings in urban areas throw sharp shadows. The sniper can use the shadow to aid in concealment during movement. He will avoid lighted areas around windows and loopholes. A lace curtain or a piece of cheesecloth provides additional concealment to snipers in interiors of rooms, if curtains are common in the area.
- *Color and Texture.* The need to break up the silhouette of helmets and individual equipment exists in urban areas as elsewhere. However, burlap or canvas strips are a more effective camouflage garnish than foliage. Predominant colors are normally browns, tans, and sometimes grays, rather than greens; but the sniper should evaluate each camouflage location separately.
- *Dust.* In weapons emplacements, the sniper should use a wet blanket, canvas, or type of cloth to keep dust from rising when the weapons are fired.
- *Background.* Snipers must pay attention to the background to ensure that they are not silhouetted or skylined, but rather blend into their surroundings. Use of a neutral drop cloth to his rear will help the sniper blend with his background.
- *Common Camouflage Errors.* To defeat enemy urban camouflage, the sniper should look for errors such as tracks or other evidence of activity, shine or shadows, unnatural or peculiar colors or textures,

muzzle flash smoke or dust, unnatural sounds and smells, and finally, movements. Things to remember when camouflaging include—

- Use dummy positions to distract the enemy and make him reveal his position by firing.
- Use the terrain and alter camouflage habits to suit the surroundings.
- Do not forget deceptive camouflage of buildings.
- Continue to improve positions. Reinforce fighting positions with sandbags or other shrapnel and blast absorbing material.
- Do not upset the natural look of the area.
- Do not make positions obvious by clearing away too much debris for fields of fire.
- Choose firing ports in inconspicuous spots when available.

INFILTRATION AND EXFILTRATION IN URBAN TERRAIN

6-34. A sniper can more easily infiltrate into the outskirts of a town because the outskirts are usually not strongly defended. Its defenders may only have a series of antitank positions, security elements on the principal approach, or positions blocking the approaches to key features in the town. The strong points and reserves are deeper in the city.

6-35. As part of a larger force, the sniper moves by stealth on secondary streets using cover and concealment of back alleys and buildings. These moves enable him to assist in seizing key terrain features and isolating enemy positions, thus aiding following units' entry into the urban area. Sniper teams may also infiltrate into the city after the initial force has seized a foothold and move into their respective sniper positions.

6-36. Snipers may use mortar and artillery fire to attract the enemy's attention and cover the sound of infiltrating troops. They should infiltrate when visibility is poor; chances of success are greater if there are no civilians in the area. Snipers may also infiltrate into a city (as part of a larger force) during an airborne or airmobile operation.

6-37. During exfiltration, snipers must be extremely careful to avoid detection. As in infiltration, snipers must use stealth and all available cover and concealment when leaving their positions. Snipers should always try to exfiltrate during darkness.

MOVEMENT TECHNIQUES IN URBAN TERRAIN

6-38. Movement in urban areas is one of the first fundamental skills that a sniper must master. He must practice movement techniques until they become second nature. To minimize exposure to enemy fire, the urban sniper must move so that he—

- Does not silhouette himself, but keeps low at all times.
- Avoids open areas (streets, alleys, parks).
- Selects the next covered position before moving.
- Conceals movement by using buildings, rubble, foliage, smoke, or limited visibility.

- Advances rapidly from one position to another, but not so rapidly that he creates dust clouds or noise that will help the enemy locate him.
- Does not mask his covering fire.
- Remains alert, ready for the unexpected.

6-39. Specific movement techniques used frequently in urban operations must be learned by all snipers. They are—

- *Crossing a Wall*. After the sniper has reconnoitered the other side, he quickly rolls over the wall, keeping a low silhouette. The speed and the low silhouette will deny the enemy a good target.
- *Moving Around a Corner*. Corners are dangerous. The sniper must observe the area around the corner before he moves beyond the corner. The most common mistake that a sniper makes at a corner is allowing his weapon to extend beyond the corner, exposing his position (flagging). Also, a sniper should not show his head at the height that an enemy soldier would expect to see it. When using the correct technique for looking around a corner, the sniper lies flat on the ground and does not extend his weapon beyond the corner of the building. He exposes his head or a hand-held mirror (at ground level) only enough to permit observation around the corner.
- *Moving Past Windows*. When using the correct technique for passing a window, the sniper stays below the window level, taking care not to silhouette himself in the window. He hugs the side of the building. An enemy gunner inside the building would have to expose himself to fire from another position if he wished to engage the sniper.
- *Moving Past Basement Windows*. When using the correct procedure of negotiating a basement window, the sniper stays close to the wall of the building and steps or jumps over the window without exposing his legs.
- *Using Doorways*. The sniper should not use doorways as entrances or exits. If he must use a doorway as an exit, he should move quickly through it to his next covered position, staying as low as possible to avoid silhouetting himself.
- *Moving Parallel to a Building*. At times, it may not be possible to use interiors of buildings for a route of advance. To correctly move along the outside of a building, the sniper moves along the side of the building, staying in the shadows, presenting a low silhouette, and moves deliberately to his next position. He must plan one position ahead of his next position. This will prevent getting into a dead-end position with nowhere to go.
- *Crossing Open Areas*. Snipers should avoid open areas such as streets, alleys, and parks whenever possible. However, they can be crossed safely if the sniper applies certain fundamentals. Even using the correct method for crossing, the sniper may employ a distraction or limited visibility to conceal his movement. He crosses the open area at the shortest distance between two points.

6-40. Before moving from one position to another, a sniper should make a visual reconnaissance and select the position that will give him the best cover and concealment. At the same time, he should select the route that he will take to that position.

NOTE: The sniper team should not move together when crossing from one building to another or across an open area.

BUILDING ENTRY TECHNIQUES

6-41. When entering a building, a sniper may be required to enter by means other than through doorways or reach top levels of buildings by means other than stairs.

6-42. The sniper team can use various means, such as ladders, drainpipes, vines, helicopters, or the roofs and windows of adjoining buildings, to reach the top floor or roof of a building. Additional aids and methods to reach higher levels include—

- The two-man lift, supported and unsupported; the two-man lift with heels raised; the one-man lift; the two-man pull; and individual climbing techniques. These techniques are more commonly used to gain entry into areas at lower levels.
- Ladders or grappling hooks with knotted ropes. By attaching a grappling hook to the end of a scaling rope, a sniper can scale a wall, swing from one building to another, or gain entry to an upstairs window.
- Rappelling. The sniper can use this combat technique to descend from the roof of a tall building to other levels or to a window.

SNIPER SUPPORT IN URBAN OPERATIONS

6-43. A sniper should be given general areas (buildings or a group of buildings) in which to position himself, but he selects the best positions for engagements. Sniper positions should cover obstacles, roofs, gaps in the final protective fires, and dead space. The sniper also selects numerous secondary and supplementary positions to cover his areas of responsibility. He should think three-dimensionally.

6-44. The sniper determines his engagement priorities by the relative importance of the targets to the effective operations of the enemy. The following are normally sniper targets:

- Enemy snipers.
- Key leaders.
- Tank commanders.
- Direct fire-support weapons crewmen.
- Crew-served weapons personnel.

- Forward observers.
- Radiotelephone operators.
- Protected equipment.

6-45. The characteristics of built-up areas and the nature of urban warfare impact on both the effectiveness of the SWS and how the sniper can use it. The sniper must consider the following basic factors during urban operations:

- *Relative Location of the Shooter and the Target.* Both the target and the shooter may be inside or outside of buildings, or either one may be inside a building while the other is outside.
- *Structural Configuration of Buildings.* The basic classes of structures encountered in a built-up area can generally be classified as concrete, masonry, or wooden. However, any one building may include a combination of these materials. All buildings offer concealment, although the degree of protection varies with the material used.
- *Firing Ranges and Angles.* Engagement ranges may vary from distances of less than 100 meters up to the maximum effective range of a sniper system. Depression and elevation limits may create dead space. Target engagement from oblique angles, either vertical or horizontal, demands increased marksmanship skills. Urban areas often limit snipers to firing down or across streets, but open spaces of urban areas permit engagements at long ranges.
- *Visibility Limitations.* Added to the weather conditions that limit visibility are the urban factors of target masking and increased dead space caused by buildings and rubble. Observation through smoke, dust, and concealment offered by shaded areas, rubble, and man-made structures influence visibility.

DURING AN ATTACK IN URBAN TERRAIN

6-46. Snipers employed during the attack of a built-up area are usually divided into three phases:

- Phase I should allow snipers to isolate the battle area by seizing terrain features that dominate the avenues of approach. Snipers deliver long-range precision fire at targets of opportunity.
- Phase II consists of the advance to the built-up area and seizure of a foothold on its edge. It is during this period that snipers displace forward and assume their initial position from which to support continuation of the attack.
- Phase III consists of the advance through the built-up area IAW the plan of attack. Sniper teams should operate in each zone of action, moving with and supporting the infantry units. They should operate at a sufficient distance from the riflemen to keep from getting involved in firefights but close enough to kill more distant targets that threaten the advance. Some sniper teams can operate independently of the infantry on missions of search for targets of opportunity, particularly the search for enemy snipers.

6-47. Snipers that are in a defensive posture should place themselves in buildings that offer the best long-range fields of fire and all-around observation. They are assigned various missions, such as—

- Providing countersniper fire.
- Firing at targets of opportunity.
- Denying the enemy access to certain areas or avenues of approach.
- Providing fire support over barricades and obstacles.
- Observing the flank and rear areas.
- Supporting counterattacks.
- Preventing enemy observation.

INTERNAL SECURITY OPERATIONS

6-48. Commanders can use snipers in internal security operations during urban guerrilla warfare and hostage situations. The following paragraphs explain each situation.

Urban Guerrilla Warfare

6-49. In this type of environment, the sniper dominates the AO by delivery of selective, aimed fire against **specific** targets as authorized by local commanders. Usually this authorization comes when targets are about to employ firearms or other lethal weapons against the peacekeeping force or innocent civilians. The sniper's other role, almost equally as important as his primary role, is the gathering and reporting of intelligence. Within the above roles, some specific tasks that may be assigned include—

- When authorized by local commanders, engaging dissidents or urban guerrillas who are involved in hijacking, kidnapping, or holding hostages.
- Engaging urban guerrilla snipers as opportunity targets or as part of a deliberate clearance operation.
- Covertly occupying concealed positions to observe selected areas.
- Recording and reporting all suspicious activities in the area of observations.
- Assisting in coordinating the activities of other elements by taking advantage of hidden observation posts.
- Providing protection for other elements of the peacekeeping force, including firemen and repair crews.

6-50. In urban guerrilla operations, there are several limiting factors that snipers would not encounter in unconventional warfare. Some of these limitations follow:

- There is no forward edge of the battle area (FEBA) and therefore no "no man's land" in which to operate. Snipers can therefore expect to operate in entirely hostile surroundings in most circumstances.
- The enemy is hidden from or perfectly camouflaged among the everyday populace that surrounds the sniper. The guerrilla force usually uses an identifying clothing code each day to distinguish

themselves from civilians. This code is a PIR each day. The sooner the sniper can begin to distinguish this code, the easier his job will be.

- In areas where confrontation between peacekeeping forces and the urban guerrillas takes place, the guerrilla dominates the ground entirely from the point of view of continued presence and observation. He knows every yard of ground; it is ground of his own choosing. Anything approximating a conventional stalk to and occupation of a hide is doomed to failure.

- Although the sniper is not subject to the same difficult conditions as he is in conventional war, he is subject to other pressures. These include not only legal and political restraints but also requirements to kill or wound without the motivational stimulus normally associated with the battlefield.

- In conventional war, the sniper normally needs no clearance to fire his shot. In urban guerrilla warfare, the sniper must make every effort possible to determine the need to open fire, and that doing so constitutes reasonable or minimum force under the circumstances.

Hostage Situations

6-51. Snipers and commanding officers must appreciate that even a well-placed shot **may not always** result in the instantaneous incapacitation of a terrorist. Even the best sniper, armed with the best weapon and bullet combination, **cannot** guarantee the desired results. Even an instantly fatal shot **may not** prevent the death of a hostage when muscle spasms in the terrorist's body trigger his weapon. As a rule then, the commander should use a sniper only when all other means of solving a hostage situation have been exhausted.

6-52. **Accuracy Requirements.** The sniper must consider the size of the target in a hostage situation. The head is the only place on the human body where a bullet strike can cause instantaneous death. (Generally, the normal human being will live 8 to 10 seconds after being shot directly in the heart.) The entire head of a man is a relatively large target, measuring approximately 7 inches wide and 10 inches high. But the area where a bullet strike can cause instantaneous death is a much smaller target. The portion of the brain that controls all motor reflex actions is the medulla. When viewed at eye level, it is located directly behind the eyes, runs generally from ear lobe to ear lobe, and is roughly 2 inches wide. In reality then, the size of the sniper's target is 2 inches, not 7 inches. The easiest way for the sniper to view this area under all circumstances is to visualize a 2-inch ball (the medulla) directly in the middle of the 7-inch ball (the head).

6-53. Application of the windage and elevation rule makes it clear that the average sniper cannot and should not attempt to deliver an instantly killing head shot beyond 200 meters. To ask him to do so requires him to do something that the rifle and ammunition combination available to him will not do.

6-54. **Position Selection.** Generally, the selection of a firing position for a hostage situation is not much different from selecting a firing position for any other form of combat. The same guidelines and rules apply. The terrain and situation will dictate the choice of firing positions.

6-55. Although the commander should use the sniper only as a last resort, he should place the sniper into position as early as possible. Early positioning will enable him to precisely estimate his ranges, positively identify both the hostages and the terrorists, and select alternate firing positions for use if the situation should change. He is also the main HUMINT asset to the command element.

Command and Control

6-56. Once the commander decides to use the sniper, all C2 of his actions should pass to the sniper team leader. At no time should the sniper receive the command to fire from someone not in command. When he receives clearance to fire, then he and the sniper team leader alone will decide exactly when.

6-57. If the commander uses more than one sniper team to engage one or more targets, it is imperative that the same ROE apply to all teams. However, it will be necessary for snipers to communicate with each other. The most reliable method is to establish a "land line" or TA-312 telephone loop much like a gun loop used in artillery battery firing positions. This loop enables all teams to communicate with all the others without confusion about frequencies or radio procedures.

SNIPER AMBUSH IN URBAN TERRAIN

6-58. In cases where intelligence is forthcoming that a target will be in a specific place at a specific time, a sniper ambush is frequently a better alternative than a more cumbersome cordon operation.

6-59. Close reconnaissance is easier than in normal operations. The sniper can carry it out as part of a normal patrol without raising any undue suspicion. The principal difficulty is getting the ambush party to its hide undetected. To place snipers in positions that are undetected will require some form of deception plan. The team leader often forms a routine search operation in at least platoon strength. During the course of the search, the snipers position themselves in their hide. They remain in position when the remainder of the force withdraws. This tactic is especially effective when carried out at night.

6-60. Once in position, the snipers must be able to remain for lengthy periods in the closest proximity to the enemy and their sympathizers. Their security is tenuous at best. Most urban OPs have "dead spots." This trait, combined with the fact that special ambush positions are frequently out of direct observation by other friendly forces, makes them highly susceptible to attack, especially from guerrillas armed with explosives. The uncertainty about being observed on entry is a constant worry to the snipers. This feeling can and does have a most disquieting effect on the sniper and underlines the need for highly trained men of stable character.

6-61. If the ambush position cannot be directly supported from a permanent position, the commander must place a "backup" force on immediate notice to extract the snipers after the ambush or in case of compromise. Commanders normally assume that during the ambush the snipers cannot make their exit without assistance. They will be surrounded by large, extremely hostile crowds. Consequently, backup forces must not only be nearby but also be sufficient in size to handle the extraction of the snipers.

URBAN HIDES

6-62. A sniper team's success or failure greatly depends on each sniper's ability to place accurate fire on the enemy with the least possible exposure to enemy fire. Consequently, the sniper must constantly seek firing positions and use them properly when he finds them. Positions in urban terrain are quite different from positions in the field. The sniper team can normally choose from inside attics to street-level positions in basements. This type of terrain is ideal for a sniper and can provide the team a means of stopping an enemy's advance through its area of responsibility. However, one important fact for the team to remember is that in this type of terrain the enemy will use every asset it has to detect and eliminate them. The following paragraphs explain the two categories of urban hide positions.

HASTY HIDE

6-63. The sniper normally occupies a hasty hide in the attack or the early stages of the defense. This position allows the sniper to place fire upon the enemy while using available cover to gain protection from enemy fire. There are some common hasty firing positions in a built-up area and techniques for occupying them are as follows:

- *Firing From Corners of Buildings*. The corner of a building, used properly, provides cover for a hasty firing position. A sniper must be capable of firing his weapon from either shoulder to minimize body exposure to the enemy. A common mistake when firing around corners is firing from the standing position. The sniper exposes himself at the height the enemy would expect a target to appear and risks exposing the entire length of his body as a target.

- *Firing From Behind Walls*. When firing from behind a wall, the sniper should attempt to fire around cover rather than over it.

- *Firing From Windows*. In a built-up area windows provide readily accessible firing ports. However, the sniper must not allow his weapon to protrude beyond the window. It is an obvious sign of the firer's position, especially at night when the muzzle flash can easily be seen. A sniper should position himself as far into the room as possible to prevent the muzzle flash from being seen. He should fire from a supported position (table and sandbag) low enough to avoid silhouetting himself. He should use room shadow during darkness and leave blinds or shades drawn to a maximum to avoid being seen. The sniper must be careful when firing to prevent the drapes or curtains from moving due to the muzzle blast. He can do this by tacking them down or using sufficient standoff. He should also use drop clothes behind himself to cut down on silhouetting.

- *Firing From an Unprepared Loophole.* The sniper may fire through a hole torn in the wall, thus avoiding the windows. He should stay as far from the loophole as possible so the muzzle does not protrude beyond the wall, thus concealing the muzzle flash. If the hole is natural damage, he should ensure that it is not the only hole in the building. If the sniper constructs it, then the hole must blend with the building or he should construct multiple holes. There are several openings in a building that naturally occur and the sniper can enlarge or use them.

- *Firing From the Peak of a Roof.* This position provides a vantage point for snipers that increases their field of vision and the ranges at which they can engage targets. A chimney, a smokestack, or any other object protruding from the roof of a building can reduce the size of the target exposed, and the sniper should use it. However, his head and weapon breaks the clean line of a rooftop and this position is a "last choice" position.

- *Firing When No Cover Is Available.* When no cover is available, target exposure can be reduced by firing from the prone position, firing from shadows, presenting no silhouette against buildings or skyline, and using tall grass, weeds, or shrubbery for concealment if available.

PREPARED HIDE

6-64. A prepared hide is one built or improved to allow the sniper to engage a particular area, avenue of approach, or enemy position while reducing his exposure to return fire. Common sense and imagination are the sniper team's only limitation in the construction of urban hides. The sniper must follow several principles in urban and field environments. In urban environments, the sniper must still avoid silhouetting, consider reflections and light refraction, and be sure to minimize muzzle blast effects on dust, curtains, and other surroundings. The team constructs and occupies one of the following positions or a variation thereof:

- *Chimney Hide.* The sniper can use a chimney or any other structure that protrudes through the roof as a base to build his sniper position. Part of the roofing material is removed to allow the sniper to fire around the chimney while standing inside the building on beams or a platform with only his head and shoulders above the roof (behind the chimney). He should use sandbags on the sides of the position to protect his flanks.

- *Roof Hide.* When preparing a sniper position on a roof that has no protruding structure to provide protection, the sniper should prepare his position underneath on the enemy side of the roof (Figure 6-1, page 6-16). He should remove a small piece of roofing material to allow him to engage targets in his sector. He then reinforces the position with sandbags and prepares it so that the only sign that a position exists is the missing piece of roofing material. The sniper should also remove other pieces of roofing to deceive the enemy as to the true sniper position. The sniper should not be visible from outside the building. Care must be taken to hide the muzzle flash from outside the building.

Figure 6-1. Roof Hide

- *Room Hide*. In a room hide, the sniper team uses an existing room and fires through a window or loophole (Figure 6-2). It can use existing furniture, such as desks or tables to establish weapon support. When selecting a position, teams must notice both front and back window positions. To avoid silhouetting, they may need to use a backdrop, such as a dark-colored blanket, canvas, carpet, and a screen. Screens (common screening material) are important since they allow the sniper teams maximum observation and deny observation by the enemy. They must not remove curtains; however, they can open windows or remove single panes of glass. Remember, teams can randomly remove panes in other windows so the position is not obvious.

Figure 6-2. Internal View of a Room Hide

- *Crawl Space Hide.* The sniper team builds this position into the space between floors in multistory buildings (Figure 6-3). Loopholes are difficult to construct, but a damaged building helps considerably. Escape routes can be holes knocked into the floor or ceiling. Carpet or furniture placed over escape holes or replaced ceiling tiles will conceal them until needed.

Figure 6-3. Crawl Space Hide

PRINCIPLES FOR SELECTING AND OCCUPYING SNIPER FIRING POSITIONS

6-65. Upon receiving a mission, the sniper team locates the target area and then determines the best location for a tentative position by using various sources of information. The team ensures the position provides optimum balance between the following principles:

- Avoid obvious sniper positions.
- Make maximum use of available cover and concealment.
- Carefully select a new firing position before leaving an old one.
- Avoid setting a pattern. The sniper should fire from both barricaded and unbarricaded windows.
- Never subject the sniper position to traffic of other personnel, regardless of how well the sniper is hidden. Traffic invites observation and the sniper may be detected by optical devices. He should also be aware of backlighting that might silhouette him to the enemy.
- Abandon a position from which two or three misses have been fired; detection is almost certain.

- Operate from separate positions. In built-up areas, it is desirable that sniper teams operate from separate positions. Detection of two teams in close proximity is very probable, considering the number of positions from which the enemy may be observing. The snipers should position themselves where they can provide mutual support.
- Select alternate positions as well as supplementary positions to engage targets in any direction.
- Always plan the escape route ahead of time.
- Minimize the combustibility of selected positions (fireproofing).
- Select a secure and quiet approach route. This route should, if possible, be free of garbage cans, crumbling walls, barking dogs, and other impediments.
- Select a secure entry and exit point. The more obvious and easily accessible entry and exit points are not necessarily the best, as their constant use during subsequent relief of sniper teams may more readily lead to compromise.
- Pick good arcs of observation. Restricted arcs are inevitable, but the greater the arc, the better.
- Ensure the least impedance of communications equipment.
- Consider all aspects of security.
- Try to pick positions of comfort. This rule is important but should be the lowest priority. Uncomfortable observation and firing positions can be maintained only for short periods. If there is no adequate relief from observation, hides can rarely remain effective for more than a few hours.
- Never return to a sniper position that the sniper has fired from, no matter how good it is.

CHARACTERISTICS OF URBAN HIDES

6-66. The overriding requirement of a hide is that it must dominate its area of responsibility and provide maximum observation of the target area.

6-67. When selecting a suitable location, there is always a tendency to go for height. In an urban operation, this can be a mistake. The greater the height attained, the more the sniper has to look out over an area and away from his immediate surroundings. For example, if a hide were established on the tenth floor of an apartment building to see a road beneath, the sniper would have to lean out of the window, which does little for security. The sniper should get only close enough to provide observation and fire without compromise. The sniper should stay at the second and third floor levels unless his area of interest is on a higher floor in another building. He would then want to be slightly above that floor if possible

6-68. The locations of incidents that the sniper might have to deal with are largely unpredictable, but the ranges are usually relatively short. Consequently, a hide must cover its immediate surroundings, as well as middle and far distances. In residential areas, this goal is rarely possible, as hides are forced off ground floor levels by passing pedestrians. However, it is not advisable to go above the second floor because to go higher greatly increases the dead space in front of the hide. This practice is not a cardinal rule, however. Local conditions, such as being on a bus route, may force the sniper to go higher to avoid direct observation by passengers.

6-69. In view of this weakness in local defense of urban hides, the principle of mutual support between hides assumes even greater importance and is one reason why coordination and planning must take place at battalion level.

CONSTRUCTING AN URBAN POSITION

6-70. Positions in urban terrain are quite different from in the field. When the sniper team must construct an urban position, it should consider the following factors:

- Use a backdrop to minimize detection from the outside of the structure.
- Position the weapon to ensure adequate observation and engagement of the target area and mark the vertical and horizontal limits of observation.
- If adequate time and materials are available, hang drop cloths to limit the possibility of observation from the outside of the structure. Cut loopholes in the drop cloth fabric to allow observation of the target area.
- Always be aware of the outside appearance of the structure. Firing through loopholes in barricaded windows is preferred, but the team must also barricade all other windows.
- Build loopholes in other windows to provide more than one firing position. When building loopholes, the team should make them different shapes (not perfect squares or circles). Dummy loopholes also confuse the enemy.
- Establish positions in attics. The team removes the shingles and cuts out loopholes in the roof; however, they must make sure there are other shingles missing from the roof so that the firing position loophole is not obvious.
- Do not locate the position against contrasting background or in prominent buildings that automatically draw attention. The team must stay in the shadows while moving, observing, and engaging targets. **AVOID** obvious locations.
- Never fire close to a loophole. The team must always back away from the hole as far as possible to hide the muzzle flash and to muffle the sound of the weapon when it fires.
- Locate positions in a different room than the one the loophole is in by making a hole through a wall to connect the two and fire from inside the far room. Thus, the sniper is forming a "double baffle" with his loopholes by constructing two loopholes in succession. This method will further reduce his muzzle flash and blast and improve his concealment from enemy observation.
- Do not fire continually from one position.

FM 3-05.222

NOTE: These factors are why the sniper should construct more than one position if time and the situation permit. When constructing other positions, the team should make sure it can observe the target area. Sniper team positions should never be used by any personnel other than a sniper team.

POSSIBLE HIDE AND OBSERVATION POST LOCATIONS

6-71. Common sense and imagination are the sniper team's only limitation in determining urban hide or OP locations. Below are just a few options that the team can use to maximize cover and meet mission requirements:

- *Old Derelict Buildings.* The team should pay special attention to the possibility of encountering booby traps. One proven method of detecting guerrilla booby traps is to notice if the locals (especially children) move in and about the building freely.
- *Occupied Houses.* After carefully observing the inhabitants' daily routine, snipers can move into occupied homes and establish hides or OPs in basements and attics. This method was used very successfully by the British in Northern Ireland. However, these locations cannot be occupied for extended periods due to the strict noise discipline required.
- *Shops.*
- *Schools and Churches.* When using these buildings, the snipers risk possible damage to what might already be strained public relations. They should not use these positions if they are still active buildings in the community.
- *Factories, Sheds, and Garages.*
- *Basements and Between Floors in Buildings.* It is possible for the sniper team to locate itself in these positions, although there may be no window or readily usable firing port available. These locations require the sniper to remove bricks or stones without leaving any noticeable evidence outside the building. The sniper should try to locate those crawl spaces that already vent to the outside.
- *Rural Areas From Which Urban Areas Can Be Observed.*

MANNING THE SNIPER HIDES AND OBSERVATION POSTS

6-72. Before moving into the hide or OP, the snipers **must** have the following information:

- The exact nature of the mission (observe, fire).
- The length of stay.
- The local situation.
- Procedure and timing for entry.
- Emergency recall code and procedures
- Emergency evacuation procedures.
- Radio procedures.

- Movement of any friendly troops.
- Procedure and timing for exit.
- Any special equipment needed.

6-73. The well-tried and understood principle of remaining back from windows and other apertures when in buildings has a marked effect on the manning of hides or OPs. The field of view from the back of a room through a window is limited. To enable a worthwhile area to be covered, two or even three men may have to observe at one time from different parts of the room.

SNIPER TECHNIQUES IN URBAN HIDES

6-74. Although the construction of hide positions may differ, the techniques or routines while in position are the same. Sniper teams use the technique best suited for the urban position. These may include any of the following:

- The second floor of a building is usually the best location for the position. It presents minimal dead space but provides the team more protection since passersby cannot easily spot it.
- Normally, a window is the best viewing aperture or loophole.
    - If the window is dirty, do not clean it for better viewing.
    - If curtains are prevalent in the area, do not remove those in the position. Lace or net-type curtains can be seen through from the inside, but they are difficult to see through from the outside.
    - If strong winds blow the curtains open, staple, tack, or weigh them down. However, do the same with all other curtains in open windows or the nonmovement of the curtains will attract attention.
    - Firing a round through a curtain has little effect on accuracy; however, ensure the muzzle is far enough away to avoid muzzle blast.
    - When area routine indicates open curtains, follow suit. Set up well away from the viewing aperture; however, ensure effective coverage of the assigned target area, or place a secondary drop cloth behind the open curtain where it would not be noticeable. With the sniper sandwiched between the two drop cloths, his movement and activities will be more difficult to observe with open curtains.
- Firing through glass should be avoided since more than one shot may be required. The copper jacket of the M118 round is usually stripped as the round passes through the glass. However, the mass of the core will continue and should stay on target for approximately 5 feet after penetrating standard house pane glass. The sniper should consider the following variables when shooting through glass:
    - Type and thickness of glass (tempered or safety glass reacts much differently from pane glass).
    - Distance of weapon to glass.
    - Type of weapon and ammunition.

- Distance of glass to target.
- Angle of bullet path to glass; if possible, he should fire at a 90-degree angle to the glass.
• If firing through glass, the team should also consider the following options:
  - Break or open several windows throughout the position before occupation. This can be done during the reconnaissance phase of the operation; however, avoid drawing attention to the area.
  - Remove or replace panes of glass with plastic sheeting of the heat-shrink type. The sheeting will not disrupt the bullet but will deceive the enemy into believing that the glass is still in place.
• Other loopholes or viewing apertures are nearly unlimited, such as—
  - Battle damage.
  - Drilled holes (hand drill).
  - Brick removal.
  - Loose boards or derelict houses.
• Positions can also be set up in attics or between the ceiling and roof:
  - Gable ends close to the eaves (shadow adding to concealment).
  - Battle damage to gables or roof.
  - Loose or removed tiles, shingles, or slates.
  - Skylights.
• The sniper makes sure the bullet clears the loophole. The muzzle must be far enough from the loophole and the rifle boresighted to ensure the bullet's path is not in line with the bottom of the loophole. The observer and sniper must clear the muzzle before firing.
• Front drops, usually netting, may have to be changed (if the situation permits) from dark to light colors at beginning morning nautical twilight or ending evening nautical twilight due to sunlight or lack of sunlight into the position.
• If the site is not multiroomed, partitions can be made by hanging blankets or nets to separate the operating area from the rest and administrative areas.
• If sandbags are required, the team can fill and carry them inside of rucksacks, or fill them in the basement, depending on the situation or location of the position site.
• There should always be a planned escape route that leads to the ORP. When forced to vacate the position, the team meets the reaction force at the ORP. Normally, the team will not be able to leave from the same point at which it gained access; therefore, a separate escape point may be required in emergencies. The team must consider windows (other than the viewing apertures), anchored ropes to climb down the building, or a small, preset explosive charge situation on a wall or floor for access into adjoining rooms, buildings, or the outside.

- The type of uniform or camouflage that the team will wear is dictated by the tactical situation, the rules of engagement, the team's mission, and the AO. The following applies:
  - Most often, the normal BDU and required equipment are worn.
  - Urban-camouflaged uniforms can be made or purchased. Urban areas vary greatly in color (mostly gray [cinder block]; red [brick]; white [marble]; dark gray [granite]; or stucco, clay, or wood). Regardless of area color, uniforms should include angular-lined patterns.
  - When necessary, most woodland-patterned BDUs can be worn inside out, as they are a green-gray color underneath.
  - Soft-soled shoes or boots are the preferred footwear in the urban environment.
  - The team can reduce its visual profile during movement by using nonstandard uniforms or a mixture of civilian clothes as part of a deliberate deception plan. With theater approval, civilian clothing can be worn (native or host country populace).
  - Tradesmen's or construction workers' uniforms and accessories can aid in the deception plan.

WEAPONS CHARACTERISTICS IN URBAN TERRAIN

6-75. The characteristics of built-up areas and the nature of urban warfare influence the effectiveness of sniper systems and how they may be employed. The sniper must consider the following basic factors during all urban operations:

STRUCTURAL CONFIGURATION OF BUILDINGS

6-76. The basic classes of structures encountered in a built-up area can generally be classified as concrete, masonry, or wooden. However, any one building may include a combination of these materials. All buildings offer concealment, although the degree of protection varies with the material used. The 7.62- x 51-mm NATO ball cartridge will penetrate at 200 meters—

- Fifty inches of pinewood boards.
- Ten inches of loose sand.
- Three inches of concrete.

GLASS PENETRATION

6-77. If the situation should require firing through glass, the sniper should know—

- When the M118 ammunition penetrates glass, in most cases, the copper jacket is stripped of its lead core and the core fragments. These fragments will injure or kill should they hit either the hostage or the terrorist. The fragments show no standard pattern, but randomly fly in a cone-shaped pattern, much like shot from a shotgun. Even when the glass is angled to as much as 45 degrees, the lead core will not show minimum signs of deflection up to approximately **5 feet** past the point of impact with standard house pane glass.

- When the bullet impacts with the glass, the glass will shatter and explode back into the room. The angle of the bullet impact with the glass has no bearing on the direction of the shattered glass. The shattered glass will always fly perpendicular to the pane of the glass.

6-78. The U.S. Secret Service tested the efficiency of Federal 168 grain Sierra hollow-point boattail ammunition on several types of glass and found that—

- Targets placed up to 20 feet behind the glass were neutralized when the weapon was fired from 100 meters away, from a 0- to 45-degree angle of deflection.

NOTE: These results do not fit with the tests run by the Marine Corps, U.S. Army, or the FBI. Their tests showed a deviation that was acceptable to 5 or 7 feet.

- Glass fragmentation formed a cone-shaped hazard area 10 feet deep and 6 feet in diameter; the axis of which is perpendicular to the line and angle of fire.
- The jacket separated from the round but both jacket and round maintained an integrated trajectory.

ENGAGEMENT TECHNIQUES

6-79. Engaging targets not only requires the sniper to determine specific variables, but also to be trained and proficient in the methods below.

SIMULTANEOUS SHOOTING

6-80. Shooting simultaneously by command fire with another sniper is a very important skill to develop and requires much practice. The senior man in the command post (CP) will usually give this command. He may delegate the actual firing decision to the assault element team leader, so that the sniper fire may be better coordinated with the rescue effort. The actual command "standby, (pause) ready, ready, fire" must be given clearly, without emotion, or tonal change. Procedure is as follows:

- Team leader requests, "SNIPER STATUS."
- Snipers respond by numbers on availability of targets, "ONE ON," "TWO ON," "THREE OFF," "FOUR ON."
- Team leader will respond with "STANDBY," or "HOLD," depending on availability of targets.
- If assault is a GO, the command "READY, READY FIRE" is given." All snipers with targets will shoot simultaneously. (This action should sound as one shot.)
- Or the team leader may indicate specific snipers to fire.
- Alert commands should be repeated twice, "READY, READY, FIRE."
- After shooting the sniper will acknowledge, "SHOT OUT." He will then confirm results.

6-81. Reactive targets give a positive visual indication of simultaneous impact and should be used whenever possible for this exercise.

COUNTDOWN SYSTEM

6-82. During a multiple action engagement or a sniper-initiated assault, a countdown technique will be used. The CP or team leader gives a verbal countdown as follows:

STANDBY

(PAUSE)

5-

4-

3-

2- SNIPERS FIRE.

1- BLAST FROM GRENADES OR BREACHING CHARGE.

6-83. If glass must be shattered to provide the primary sniper a clear shot at his target, it is best if the support sniper also aims his "window breaking" bullet at the target. This way, the sniper team has **two** projectiles aimed at the target, increasing the likelihood of a hit.

Appendix A
Weights, Measures, and Conversion Tables

Tables A-1 through A-5, pages A-1 and A-2, show metric units and their U.S. equivalents. Tables A-6 through A-15, pages A-2 through A-5, are conversion tables.

Table A-1. Linear Measure

Unit	Other Metric Equivalent	U.S. Equivalent
1 centimeter	10 millimeters	0.39 inch
1 decimeter	10 centimeters	3.94 inches
1 meter	10 decimeters	39.37 inches
1 decameter	10 meters	32.8 feet
1 hectometer	10 decameters	328.08 feet
1 kilometer	10 hectometers	3,280.8 feet

Table A-2. Liquid Measure

Unit	Other Metric Equivalent	U.S. Equivalent
1 centiliter	10 milliliters	0.34 fluid ounce
1 deciliter	10 centiliters	3.38 fluid ounces
1 liter	10 deciliters	33.81 fluid ounces
1 decaliter	10 liters	2.64 gallons
1 hectoliter	10 deciliters	26.42 gallons
1 kiloliter	10 hectoliters	264.18 gallons

Table A-3. Weight

Unit	Other Metric Equivalent	U.S. Equivalent
1 centigram	10 milligrams	0.15 grain
1 decigram	10 centigrams	1.54 grains
1 gram	10 decigrams	0.035 ounce
1 decagram	10 grams	0.35 ounce
1 hectogram	10 decigrams	3.52 ounces
1 kilogram	10 hectograms	2.2 pounds
1 quintal	100 kilograms	220.46 pounds
1 metric ton	10 quintals	1.1 short tons

Table A-4. Square Measure

Unit	Other Metric Equivalent	U.S. Equivalent
1 square centimeter	100 square millimeters	0.155 square inch
1 square decimeter	100 square centimeters	15.5 square inches
1 square meter (centaur)	100 square decimeters	10.76 square feet
1 square decameter (are)	100 square meters	1,076.4 square feet
1 square hectometer (hectare)	100 square decameters	2.47 acres
1 square kilometer	100 square hectometers	0.386 square mile

Table A-5. Cubic Measure

Unit	Other Metric Equivalent	U.S. Equivalent
1 cubic centimeter	1,000 cubic millimeters	0.06 cubic inch
1 cubic decimeter	1,000 cubic centimeters	61.02 cubic inches
1 cubic meter	1,000 cubic decimeters	35.31 cubic feet

Table A-6. Temperature

Convert From	Convert To
Fahrenheit	Celsius Subtract 32, multiply by 5, and divide by 9
Celsius	Fahrenheit Multiply by 9, divide by 5, and add 32

Table A-7. Approximate Conversion Factors

To Change	To	Multiply By	To Change	To	Multiply By
Inches	Centimeters	2.540	Ounce-inches	Newton-meters	0.007062
Feet	Meters	0.305	Centimeters	Inches	3.94
Yards	Meters	0.914	Meters	Feet	3.280
Miles	Kilometers	1.609	Meters	Yards	1.094
Square inches	Square centimeters	6.451	Kilometers	Miles	0.621
Square feet	Square meters	0.093	Square centimeters	Square inches	0.155
Square yards	Square meters	0.836	Square meters	Square feet	10.76
Square miles	Square kilometers	2.590	Square meters	Square yards	1.196
Acres	Square hectometers	0.405	Square kilometers	Square miles	0.386

Table A-7. Approximate Conversion Factors (Continued)

To Change	To	Multiply By	To Change	To	Multiply By
Cubic feet	Cubic meters	0.028	Square hectometers	Acres	2.471
Cubic yards	Cubic meters	0.765	Cubic meters	Cubic feet	35.315
Fluid ounces	Millimeters	29.573	Cubic meters	Cubic yards	1.308
Pints	Liters	0.473	Millimeters	Fluid ounces	0.034
Quarts	Liters	0.946	Liters	Pints	2.113
Gallons	Liters	3.785	Liters	Quarts	1.057
Ounces	Grams	28.349	Liters	Gallons	0.264
Pounds	Kilograms	0.454	Grams	Ounces	0.035
Short tons	Metric tons	0.907	Kilograms	Pounds	2.205
Pounds-feet	Newton-meters	1.356	Metric tons	Short tons	1.102
Pounds-inches	Newton-meters	0.11296	Nautical Miles	Kilometers	1.852

Table A-8. Area

To Change	To	Multiply By	To Change	To	Multiply By
Square millimeters	Square inches	0.00155	Square inches	Square millimeters	645.16
Square centimeters	Square inches	9.155	Square inches	Square centimeters	6.452
Square meters	Square inches	1,550	Square inches	Square meters	0.00065
Square meters	Square feet	10.764	Square feet	Square meters	0.093
Square meters	Square yards	1.196	Square yards	Square meters	0.836
Square kilometers	Square miles	0.386	Square miles	Square kilometers	2.59

Table A-9. Volume

To Change	To	Multiply By	To Change	To	Multiply By
Cubic centimeters	Cubic inches	0.061	Cubic inches	Cubic centimeters	16.39
Cubic meters	Cubic feet	35.31	Cubic feet	Cubic meters	0.028
Cubic meters	Cubic yards	1.308	Cubic yards	Cubic meters	0.765
Liters	Cubic inches	61.02	Cubic inches	Liters	0.016
Liters	Cubic feet	0.035	Cubic feet	Liters	28.32

Table A-10. Capacity

To Change	To	Multiply By	To Change	To	Multiply By
Milliliters	Fluid drams	0.271	Fluid drams	Milliliters	3.697
Milliliters	Fluid ounces	0.034	Fluid ounces	Milliliters	29.57
Liters	Fluid ounces	33.81	Fluid ounces	Liters	0.030
Liters	Pints	2.113	Pints	Liters	0.473
Liters	Quarts	1.057	Quarts	Liters	0.946
Liters	Gallons	0.264	Liters	Gallons	3.785

Table A-11. Statute Miles to Kilometers and Nautical Miles

Statute Miles	Kilometers	Nautical Miles	Statute Miles	Kilometers	Nautical Miles
1	1.61	0.869	60	96.60	52.14
2	3.22	1.74	70	112.70	60.83
3	4.83	2.61	80	128.80	69.52
4	6.44	3.48	90	144.90	78.21
5	8.05	4.35	100	161.00	86.92
6	9.66	5.21	200	322.00	173.80
7	11.27	6.08	300	483.00	260.70
8	12.88	6.95	400	644.00	347.60
9	14.49	7.82	500	805.00	434.50
10	16.10	8.69	600	966.00	521.40
20	32.20	17.38	700	1127.00	608.30
30	48.30	26.07	800	1288.00	695.20
40	64.40	34.76	900	1449.00	782.10
50	80.50	43.45	1000	1610.00	869.00

Table A-12. Nautical Miles to Kilometers and Statute Miles

Nautical Miles	Kilometers	Statute Miles	Nautical Miles	Kilometers	Statute Miles
1	1.85	1.15	60	111.00	69.00
2	3.70	2.30	70	129.50	80.50
3	5.55	3.45	80	148.00	92.00
4	7.40	4.60	90	166.50	103.50
5	9.25	5.75	100	185.00	115.00
6	11.10	6.90	200	370.00	230.00
7	12.95	8.05	300	555.00	345.00
8	14.80	9.20	400	740.00	460.00
9	16.65	10.35	500	925.00	575.00
10	18.50	11.50	600	1110.00	690.00
20	37.00	23.00	700	1295.00	805.00
30	55.50	34.50	800	1480.00	920.00
40	74.00	46.00	900	1665.00	1033.00
50	92.50	57.50	1000	1850.00	1150.00

Table A-13. Kilometers to Statute and Nautical Miles

Kilometers	Statute Miles	Nautical Miles	Kilometers	Statute Miles	Nautical Miles
1	0.62	0.54	60	37.28	32.38
2	1.24	1.08	70	43.50	37.77
3	1.86	1.62	80	49.71	43.17
4	2.49	2.16	90	55.93	48.56
5	3.111	2.70	100	62.14	53.96
6	3.73	3.24	200	124.28	107.92
7	4.35	3.78	300	186.42	161.88
8	4.97	4.32	400	248.56	215.84
9	5.59	4.86	500	310.70	269.80
10	6.21	5.40	600	372.84	323.76
20	12.43	10.79	700	434.98	377.72
30	18.64	16.19	800	497.12	431.68
40	24.86	21.58	900	559.26	485.64
50	31.07	26.98	1000	621.40	539.60

Table A-14. Yards to Meters

Yards	Meters	Yards	Meters	Yards	Meters
100	91	1000	914	1900	1737
200	183	1100	1006	2000	1828
300	274	1200	1097	3000	2742
400	366	1300	1189	4000	3656
500	457	1400	1280	5000	4570
600	549	1500	1372	6000	5484
700	640	1600	1463	7000	6398
800	732	1700	1554	8000	7212
900	823	1800	1646	9000	8226

Table A-15. Meters to Yards

Meters	Yards	Meters	Yards	Meters	Yards
100	109	1000	1094	1900	2078
200	219	1100	1203	2000	2188
300	328	1200	1312	3000	3282
400	437	1300	1422	4000	4376
500	547	1400	1531	5000	5470
600	656	1500	1640	6000	6564
700	766	1600	1750	7000	7658
800	875	1700	1860	8000	8752
900	984	1800	1969	9000	9846

Appendix B
Mission-Essential Tasks List

SPECIAL OPERATIONS TARGET INTERDICTION COURSE
MOS 18—SKILL LEVEL 3
CRITICAL INDIVIDUAL TASKS

Subject Area 1: Special Operations Target Interdiction

Task No.	Title
331-202-4200	Detect Targets Based on Target Indicators
331-202-4201	Produce a Panoramic Sketch
331-202-4202	Prepare a Sniper's Observation Log
331-202-4204	Prepare a Sniper Mission Operation Order
331-202-4205	Conduct Training of Special Operations Snipers
331-202-4206	Employ the Methods Used for Indicating Targets
331-202-4208	Determine Sniper Assessment and Selection Procedures
331-202-4209	Determine the Capabilities and Roles of Special Operations Snipers
331-202-4214	Observe an Arc of Observation
331-202-4215	Perform Selected Special Operations Sniper/Observer Team Functions/Tasks in Support of Special Operations Forces Mission/Combat Operations

Subject Area 2: Sniper Weapon System

Task No.	Title
331-202-4210	Maintain Personal and Team Optical Equipment
331-202-4211	Mount a Night Vision Device on the Sniper Weapon System
331-202-4212	Maintain the Sniper Weapon System
331-202-4213	Prepare the Sniper Weapon System for Infiltration

Subject Area 3: Ballistics

Task No.	Title
331-202-4220	Determine Distance With the Unaided Eye
331-202-4221	Determine Distance With Mechanical and/or Optical Aids
331-202-4222	Engage Targets Applying Sniper System Ballistic Theory
331-202-4223	Apply Sight Corrections to Compensate for Wind and Meteorological Conditions

331-202-4224	Determine the Point of Impact on the Target by Reading the Bullet Trace

Subject Area 4: Tracking

331-202-4231	Employ the Observation Techniques/Categories Needed to Enhance the Recall of Details
331-919-0161	Evade Dog/Visual Tracker Teams
331-919-0162	Demonstrate Visual Tracking Techniques

Subject Area 5: Concealment

331-202-4203	Select a Line of Advance
331-202-4230	Employ Stealth Movement Methods
331-202-4232	Conceal Yourself and Your Equipment
331-202-4233	Construct a Ghillie Suit
331-202-4234	Construct Sniping Hides
331-202-4235	Camouflage Yourself and Your Equipment

Subject Area 6: Marksmanship

331-202-4240	Employ the Four Fundamentals of Shooting
331-202-4241	Employ Supported Shooting Positions
331-202-4242	Adjust the Iron and Telescopic Sights for the M24 Sniper Weapon System
331-202-4243	Engage Stationary Targets With the Sniper Weapon System
331-202-4244	Engage Moving Targets With the Sniper Weapon System
331-202-4245	Engage Snap Targets With the Sniper Weapon System
331-202-4246	Zero the Sniper Weapon System
331-202-4247	Prepare a Sniper's Range Card
331-202-4248	Engage Targets Using the Night Vision Device on the Sniper Weapon System
331-202-4249	Engage Targets During Time of Limited Visibility With the Telescopic Sight
331-202-4250	Engage Targets With Selected U.S., Foreign, Special Purpose, and Obsolete Sniper Weapon Systems
331-202-4251	Zero a Night Vision Device on a Sniper Weapon System During Daylight
331-202-4252	Engage Targets at the Maximum Effective Range of the Sniper Weapon System
331-202-4253	Demonstrate Planning Considerations for Operations on Urban Terrain

331-202-4254 Engage Targets Over Uneven Ground With the Sniper Weapon System

COLLECTIVE TASKS

Task No.	Title
7-5-1825	Move Tactically (Sniper)
7-5-1869	Select/Engage Targets (Sniper)
7-5-1871	Select/Occupy Firing Position (Sniper)
7-5-1872	Estimate Range (Sniper)
7-5-1809	Debrief (Sniper)

ELEMENT: Sniper Team

TASK: Move Tactically (7-5-1825) (FM 7-8, TC 23-10)

ITERATION: 1   2   3   4   5   M   (Circle)

COMMANDER/LEADER ASSESSMENT: T   P   U   (Circle)

CONDITIONS: The sniper team is given a mission to move with a security element. Both friendly and opposing forces units have indirect fire and combat air support (CAS) available.

TASK STANDARDS:
1. The sniper team moves undetected.
2. The sniper team moves tactically based on METT-TC.
3. The sniper team complies with all graphic control measures.
4. The sniper team moves along the route specified in the order.
5. The sniper team arrives at the destination specified in the order.
6. The sniper team arrives at the specified time.
7. The sniper team sustains no casualties.

TASK STEPS AND PERFORMANCE MEASURES	GO	NO-GO

*+1. The sniper team leader selects the movement routes that—
  a. Avoid known opposing force (OPFOR) positions and obstacles.
  b. Offer cover and concealment.
  c. Take advantage of difficult terrain, swamp, and dense woods.
  d. Avoid natural lines of drift.
  e. Avoid trails, roads, footpaths, or built-up or populated areas unless required by the mission.
2. The sniper team uses the proper movement techniques: sniper low crawl, medium crawl, high crawl, hand-and-knee crawl, and walk.
  a. The observer is the point man; the sniper follows.
  b. The observer's sector is from 9 o'clock to 3 o'clock; the sniper's sector is from 3 o'clock to 9 o'clock.		

FM 3-05.222

c. The observer and the sniper must maintain visual contact even when lying on the ground.
d. The interval between the observer and the sniper is not more than 20 meters.
e. The sniper reacts to the point man's actions.
f. The sniper and the point man cross danger areas. (See T&EO 7-3/4-1028, Cross Danger Area, ARTEP 7-8-MTP, *Mission Training Plan for Infantry Rifle Platoon and Squad*).
3. The sniper team maintains operations security.
   a. Moves slowly and cautiously.
   b. Uses camouflage.
   c. Avoids making sounds.
4. The sniper team maintains proper communication procedures.
   a. Maintains radio listening silence.
   b. Uses visual signals.		

TASK PERFORMANCE SUMMARY BLOCK							
ITERATION	1	2	3	4	5	M	TOTAL
TOTAL TASK STEPS EVALUATED							
TOTAL TASK STEPS "GO"							

"*" indicates a leader task step.
"+" indicates a critical task step.

OPFOR TASK: Engage Sniper Team
STANDARDS:
1. The OPFOR detects the moving sniper team.
2. The OPFOR delays the team beyond its allotted time (leader evaluation).
3. The OPFOR prevents the team from moving to its assigned destination or along its prescribed route (leader evaluation).
4. The OPFOR inflicts one casualty on the sniper team.

FM 3-05.222

ELEMENT: Sniper Team

TASK: Engage Targets (7-5-1869) (TC 23-10)

ITERATION: 1 2 3 4 5 M (Circle)

COMMANDER/LEADER ASSESSMENT: T P U (Circle)

CONDITIONS: The sniper team has a specific sniper mission (target criteria and priority), either by supporting a unit or acting independently. The sniper team observes the targets. Both friendly and OPFOR units have indirect fire and CAS available.

TASK STANDARDS:

1. The sniper team selects the priority target and destroys it with no more than two rounds.
2. The sniper team sustains no casualties.

TASK STEPS AND PERFORMANCE MEASURES	GO	NO-GO

1. The sniper team identifies the following priority targets that will limit the OPFOR's fighting ability:
  a. OPFOR sniper.
  b. Officers, both military and political.
  c. NCOs.
  d. Scout or dog team.
  e. Crew-served weapons personnel.
  f. Vehicle commanders and drivers.
  g. Communications personnel.
  h. Forward observers.
  i. Critical equipment such as optical sights or radios.		

*2. The sniper team leader selects the priority targets to be engaged.
  a. The sniper team selects the target that is critical to the mission.
  b. The sniper team does not become a target while searching for or firing on an OPFOR target.
  c. The sniper team estimates its range from the target. (See T&EO 7-5-1872, Estimate Range). The range must be within 300 to 800 meters.
  d. The sniper team leader chooses to engage targets or continues the observation of the targets.		

3. The sniper team engages the target.
  a. The observer gives the wind adjustment.
  b. The sniper adjusts the scope on the target and informs the observer when completed.
  c. The observer reconfirms the wind adjustment and notifies the sniper of any changes.
  d. The sniper fires.
  e. The observer watches the vapor trail and the strike of the round. He then prepares to give an adjustment if the sniper misses.
  f. If the sniper misses, he checks the scope and fires again, or he may engage a second target.		

TASK PERFORMANCE SUMMARY BLOCK							
ITERATION	1	2	3	4	5	M	TOTAL
TOTAL TASK STEPS EVALUATED							
TOTAL TASK STEPS "GO"							

"*" indicates a leader task step.

OPFOR TASK: React to Sniper Fire

STANDARDS:

1. The OPFOR assumes covered and concealed positions within 3 seconds of receiving sniper fire.
2. The OPFOR detects the sniper team's location within 5 seconds.
3. The OPFOR returns fire within 5 seconds of receiving sniper fire.
4. The OPFOR inflicts one casualty on the sniper team.
5. The OPFOR sustains no more than one casualty.

ELEMENT: Sniper Team

TASK: Occupy Firing Position (7-5-1871) (TC 23-10)

ITERATION: 1 2 3 4 5 M (Circle)

COMMANDER/LEADER ASSESSMENT: T P U (Circle)

CONDITIONS: The sniper team is given a mission to engage a target and an area of operations. Both friendly and OPFOR units have indirect fire and CAS available.

TASK STANDARDS:
1. The sniper team selects a final firing position within 300 to 600 meters of the target area.
2. The sniper team is not detected while occupying the position.
3. The sniper team sustains no casualties.

TASK STEPS AND PERFORMANCE MEASURES	GO	NO-GO

*1. The sniper team leader selects a final firing position that has—
   a. Maximum fields of fire and observation of the target area.
   b. Maximum concealment from the OPFOR's observation.
   c. Covered routes into and out of the position.
   d. A position no closer than 300 meters to the target area.
   e. A natural or man-made obstacle (if available) between the sniper team's position and the target.		

2. The sniper team maintains operations security by avoiding—
   a. Prominent, readily identifiable objects and terrain features.
   b. Roads and trails.
   c. Objects that may make noise.
   d. Optical devices that may reflect light.
   e. Leaving a path that leads to its position.
   f. Firing position(s).		

3. The sniper team operates from a position by—
   a. Using shadows (if available).
   b. Using camouflage that does not contrast with the surrounding area.		

4. The sniper team occupies the position.
   a. Moves into the position undetected.
   b. Scans ahead and watches for overhead movement.
   c. Keeps the body outline low to the ground.		

5. The sniper team sustains the firing position.
   a. Organizes the equipment.
   b. Establishes a system of observation and relief. (See T&EO 7-3/4-1058, Sustain, ARTEP 7-8-MTP).		

TASK PERFORMANCE SUMMARY BLOCK							
ITERATION	1	2	3	4	5	M	TOTAL
TOTAL TASK STEPS EVALUATED							
TOTAL TASK STEPS "GO"							

"*" indicates a leader task step.

OPFOR TASK: Detect Snipers

STANDARDS:

1. The OPFOR detects movement of the snipers moving into the firing position.
2. The OPFOR inflicts more than one casualty.
3. The OPFOR engages the sniper team within 5 seconds.
4. The OPFOR sustains no more than one casualty.

_____ FM 3-05.222

ELEMENT: Sniper Team

TASK: Estimate Range (7-5-1872) (TC 23-10)

    ITERATION: 1   2   3   4   5   M   (Circle)

    COMMANDER/LEADER ASSESSMENT: T   P   U   (Circle)

CONDITIONS: The sniper team has to employ range estimation throughout the target area to engage targets. Both friendly and OPFOR units have indirect fire and CAS available.

TASK STANDARDS:

1. The sniper team agrees on range estimation.
2. The averaged range estimation must be within 10 percent of the actual distance.
3. The sniper team sustains no casualties.

TASK STEPS AND PERFORMANCE MEASURES	GO	NO-GO

1. Each member of the sniper team estimates the range to the target by selecting one or more of the following methods:
  a. The use of maps.
  b. A 100-meter increment.
  c. The appearance of objects.
  d. The mil-scale formula.
  e. The use of the SWS.
  f. The use of the range card.
  g. The bracketing method.
  h. A combination of methods.
2. The snipers estimate the range throughout the target area.
  a. Each sniper estimates the range to the target(s).
  b. The estimated range by individuals is averaged within 10 percent, plus or minus, of the true range.
*+3. The team leader determines the estimated range to be used.
  a. Each sniper estimates the range to the target(s).
  b. The team leader compares the estimates.
  c. The team leader makes the final determination of the range to the target(s).
  c. The range to the target(s) is within 10 percent, plus or minus, of the true range.		

TASK PERFORMANCE SUMMARY BLOCK							
ITERATION	1	2	3	4	5	M	TOTAL
TOTAL TASK STEPS EVALUATED							
TOTAL TASK STEPS "GO"							

"*" indicates a leader task step.

"+" indicates a critical task step.

FM 3-05.222

ELEMENT: Sniper Team
TASK: Debrief (7-5-1809) (TC 23-10)

    ITERATION: 1   2   3   4   5   M   (Circle)

    COMMANDER/LEADER ASSESSMENT: T   P   U   (Circle)

CONDITIONS: The sniper team completes the mission and conducts a debriefing.

TASK STANDARDS:
1. All team members and the sniper employment officer are present.
2. All information is collected and recorded in the correct format.

TASK STEPS AND PERFORMANCE MEASURES	GO	NO-GO

*1. The sniper employment officer designates an area for debriefing.
   a. The size of the area is large enough for the personnel.
      (1) S-2.
      (2) Sniper employment officer.
      (3) Sniper team.
      (4) Battalion commander or his representative.
   b. The area is equipped with the necessary maps.
   c. The debriefing is free from all distractions.
2. The sniper team links up with the sniper employment officer.
   a. The sniper team links up with the sniper employment officer at the time specified in the patrol order.
   b. The location is in a secure area behind the FLOT.
3. The team members and the sniper employment officer conduct the debriefing.
   a. All members are present.
   b. The sniper team has all recorded information.
      (1) Range card.
      (2) Field sketch.
      (3) Log book.
   c. The team leader conducts the debriefing in chronological order.		

TASK PERFORMANCE SUMMARY BLOCK							
ITERATION	1	2	3	4	5	M	TOTAL
TOTAL TASK STEPS EVALUATED							
TOTAL TASK STEPS "GO"							

"*" indicates a leader task step.

NO OPFOR TASK

Appendix C

Sustainment Program

The sustainment program enables the sniper to maintain the high degree of skill and proficiency required to complete SOF missions. The sniper's training program should emphasize marksmanship and stalking because they are the most perishable of sniper skills.

TRAINING

C-1. The frequency of training is important to maintain sniper proficiency. The sniper should be tested or evaluated on all sniper skills at least annually; semiannually is better. Marksmanship qualification should occur at least quarterly to the standards outlined in the SOTIC program of instruction (POI). Figure C-1, pages C-2 through C-11, provides a sample POI for a Level II Program.

TIME DEVOTED TO TRAINING

C-2. The time the unit allows the sniper to devote to sustainment training determines the sniper's overall proficiency. Experience has shown that to maintain the degree of weapon familiarity needed to engage targets at unknown distances, the sniper should devote at least 8 hours a week in sniper marksmanship training. This amount of time spent in **quality** marksmanship training will sustain the sniper's proficiency in the art of precision long-range rifle fire.

BASIC AMMUNITION REQUIREMENTS

C-3. Basic ammunition requirements for sustainment-type firing can be found in DA Pam 350-39, *M21/24 Sniper Rifle (Category I) Ammunition/Training Strategy* (Table C-1, page C-11). This amount is the **minimum** ammunition requirement, not the maximum.

PROGRAM OF INSTRUCTION (RECOMMENDED)

COURSE: SPECIAL OPERATIONS TARGET INTERDICTION COURSE, LEVEL II (SOTIC II)

TRAINING LOCATION: UNIT TRAINING AREA

PURPOSE: To train selected personnel in the technical skills and operational procedures necessary to deliver precision rifle fire from concealed positions to select targets in support Special Operations Forces (SOF) missions at ranges to 600 meters. This course also prepares personnel to train foreign personnel in target interdiction techniques within established U.S. policy. In addition, this course prepares personnel for the Level I SOTIC conducted at the United States Army John F. Kennedy Special Warfare Center and School, Fort Bragg, North Carolina.

SCOPE: Maximum hands-on training in advanced rifle marksmanship, techniques of observation, judging distance, advanced methods of concealment, camouflage, stalking, target selection, and interdiction mission planning.

Prerequisites: As established by the local unit commander responsible for the course.

SPECIAL INFORMATION: Instructors for this course must all be graduates of the Level I SOTIC conducted at USAJFKSWCS, Fort Bragg, NC. The student/instructor ratio must be maintained so that each team on the firing line will have an assigned instructor behind them on a mentor basis. This ratio may be 2:1 or 4:1 depending on the pit support for the course. The mentoring relationship should begin at the onset of the course and maintained throughout the course.

NOTE: This recommended course POI can be established as a two-week or a five-week POI. The unit commander may decide to add to this POI dependent upon his mission requirements. The course materials should be left as intact as possible and taught as a baseline for the unit Level II course.

COURSE LENGTH: 2 Weeks to 5 Weeks

COURSE HOURS: 140.5 - 200 Hours

CLASS SIZE: 4 minimum with size maximum dependent upon maintaining instructor to student ratio and available ranges, training areas, and classroom size.

Hours shown do not reflect maintenance that must be performed daily, transportation to and from ranges and training areas, nor breaks for meals.

Figure C-1. Sample Sniper Sustainment Program of Instruction

COURSE SUMMARY

GENERAL SUBJECTS

		HOURS	
		5 WK	2 WK
G1	Inprocessing of Students	2.0	2.0
G2	Introduction to SOTIC II	1.0	1.0
G3	Mission Planning	2.0	1.0
	TOTAL HOURS	5.0	4.0

MARKSMANSHIP

M1	The M24 SWS and Equipment	3.0	3.0
M2	Advanced Rifle Marksmanship	4.0	4.0
M3	Sniper Marksmanship	2.5	2.5
M4	Sight Adjustment and Zero	4.0	4.0
M5	Correcting for Meteorological Conditions	2.0	2.0
M6	Reading Wind and Spotting	2.0	1.0
M7	Range Exercise #1 (Position Shooting)	16.0	8.0
M8	Range Exercise #2 (Grouping and Zeroing Telescopic Sight)	8.0	4.0
M9	Range Exercise #3 (Snap Shooting; 200, 300, and 400 Meters)	12.0	12.0
M10	Range Exercise #4 (Moving Targets; 200 and 300 Meters)	12.0	12.0
M11	Range Exercise #5 (Dusk Shoot)	4.0	2.0
M12	Range Exercise #6 (Deliberate Targets; 400, 500, and 600 Meters)	12.0	12.0
M13	Range Exercise #7 (NVD Shoot–Use Unit Organic NVDs)	6.0	3.0

Figure C-1. Sample Sniper Sustainment Program of Instruction (Continued)

MARKSMANSHIP (Continued)		HOURS	
		5 WK	2 WK
M14	Application of Fire (Ballistics)	3.0	3.0
M15	Field Shooting	32.0	16.0
M16	Judging Distance	3.0	3.0
M17	Judging Distance Exercises	5.0	2.0
	TOTAL HOURS	**130.5**	**93.5**
OBSERVATION			
O1	Observation of Ground	2.0	2.0
O2	Observation Exercises	5.0	2.0
O3	Observer's Log and Range Card	1.0	1.0
O4	Panoramic Sketching and Electronic Reporting	2.0	2.0
O5	Kim's Game	5.0	2.0
	TOTAL HOURS	**15.0**	**7.0**
CONCEALMENT			
C1	Individual Camouflage and Concealment	2.0	2.0
C2	Ghillie Suit Construction	1.0	1.0
C3	Individual Movement (Stalking)	2.0	2.0
C4	Selecting Lines of Advance	2.0	2.0
C5	Stalking Exercise	17.5	7.0
C6	Sniper Hides and Loopholes	2.0	2.0
C7	Sniper Hide Construction	6.0	0.0
	TOTAL HOURS	**32.5**	**19.0**

Figure C-1. Sample Sniper Sustainment Program of Instruction (Continued)

EXAMINATIONS		HOURS	
		5 WK	2 WK
E1	Written Examination	1.0	1.0
E2	Sniper Marksmanship	8.0	8.0
E3	Field Shoot Examination	8.0	8.0
	TOTAL HOURS	**17.0**	**17.0**

NOTE: It is not mandatory that the above listed examinations are "must pass" but it is highly recommended that they are considered critical tasks and thus are "must pass" events.

The sponsoring unit commander may add to this list or adjust between the two as he sees fit, according to his mission perimeters and requirements. Some of the course that may be added according to the command requirements would be:

Urban Operations—Urban Hides, Urban Mission Considerations, Building Surveys, Advance Surveys, Sniper in Support of Close Combat, Urban Shooting Positions, Shooting Through Medium, Aerial Platform Shooting High Angle Shooting, Video Surveillance and Reporting.

Mountain Shooting—Wind Tunnel Effect of Terrain, High Angle Shooting, High Altitude Shooting and Environmental Effects of High Altitude.

Desert Shooting—High Temperature Effects on Shooting and Weapons, Mirage Problems of Desert, Temperature Inversion Effect on Observation.

This list is limited only by the commander's imagination and requirements based on his mission, equipment available, training terrain available, time allowed, and troops available for the giving instruction.

TRAINING ANNEXES

COURSE: SOTIC II

ANNEX A: GENERAL SUBJECTS

PURPSE: To provide the student with the background information and perspective necessary to understand the role of snipers in the SOF.

TOTAL HOURS: 5.0 (4.0)

SUBJECT	TITLE	HOURS	
		5 WK	2 WK
G1	Inprocessing of students	2.0	2.0

SCOPE: The student will inprocess into the course and all paperwork will be completed to establish the student's record in the course. Weapons and equipment will be issued and inspected for serviceability.

Figure C-1. Sample Sniper Sustainment Program of Instruction (Continued)

G2	Introduction to SOTIC II	1.0	1.0

SCOPE: The student will know and describe the course content, conduct on the range, range procedures, safety procedures, pass requirements, retrain and retest procedures, and examination perimeters.

G3	Mission Planning	2.0	1.0

SCOPE: The student will analyze mission requirements and operational perimeters and prepare an operations order for a mission. Due consideration will be taken of the unique characteristics of the sniper mission.

ANNEX B: MARKSMANSHIP

PURPOSE: To provide the student with the necessary skills to qualify him in sniper marksmanship to 600 meters with the M24 Sniper Marksmanship System.

TOTAL HOURS: 130.5 (93.5)

SUBJECT	TITLE	HOURS	
		5 WK	2 WK
M1	The M24 SWS and Equipment	3.0	3.0

SCOPE: The student will describe the functioning, inspection, disassembly, and reassembly of the M24; and the maintenance procedures for the M24 SWS and related sniper equipment used by their unit.

M2	Advanced Rifle Marksmanship	4.0	4.0

SCOPE: The student will demonstrate the four fundamentals of shooting to include the use of the sling and iron sights. The student will demonstrate the positions of standing off-hand, sitting rapid, prone rapid, and prone slow. NOTE: It is understood that most known distance (KD) ranges are established in yards, and yards may be substituted for meters in these exercises; see Range Exercise #1.

M3	Sniper Marksmanship	2.5	2.5

SCOPE: The student will demonstrate sniper marksmanship skills, to include field shooting positions and engagement techniques of snap targets, moving targets, and deliberate targets at known ranges.

M4	Sight Adjustment and Zero	4.0	4.0

SCOPE: The student will descrbe the principles of minute of angle and mils and how these measurements relate to zeroing the M24. He will demonstrate the zeroing of both the iron sights and the M3A scope, sight adjustment, and use of the sniper data logbook.

M5	Correcting for Meteorological Conditions	2.0	2.0

SCOPE: The student will calculate and correct for zero change and bullet deflection caused by meteorological conditions.

Figure C-1. Sample Sniper Sustainment Program of Instruction (Continued)

ANNEX B: MARKSMANSHIP (Continued)	HOURS	
	5 WK	2 WK
M6 Reading and Spotting	2.0	1.0

SCOPE: The student will demonstrate the use of the spotting scope to determine wind velocity and direction. He will read bullet trace to determine round impact. This class is conducted on a known distance range.

M7 Range Exercise #1 (Position Shooting)	16.0	8.0

SCOPE: The student will demonstrate Advanced Rifle Marksmanship skills while firing the NRA course of fire on a known distance range. The student will first zero his iron sights at 200 meters (yards) in the prone supported position. This exercise will also be the first exercise to utilize the sniper—observer pair for reading wind and spotting. Exercise will be conducted on a known distance range and consists of 200 meter (yard) standing off-hand, 200 meter (yard) sitting rapid, 300 meter (yard) prone rapid, and 600 meter (yard) slow fire.

NOTE: Two-week course may wish to substitute 200 meter (yard) prone slow and rapid for 200-meter exercises, drop 300 and fire the 600 prone slow.

M8 Range Exercise #2 (Grouping and Zeroing Telescopic Sights)	8.0	4.0

SCOPE: The student will demonstrate zeroing of his telescopic sight at 200 meters and then practice five round groups at 200, 300, 400, 500, and 600 meters (yards). This exercise will be conducted on a known distance range.

M9 Range Exercise #3 (Snap Shooting; 200, 300, and 400 Meters [Yards])	12.0	12.0

SCOPE: The student will demonstrate the techniques and ability to engage snap targets at various ranges under restricted time limits. Exercises conducted will 3 second snap target at 200 meters (yards) over a 7 meter front; 6 second snap target at 300 meters (yards) over a 7 meter front; 3 second snap target at 400 meters (yards) over a 10 meter front. The 200 and 300-meter targets are the head of an FBI target while the 400-meter target is the chest and head of an FBI target.

M10 Range Exercise #4 (Moving Targets; 200 and 300 Meters)	12.0	12.0

SCOPE: The student will demonstrate the techniques and the ability to engage targets moving at a parade ground marching cadence over a distance of 10 meters at ranges of 200 and 300 meters. This exercise will be conducted on a known distance range.

M11 Range Exercise #5 (Dusk Shoot)	4.0	2.0

SCOPE: The student will engage targets under conditions of failing light with the M3A telescopic sight. Exercise will run from sunset to EENT or until light conditions no longer permit shooting. The students will then engage targets with artificial illumination supplied by parachute flares fired from various locations. This exercise can be run on a known distance range or a machine gun range with pop-up targets.

Figure C-1. Sample Sniper Sustainment Program of Instruction (Continued)

ANNEX B: MARKSMANSHIP (Continued)	HOURS 5 WK	2 WK
M12 Range Exercise #6 (Deliberate Targets; 400, 500, and 600 Meters [Yards])	12.0	12.0

SCOPE: The student will demonstrate the ability to engage targets with precision shots at ranges to 600 meters under time restraints of 10 seconds, 400 meters; 12 seconds, 500 meters; 15 seconds, 600 meters (yards) on a known distance range.

M13 Range Exercise #7 (NVD Shoot—Use Unit Organic NVDs)	6.0	3.0

SCOPE: The student will zero the NVDs and engage targets at various ranges using the unit organic NVDs. This exercise is run on a known distance range or a machine gun range with pop-up targets.

M14 Application of Fire	3.0	3.0

SCOPE: The student will describe how to calculate the effects of internal and external ballistics on the trajectory of the round over uneven ground at targets on unknown distance, hit probability and the effects of internal and external ballistics on hit probability, target designation and selection, terminal ballistics and follow on shot calculation.

M15 Field Shooting (Unknown Distance)	32.0	16.0

SCOPE: The student will successfully engage targets at unknown distances from 200 to 600 meters under field conditions.

M16 Judging Distance	3.0	3.0

SCOPE: The student will describe and demonstrate the various techniques of judging distance by eye and with the M3A telescopic sight.

M17 Judging Distance Exercises (5 or 2 Exercises)	5.0	2.0

SCOPE: The student will conduct practical exercises in judging distance by eye and by M3A scope using the mil-relationship formula. The 5-week course will conduct 5 exercises while the 2-week course will conduct 2 exercises.

Figure C-1. Sample Sniper Sustainment Program of Instruction (Continued)

ANNEX C: OBSERVATION

PURPOSE: To provide the student with the background knowledge and skills to perform his role as an observer in the sniper observer pair.

SUBJECT	TITLE	HOURS	
		5 WK	2 WK
O1	Observation of Ground	2.0	2.0

SCOPE: The student will describe the four target indicators and how they are used to locate the enemy, demonstrate the principles and techniques of observation, and the differences in the techniques during day and night.

O2	Observation Exercises (5 or 2 Exercises)	5.0	2.0

SCOPE: The student will conduct one practical exercise and four graded, (one and one for the 2 week) where he will demonstrate his ability to draw a panoramic sketch or a sector and accurately place the 10 items in the sector through the use of observation skills and an understanding of target indicators.

O3	Observer's Log and Range Card	1.0	1.0

SCOPE: The student will record information in an observer's log and demonstrate the construction of and use of the range card.

O4	Panoramic Sketching and Electronic Reporting	2.0	2.0

SCOPE: The student will demonstrate the techniques of panoramic sketching and construction of a filed sketch including marginal information, subsketches, the use of sketching of materials, and the use of electronic reporting techniques and equipment.

O5	Kim's Game (5 or 2 Exercises)	5.0	2.0

SCOPE: The student will demonstrate observation techniques with memorization, retention, and recall of detail

ANNEX D: CONCEALMENT

PURPOSE: To provide the student with the skills to conceal himself and his equipment during sniper operations.

SUBJECT	TITLE	HOURS	
		5 WK	2 WK
C1	Individual Camouflage and Concealment	2.0	2.0

SCOPE: The student will describe target indicators and techniques to defeat the target indicators, principles and techniques of camouflage and concealment, and use of various materials in camouflage and concealment.

Figure C-1. Sample Sniper Sustainment Program of Instruction (Continued)

ANNEX D: CONCEALMENT (Continued)	HOURS	
	5 WK	2 WK

PURPOSE: To provide the student with the background knowledge and skills to perform his role as an observer in the sniper observer pair.

C2	Ghillie Suit Construction	1.0	1.0

SCOPE: The student will construct a Ghillie suit using the principles of camouflage and concealment.

C3	Individual Movement (Stalking)	2.0	2.0

SCOPE: The student will demonstrate the 5 techniques of stealth movement over ground with the M24 SWS and equipment.

C4	Selecting Lines of Advance	2.0	2.0

SCOPE: The student will demonstrate terrain analysis, route selection using vantage points, dead space and masking techniques to facilitate stealth movement to an objective and return.

C5	Stalking Exercises (5 or 2 Exercises)	17.5	7.0

SCOPE: The student will demonstrate selecting lines of advance, stealth movement techniques and concealment by stalking to within 200 meters of an observer using binoculars to locate the student. There are 5 (2) three-hour exercises with each having a 30-minute preparation period prior to the stalk start. Each exercise is graded at 20 points, five stalk; or 50 points, 2 stalk.

C6	Sniper Hides and Loopholes	2.0	2.0

SCOPE: The student will describe site selection criteria, hasty and deliberate sniper hide construction, methods of loophole construction and concealment, spoils removal, and the effects of long term operations confined in the hide, sleep deprivation, and the effects on mission completion. The student will also describe the special considerations required for urban hide construction and building construction factors effecting the hide construction.

C7	Sniper Hide Construction	6.0	0.0

SCOPE: The students will demonstrate hide construction through the construction of a class hide for sniper team. Site selection, construction materials and techniques, spoil disposal, and camouflage and concealment techniques will be demonstrated in the hide exercise. The students may be required to work on an urban hide in lieu of a rural type hide.

ANNEX E: EXAMINATIONS

PURPOSE: The student will demonstrate his grasp of the subject matter through the use of various examinations designed to test his knowledge and capabilities.

SUBJECT	TITLE	HOURS	
		5 WK	2 WK
E1	Written Examination	1.0	1.0

SCOPE: The student will demonstrate his knowledge of the subjects presented throughout the course with a multiple-choice examination consisting of 50 questions.

Figure C-1. Sample Sniper Sustainment Program of Instruction (Continued)

ANNEX E: EXAMINATIONS (Continued)	HOURS	
	5 WK	2 WK
E2 Sniper Marksmanship	8.0	8.0

SCOPE: The student will demonstrate his ability to zero the M3A telescopic sight on the M24 SWS and fire the SWS in a course of fire consisting of snaps and movers at 200 and 300 meters, snaps and deliberates at 400 meters, and deliberates at 500 and 600 meters.

E3 Field Shoot Examination	8.0	8.0

SCOPE: The student will demonstrate his ability to engage five targets over uneven terrain at unknown distances from 400 meters to 700 meters.

Figure C-1. Sample Sniper Sustainment Program of Instruction (Continued)

Table C-1. Excerpt From DA Pam 350-39, Dated 3 July 1997

Annual Ammunition Requirement and Training Strategy for the M24 Sniper Rifle (CAT I)				
EVENT	AC/RC	MATCH	.50 CAL	.300
DODIC		A171	A531	A191*
ZERO/CONFIRM ZERO	4/1	20		
FIRE KNOWN DIST 200-1000 M	4/1	100		
FIELD FIRE	4/1	100		
LFX	4/1	10		
FAM FIRE NIGHT/ZERO NVG	2/1	40		
SUSTAINMENT (7.62 MM)	2/1	40		
(.50 CAL)	2/1		40	
(.300 WIN)	2/1			40
RECORD QUALIFICATION	2/1	100		
TOTAL ROUNDS INDIVIDUAL	AC/RC	1280/420	80	80
TOTAL RDS BN	AC/RC	46,080/15,360	2880/960	
SOTIC LEVEL II TNG PROGRAM (BN)	1/.33	12,160/4,053		
Note: * USASOC procurement item.				

TRAINING EXERCISES

C-4. The following exercises may be incorporated into team training to improve every team member's skills and enhance the team's overall capabilities.

Marksmanship Exercises

C-5. Marksmanship training will take up a large amount of the sniper's overall proficiency training. The sniper must be proficient in all sniper-related skills, but without marksmanship, these other skills are useless. Some examples of marksmanship exercises are—

- *Grouping Exercises.* These are simple exercises where the sniper fires five-round shot groups at various ranges, from 100 to 800 meters. Analysis of the shot groups helps him determine firing errors and environmental effects in a more or less controlled environment. Analysis also allows the sniper to collect his cold bore shot and environmental data.
- *Moving Target Firing.* Firing at moving targets helps the sniper to maintain proficiency in this difficult skill. Targets should be engaged from 100 to 600 meters. These exercises are simple to run. Moving targets are provided by having personnel "walk" silhouette targets on a stick or board held over their head while protected in the pits of a traditional known distance range. The targets must be cut 12 inches wide to maintain realism. Snipers should not know target direction or speed during final exercises. The target will come up in the middle of a 20-meter area and then move either left or right and at a speed of either a slow walk, walk, or a run. The sniper must engage the target before it gets to the other side of the lane. Also, a second target can be moved in the same lane assigned to another sniper team. The teams must hit their assigned target. Stop-and-go targets may be engaged from 600 to 800 meters. These targets will move across the sniper's front and periodically stop for 3 to 5 seconds, then begin to move again. Each sniper will be assigned a target to engage.
- *Unknown Distance Firing.* This exercise helps the sniper to stay proficient in a variety of sniper skills. The sniper pair must fill out a range card or sector sketch and estimate the range to targets. The pair must then use the information to engage targets of unknown distance. Distances should range from 200 to 800 meters. The targets must be fully mixed and partially exposed. This will force the snipers to range on other objects and use target information for range estimation and engagement.
- *Firing Under Artificial Illumination and/or NVDs.* In this exercise, the sniper fires at both stationary and moving targets from 100 to 600 meters under artificial illumination or 100 to 600 meters (weather dependant) using NVDs. Beyond a range of 400 meters, wind plays a significant role in target engagement. Winds above 5 mph will cause a miss, regardless of the sniper's skill, if the sniper can not detect the wind and dope the wind.
- *Stress Shooting.* All previous exercises can be further enhanced with the additional application of a stress factor. Applying a time limit, stalking to the target, or physical effort before firing are but a few stresses that may be applied to the sniper.
- *Firing Air Rifles.* The sniper can effectively use match-grade air rifles (for example, RWS 75 or Daisy Gamo) for marksmanship training.

They do not require any special ranges. Any area with a minimum of 10 meters of distance can be used, indoors or outdoors. The sniper can use this range, in conjunction with caliber .22 bullet traps and standing off-hand (unsupported), to reinforce marksmanship fundamentals. Using scaled targets and a good air-rifle scope in an outdoor area can simulate ranges of up to 1,000 meters. Ranges of 10 to 20 meters and targets, reduced in size to represent different ranges, can be used.

Stalking Exercises

C-6. These exercises enable the sniper to train and develop skills in movement, camouflage, map reading, mission planning, and position selection. Live fire may be incorporated to confirm the sniper's target engagement. Stalks should be performed by sniper teams as they must learn to work together.

Range-Estimation Exercises

C-7. There are many ways to conduct this type of exercise. The sniper estimates ranges out to 800 meters and must be within 5 percent of the correct range when miling with his scope. He should use only his binoculars and rifle telescope as aids. When using only the eye for range estimation, the sniper must be within 12 percent out to 500 meters. Range estimation beyond 600 meters by the unaided eye is very difficult and requires constant practice.

Other Exercises

C-8. The sniper's training program may include other exercises to enhance his observation, memory, and camouflage skills.

M24 SNIPER MILES TRAINING

C-9. The MILES training is an invaluable tool in realistic combat training. Other than actual combat, the sniper's best means of displaying effectiveness as a force multiplier is through the use of the M24 SWS with MILES.

CHARACTERISTICS OF THE MILES TRANSMITTER

C-10. The M24 SWS MILES transmitter is a modified M16 transmitter. A special mounting bracket attaches the laser transmitter to the right side of the barrel (looking from the butt end) of the M24 and places it parallel with the line of bore. The laser beam output has been amplified and tightened to provide precision fire capability out to 1,000 meters. (For component information and instructions on mounting, zeroing, and operation, see TM 9-1265-211-10, *Operator's Manual for Multiple Integrated Laser Engagement System (MILES) Simulator System, Firing, Laser: M89*).

TRAINING VALUE

C-11. Using the M24 with MILES, the trainer can enhance sustainment training in target engagement such as the following:

- *Selection of Firing Positions.* Due to transmitter modifications, the sniper must attain a firing position that affords clear fields of fire. Any obstruction (vegetation, terrain) can prevent a one-shot kill by deflecting or blocking the path of the laser beam. By selecting this type of position, the sniper will greatly improve his observation and firing capabilities.

- *Target Detection/Selection.* Using MILES against multiple or cluster targets requires the sniper to select the target that has the greatest effect on the enemy. The trainer provides instant feedback on the sniper's performance. Situations may be created such as bunkers, hostage situations, and MOUT firing. The hit-or-miss indicating aspects of MILES are invaluable in this type of training.
- *Marksmanship.* A target hit (kill) with MILES is the same as one with live ammunition. Proper application of marksmanship fundamentals results in a first-round kill; the training value is self-evident.

MILES TRAINING LIMITATIONS

C-12. The concept of MILES is to provide realistic training. However, MILES is limited in its capabilities as applied to the sniper's mission of long-range precision fire. These limitations are—

- *Lack of Range Estimation Training.* Due to the straight beam, once the weapon and MILES are zeroed together, the sniper cannot change the elevation knob based on range-to-target. This negates any range estimation practice during this training.
- *Lack of External Ballistics Training.* The MILES transmitter emits a concentrated beam of light. It travels from the sniper's weapon undisturbed by outside forces such as temperature, humidity, and wind. Lack of these effects may lull the sniper into a false sense of confidence. The trainers should constantly reinforce the importance of these factors. The sniper should make a mental note of changes that should be applied to compensate for these effects.
- *Engagement of Moving Targets.* The engagement of moving targets requires the sniper to establish a target lead to compensate for flight time of his bullet. Traveling in excess of 186,000 miles per second (speed of light), the MILES laser nullifies the requirement for target lead. Again, the sniper may be lulled into a false sense of confidence. The trainer should enforce the principles of moving target engagement by having the sniper note appropriate target lead for the given situation.

REDUCED-SCALE RANGE

C-13. When using air rifles for marksmanship training, the sniper can use one of several formulas to simulate distances for a reduced or subcaliber range. Listed below are several formulas and explanations on how to use them.

REDUCED-SCALE TARGET HEIGHT FORMULA

C-14. The formula to find the reduced height of a target at a given range and a simulated range is as follows:

$$\frac{R1 \times H1}{R2} = H2$$

R1 = Reduced range+

R2 = Simulated range+

H1 = Height of actual target*

H2 = Reduced height of target*

NOTES:

+ Both the reduced range and the simulated range must agree in measurement; for example, 1,000 meters and 35 meters or 1,000 yards and 35 yards.

* If the height of the real target is expressed in inches (for example, 72 inches for a 6-foot man), the answer is in inches. If the height is expressed in feet, the answer will be in feet.

EXAMPLES:

R1 = 35 meters   R2 = 500 meters   H1 = 72 inches

$$\frac{35 \times 72}{500} = 5.04 \text{ inches}$$

R1 = 25 yards   R2 = 800 yards   H1 = 6 feet

$$\frac{25 \times 6}{800} = 0.1875 \text{ feet}$$

To change the 0.1875 feet to inches, multiply by 12. 0.1875 x 12 = 2.25 inches

REDUCED-SCALE SIMULATED RANGE FORMULA

C-15. The formula to find the simulated range of a given reduced target height at a given reduced range is as follows:

$$\frac{R1 \times H1}{H2} = R2$$

EXAMPLE: A 3-inch target at 35 meters will simulate what range?

R1 = 35 meters   H1 = 72 inches   H2 = 3 inches

$$\frac{35 \times 72}{3} = 840 \text{ meters}$$

C-16. Table C-2, page C-16, simulates in inches a 6-foot man at various ranges on the given reduced ranges.

C-17. The center of the pellet strike should be the determining factor due to the size difference of the pellet diameter versus the range and simulated target size. It is also recommended that the ranges of 100 through 400 yards/meters be against a head or head and shoulders target and not the full body. For example, at 35 meters the normal human head is simulated to be 2.8 inches high by 1.75 inches wide for a simulated range of 100 meters.

C-18. An additional advantage to the reduced range targets is that when the sniper uses the above table or formulas, a reduced-scale unknown distance range for judging distance will be constructed at the same time. When he uses the mil scale in the Leupold M3A Ultra 10x scope, the targets will give the same mil readings as a real target at that distance. As an example, a 6-foot man at 500 yards is 4 mils high. A 3.60-inch target at 25 yards is 4 mils high and simulates a 6-foot man at 500 yards.

Table C-2. Reduced Range Chart

Range	15 yds/m	20 yds/m	25 yds/m	30 yds/m	35 yds/m
1,000 yds/m	1.08"	1.44"	1.80"	2.16"	2.52"
900 yds/m	1.20"	1.60"	2.00"	2.40"	2.80"
800 yds/m	1.35"	1.80"	2.25"	2.70"	3.15"
700 yds/m	1.54"	2.05"	2.57"	3.08"	3.60"
600 yds/m	1.80"	2.40"	3.00"	3.60"	4.20"
500 yds/m	2.16"	2.88"	3.60"	4.32"	5.04"
400 yds/m	2.70"	3.60"	4.50"	5.40"	6.30"
300 yds/m	3.60"	4.80"	6.00"	7.20"	8.40"
200 yds/m	5.40"	7.20"	9.00"	10.80"	12.60"
100 yds/m	10.80"	14.40"	18.00"	21.60"	25.20"

C-19. Another technique is to determine the number of MOA that the target represents at that range; 36 inches at 600 equals 6 MOA. Then use that number to simulate the reduced distance target: 6 MOA equals 1.5 inches at 25 yards. Thus, a 1.5-inch target at 25 yards will equal a 36-inch target at 600 yards.

Appendix D
Mission Packing List

The sniper team determines the type and quantity of equipment it carries by a METT-TC analysis. Some of the equipment mentioned may not be available. A sniper team, due to its unique mission requirements, carries only mission-essential equipment. This is not an inclusive list, and not all items listed will be carried on all missions.

ARMS AND AMMUNITION

D-1. Table D-1 lists mission-essential arms and ammunition carried by a sniper and observer.

Table D-1. List of Arms and Ammunition

Sniper	Observer
• M24 SWS with M3A telescope. • Rounds of M118/M852 ammunition. • Sniper's data book, mission logbook, range cards, wind tables, and slope dope. • M9 9-mm pistol. • Rounds 9-mm ball ammunition. • Each 9-mm magazines. • M9 bayonet. • M67 fragmentation grenades. • CS grenades; 2 percussion grenades (MOUT). • M18A1 mine, complete.	• M4/M16/M203 (with NVD, as appropriate). • Rounds ammunition. • Magazines for rifle. • M9 9-mm pistol. • Rounds, 9-mm ball ammunition. • Each 9-mm magazines. • M9 bayonet. • Rounds 40-mm, high-explosive ammunition if M203 is carried. • Rounds 40-mm antipersonnel ammunition. • M67 fragmentation grenades; 2 CS grenades; 2 percussion (MOUT).

SPECIAL EQUIPMENT

D-2. Table D-2 lists mission-essential special equipment carried by a sniper and observer.

Table D-2. List of Special Equipment

Sniper	Observer
• M24 SWS cleaning kit. • M24 SWS deployment kit (tools and replacement parts). • M9 pistol cleaning kit. • Extra handset for radio. • Extra batteries for radio (BA-4386 or lithium, dependent on mission length). • Signal operating instructions. • AN/PVS-5/7 series, night vision goggles. • Extra BA-1567/U or AA batteries for night vision goggles. • Pace cord. • E-tool with carrier. • 50-foot 550 cord. • 1 green and 1 red star cluster. • 2 HC smoke grenades. • Measuring tape (25-foot carpenter-type).	• M4/M16/M203 cleaning kit. • AN/PRC-77/AN-PRC-119/AN/PRC-104 radio. • Radio accessory bag, complete with long whip and base, tape antenna and base, handset, and battery (BA-4386 or lithium). • M49 20x spotting scope with M15 tripod (or equivalent 15 to 20x fixed power scope, or 15-45x spotting scope). • M19/M22 binoculars (preferably 7 x 50 power with mil scale). • Range estimation "cheat book." • 300 feet WD-1 field wire (for field-expedient antenna fabrication). • Olive-drab duct tape, olive-drab ("100 mph") tape. • Extra batteries for radio (if needed). • Extra batteries (BA-1576/U) for AN/PVS-4. • Calculator with extra battery. • Butt pack. • 10 each sandwich-sized waterproof bags. • 2 HC smoke grenades. • Lineman's tool.

UNIFORMS AND EQUIPMENT

D-3. Table D-3 lists uniforms and equipment.

Table D-3. List of Uniforms and Equipment

- Footgear (jungle/desert/cold weather/combat boots).
- 2 sets of BDUs (desert/woodland/camouflage).
- Black leather gloves.
- 2 brown T-shirts.
- 2 pairs brown underwear.
- 8 pairs olive-drab wool socks.
- Black belt.
- Headgear (BDU/jungle/desert/cold weather).
- Identification (ID) tags and ID card.
- Wristwatch (sweep-second hand with luminous dial, waterproof).
- Pocket survival knife.
- Large all-purpose, lightweight individual carrying equipment (ALICE) pack, complete with frame and shoulder straps.
- 2 waterproof bags (for ALICE pack).
- 2 2-quart canteens with covers.
- 1 bottle water purification tablets.
- Complete load-bearing equipment (LBE).
- Red-lens flashlight (angle-head type with extra batteries).
- MRE (number dependent on mission length).
- 9-mm pistol holster and magazine pouch (attached to LBE).
- 2 camouflage sticks (METT-TC-dependent).
- 2 black ink pens.
- 2 mechanical pencils with lead.
- 2 black grease pencils.
- Lensatic compass.
- Map(s) of operational area.
- Protractor.
- Poncho.
- Poncho liner.
- 2 ghillie suits, complete.
- 2 protective masks/MOPP suits.
- Foot powder.
- Toiletries.

OPTIONAL EQUIPMENT

D-4. Table D-4 lists optional equipment.

Table D-4. List of Optional Equipment

- M203 vest.
- Desert camouflage netting.
- Natural-colored burlap.
- Glitter tape.
- VS-17 panel.
- Strobe light with filters.
- Special patrol insertion/extraction system harness.
- 12-foot sling rope.
- 2 each snap links.
- 120-foot nylon rope.
- Lip balm or sunscreen.
- Signal mirror.
- Pen gun with flares.
- Chemical lights (to include infrared).
- Body armor/flak jacket.
- Sniper veil.
- Sewing kit.
- Insect repellent.
- Sleeping bag.
- Knee and elbow pads.
- Survival kit.

- Rifle drag bag.
- Pistol silencer/suppressor.
- 2.5 pounds C4 with caps, cord, fuse, and igniter.
- Rifle bipod/tripod.
- Empty sandbags.
- Hearing protection (ear muffs).
- Thermometer.
- Laser range finder.
- Thermal imager.
- KN-200-KN-250 image intensifier.
- Pocket binoculars.
- 35-mm automatic loading camera with appropriate lenses and film.
- 1/2-inch camcorder with accessories.
- Satellite communication equipment.
- Short-range radio with earphone and whisper microphone.
- Field-expedient antennas.
- Information reporting formats.
- Encryption device for radio.
- SO sniper training/employment manual.

SPECIAL TOOLS AND EQUIPMENT

D-5. Table D-5 lists special tools and equipment (MOUT).

Table D-5. List of Special Tools and Equipment

• Pry bar.	• Power saw.
• Pliers.	• Cutting torch.
• Screwdriver.	• Shotgun.
• Rubber-headed hammer.	• Spray paint.
• Glass cutter.	• Stethoscope.
• Masonry drill and bits.	• Maps or street plans.
• Metal shears.	• Photographs, aerial and panoramic.
• Chisel.	• Whistle.
• Auger.	• Luminous tape.
• Lock pick, skeleton keys, cobra pick.	• Flex cuffs.
• Bolt cutters.	• Padlocks.
• Hacksaw or handsaw.	• Intrusion detection system (booby traps).
• Sledgehammer.	• Portable spotlight(s).
• Ax.	• Money (U.S. and indigenous).
• Ram.	• Civilian attire.

ADDITIONAL EQUIPMENT TRANSPORT

D-6. The planned use of air and vehicle drops and caching techniques eliminates the need for the sniper team to carry extra equipment. Another method is to use the stay-behind technique when operating with a security patrol (Chapter 5). Through coordination with the security patrol leader, the team's equipment may be broken down among the patrol members. On arrival at the ORP, the security patrol may leave behind all mission-essential equipment. After completing the mission, the team may cache the equipment for later pickup, or it may be returned the same way it was brought in.

Appendix E

M82A1 Caliber .50 Sniper Weapon System

Changes in modern warfare required an expansion of the sniper's role. On the fluid, modern battlefield, the sniper must be prepared to engage a wide range of targets at even greater distances. After years of research and development, the military adopted the M82A1 caliber .50 SWS. However, even after having been deployed to operational units, no comprehensive training plan has been developed to train snipers on this new role. The basic approach to the large-bore sniper rifle has been that it is nothing more than a big M24 (7.62-mm sniper rifle). This logic has its obvious flaws. Many of the techniques learned by the sniper need to be modified to compensate for this new weapon system. Some of these changes include movement techniques, maintenance requirements, sniper team size and configuration, support requirements, and the marksmanship skills necessary to engage targets at ranges in excess of 1,800 meters. To keep up with the battles fought in-depth, as well as smaller-scale conflicts, the need for a sniper trained and equipped with a large-bore rifle is apparent.

ROLE OF THE M82A1 CALIBER .50 SWS

E-1. The military can use the M82A1 in several different roles—as the long-range rifle, the infantry support rifle, and the explosive ordnance disposal tool. Personnel use the M82A1 as—

- A long-range rifle to disable valuable targets that are located outside the range or the capabilities of conventional weapons, many times doing so in situations that may preclude the use of more sophisticated weapons.
- An infantry support rifle to engage lightly armored vehicles and to penetrate light fortifications that the 5.56 mm and the 7.62 mm cannot defeat.
- An explosive ordnance disposal tool to engage and disrupt several types of munitions at ranges from 100 to 500 meters. In most cases, the munitions are destroyed or disrupted with a single hit and without a high-order detonation.

E-2. When used in any of its roles, the caliber .50 SWS and personnel trained to use it are vital assets to the commander. In light of this, a training program is necessary to maximize their potential.

M82A1 CALIBER .50 SWS CHARACTERISTICS

E-3. The Barrett Caliber .50 Model 82A1 is a short recoil-operated, magazine-fed, air-cooled, semiautomatic rifle (Figure E-1). Its specifications are as follows:

- Caliber: .50 Browning machine gun cartridge (12.7 x 99 mm).
- Weight: 30 lbs (13.6 kg).
- Overall Length: 57 inches (144.78 cm).
- Barrel Length: 29 inches (73.67 cm).
- Muzzle Velocity: 2,850 fps (M33 ball).
- Maximum Range M2 Ball: 6,800 meters (7,450 yds).
- Maximum Effective Range: 1,830 meters on an area target and 1200 to 1400 on a point target, depending on target size.
- Magazine Capacity: 10 rounds.

Figure E-1. The Barrett Caliber .50 Model 82A1

THE SWAROVSKI RANGING RETICLE RIFLE SCOPE

E-4. A qualified sniper has already been taught the fundamentals of scoped rifle fire. However, the scope that the M82A1 is equipped with differs from most of the scopes. Figure E-2, page E-3, lists the specifications for the Swarovski rifle scope.

Design Characteristics
• 30-mm main tube is compatible with existing SWS mounting rings. • Made of aluminum alloy. • NOVA ocular system filled with dry nitrogen after pressure-testing to prevent fogging. • Available add-on battery-operated reticle illuminator for low-light conditions. • Recoiling eyepiece offers added protection from scope "bite." • Sloped scope rail enables use of **only** the Swarovski reticle-equipped scope, rail is between 0.030 and 0.035 inches higher at rear. The sloped rail also aids in the M3A to zero at the longer ranges due to the additional slope in the rail. **NOTE:** Under no circumstances will the scope rail be removed.
Technical Data
• Magnification: 10x. • Objective Lens Diameter: 42 mm. • Field of View: 12 feet at 100 yards/4 meters at 100 meters. • Parallax-Free Distance of the Reticle: 500 meters (tolerance 250 meters to infinity). • Windage and Elevation Adjustments: 1 click: 1/5 inch at 100 yards, (1/5 MOA) maximum: 80 inches at 100 yards (80 MOA). • Operating Temperature: 131 degrees Fahrenheit/ –4 degrees Fahrenheit. • Weight: approximately 13.5 ounces.
Reticle
• Offers "ranging" capabilities from 500 to 1,800 meters. • No range estimation abilities (500 meters to 600 meters stadia lines approximately 1 mil apart, 3.5 MOA). • 5 and 10 mph wind hold-offs, also usable for moving targets. • Lack of mil dots require different hold-offs for wind and drift.

Figure E-2. Scope Specifications

USING THE M82A1 CALIBER .50 SWS

E-5. A qualified sniper has also already been taught the considerations necessary for proper employment of a sniper team. With the caliber .50 SWS, many of these have changed. The effective range, signature, weight, support requirements, and terminal performance of the round are all increased over the 7.62 SWS. As a result, the sniper must do the following to ensure proper use of this system:

- Maximize the range of the M82A1. Always engage targets at the maximum range that the weapon, target, and terrain will permit. Make sure to—
    - Select the appropriate ammunition (by effective range and terminal performance).

- Use range finder whenever possible.
- When in doubt of range estimation, aim low and adjust using the sight-to-burst method. Second shots are frowned upon with a 7.62 SWS because of two things—the target getting into a prone position (which most people do when shot at) and the sniper revealing his position. Neither of these principles apply to the caliber .50 SWS when it is used against armored or fortified targets. Armored personnel carriers (APCs) cannot "duck," nor is an enemy buttoned into a bunker or an APC as observant as he could be. The sniper needs to have carefully observed his area before assuming the above to be always true.

• Conduct movement into or occupy an FFP with the M82A1 SWS. Sniper should—
- Modify his movement techniques to accommodate for the following:
  ♦ Increased weight of the system, ammunition, and team equipment.
  ♦ Better route selection (amount of crawling is reduced).
  ♦ Better selection of withdrawal routes (after the shot, the sniper becomes a higher-priority target and must select route for quick egress).
- Occupy an FFP and adjust for the following:
  ♦ Much larger signature to front, clear area and dampen soil.
  ♦ Signature also at 65 degrees, fan to right and left of sniper.
  ♦ Size requirement for a 3-man sniper team in a permanent hide may make it unfeasible for many applications.
  ♦ FFP should prevent long-range "skylined" targets.

• Understand additional support requirements for the M82A1. Sniper should—
- Maintain an M82A1 SWS as follows:
  ♦ Clean after 10 rounds for better accuracy.
  ♦ Be aware that a chamber pressure of 55,000 copper units of pressure (CUP) could cause fatal maintenance failures.
- Modify existing training and sustainment programs as follows:
  ♦ Cannot fire the M82A1 on existing small-arms sniper ranges.
  ♦ Requires special considerations for the use of ammunition other than the standard ball and tracer (for example, multipurpose ammunition, armor-piercing incendiary [API]).
- Understand the following additional transportation requirements:
  ♦ Additional weight of system makes vehicular movement desirable.
  ♦ As there is no existing approved method for parachute or underwater infiltration with the M82A1, the sniper must plan for alternate methods of getting the SWS to the battlefield.

E-6. A sniper's ability to deploy to the battlefield with an M82A1 depends on whether he can adapt what he has learned in the past. The rules of sniper employment haven't changed, but many of the finer points have.

MAINTENANCE

E-7. Maintenance of the M82A1 SWS involves assembly and disassembly, inspection, cleaning and lubrication, and replacement of parts.

E-8. The sniper normally stores and transports the M82A1 SWS in the carrying case. The following procedure covers the initial assembly of the rifle as it would come from the case.

E-9. The sniper first removes the lower receiver from the carrying case. He extends the bipod legs by pulling them back and swinging them down to the front where they will lock into place. He then places the lower receiver on the ground.

E-10. The sniper removes the rear lock pin from its stored position in the lower receiver, found just forward of the recoil pad.

E-11. He frees the bolt carrier. The bolt carrier is held in place under tension in the lower receiver by the midlock pin. He grasps the charging handle of the bolt carrier with the right hand and pulls back against the tension of the main spring. The sniper then removes the midlock pin and allows the bolt carrier to come forward slowly until there is no more spring tension.

> **CAUTION**
> Do not pull the midlock pin without hands-on control of the bolt carrier; it can be launched from the lower receiver.

E-12. The sniper removes the upper receiver from the case. He maintains control of the barrel that is retracted in the upper receiver to prevent it from sliding and injuring his fingers. The barrel may have rotated in shipping and he will need to index it so that the feed ramp is to the bottom. The sniper then fully extends the barrel from the upper receiver.

E-13. The impact bumper that surrounds the barrel must be placed into proper position by the barrel lug. The sniper grasps the barrel key (not the springs) with the thumb and middle of the index finger. He pulls the key into place on the key slot of the barrel. This is a difficult operation, at first, because the tension of the barrel spring is approximately 70 pounds.

E-14. The sniper positions the upper receiver, rear-end up, muzzle down, over the lower receiver. He engages the front hook of the lower receiver.

NOTE: The sniper should make sure of the proper mating of the hook and bar to avoid receiver damage during the final assembly motion.

E-15. He grasps the charging handle on the bolt carrier and pulls back against the tension of the main spring until the bolt clears the barrel when the upper receiver is lowered.

E-16. The sniper lowers and closes the upper receiver onto the lower receiver. He releases the charging handle. Then he places the mid and rear lock pins into the lock pin holes in the receiver.

E-17. He places thumb safety in the "on safe" position (horizontal).

DISASSEMBLY AND ASSEMBLY

E-18. The two types of disassembly and assembly are general and detailed. General disassembly and assembly involves removing and replacing the three major weapon groups. Detailed disassembly and assembly involves removing and replacing the component parts of the major groups.

General Disassembly

NOTE: Only SOTIC personnel are authorized to perform complete disassembly.

E-19. The three major weapon groups are the upper receiver, bolt carrier, and lower receiver. As the sniper disassembles the weapon, he should note each part position, configuration, and part name. He—

- Begins by clearing the weapon and supporting the rifle on the bipod, with the magazine removed.
- Removes the mid and rear lock pins.
- Grasps the charging handle and pulls back until the bolt withdraws from and clears the barrel.
- Lifts the upper receiver at its rear. When the receiver has raised enough to clear the bolt, he slowly releases the pull on the charging handle so that the bolt carrier comes to rest.
- Continues to raise the upper receiver until the front hinge is disengaged and then lifts it from the lower receiver.
- Withdraws the barrel by resting the upper receiver group on the muzzle brake by placing it on any surface that will not damage the end of the brake.
- Withdraws the barrel key from the slot in the barrel by slowly working it out and grasping it between his thumb and the middle of the index finger (he should be prepared to assume the tension of the barrel springs upon the release of the barrel key from the slot—tension is approximately 70 pounds). Slowly lowers the key until the tension of the springs are at rest.

NOTE: The sniper should **never** pull on the barrel springs to remove the barrel key.

- Lowers the receiver down around the barrel.
- Grasps the charging handle and lifts the bolt carrier group from the lower receiver.

- Decocks the firing mechanism by depressing the sear with the rear lock pin.
- With the bolt in the left hand, uses the mid lock to depress the bolt latch against the palm. He uses the rear lock pin to lift the cam pin and frees the bolt with the right hand. The bolt will rise under the power of the bolt spring. He should never lift the cam pin more than needed to release the bolt.
- Removes the bolt and bolt spring.
- If it is necessary to remove the extractor, inserts a pin punch or paper clip through the extractor hole and slides the extractor out either side of the slot. He should be prepared to capture or contain the plunger and the plunger spring. He reverses the procedure for replacement.

Assembly

E-20. The sniper reverses the procedure for reassembly of the bolt. He—
- Replaces the bolt carrier. With the lower receiver group standing on its bipod, places the bolt carrier into the lower receiver.
- Replaces the upper receiver by—
    - Positioning the upper receiver (rear up and muzzle down) over the lower receiver so the hook of the front hinge can fully engage the hinge bar on the lower receiver.
    - **NOTE:** If not properly seated, the hinge bar can be pried off due to leverage the operator can apply when closing the receiver.
    - While positioned directly behind the rifle, preparing to close the upper receiver with the left hand.
    - Grasping the charging handle of the bolt carrier and pulling the bolt carrier back into the main spring, so the bolt clears the barrel while lowering the upper receiver.
    - Lowering and closing the upper receiver and releasing the charging handle.
    - Replacing mid and rear lock pins.

INSPECTION

E-21. Inspection begins with the weapon disassembled into its three major groups. Figure E-3, page E-8, describes each group and the steps that the sniper must perform to inspect for proper functioning.

Upper Receiver Group
• Makes sure barrel springs are not overstretched and each coil is tight with no space between the coils. • Checks to see if impact bumper is in good condition. • Ensures the muzzle brake is tight. • Inspects the upper receiver for signs of being cracked, bent, or burred. • Makes sure scope mounting rings are tight.
Bolt Carrier Group
• Checks the ejector and extractor to see that they are under spring pressure and not chipped or worn. • Decocks firing mechanism, depresses the bolt latch, and manually works the bolt in and out, feeling for any roughness. • Holding the bolt down, inspects firing pin protrusion and for any erosion of the firing pin hole. • Inspects bolt latch for deformation and free movement. • Swings cocking lever forward. The sear should capture the firing pin extension before the cocking lever is fully depressed.
Lower Receiver Group
• With the bolt carrier in place, pulls it rearwards and checks to see that the mainspring moves freely. • Holds bolt carrier under mainspring housing approximately 10 mm and checks for excessive lift that would prevent the trigger from firing. • Ensures the lower receiver is not cracked, bent, or burred. • Checks the bipod assembly to ensure it functions properly.

Figure E-3. Steps in Sniper's Inspection of Major Weapon Groups

CLEANING AND LUBRICATION

E-22. The rifle's size makes it relatively easy to clean. The sniper should clean it at the completion of each day's firing or during the day if fouling is causing the weapon to malfunction. He—

- Cleans the bore with rifle bore cleaner (RBC) or a suitable substitute. Each cleaning should include at least six passes back and forth with the bronze-bristle brush, followed by cloth patches until the patches come out clean. Immediately after using bore cleaner, he dries the bore and any parts of the rifle exposed to the bore cleaner and applies a thin coat of oil. He should always clean the bore from the chamber end.

- Cleans the rest of the weapon with a weapons cleaning toothbrush, rags, and cleaning solvent. When using cleaning solvent, he should not expose plastic or rubber parts to it. He dries and lubricates all metal surfaces when clean.

E-23. The sniper should lightly lubricate all exposed metal. These parts are as follows:

- Bolt (locking lugs and cam slot).
- Bolt carrier (receiver bearing surfaces).
- Barrel bolt locking surfaces (receiver bearing surfaces).
- Receiver (bearing surfaces for recoiling parts).

NOTE: The sniper lubricates according to the conditions in the AO.

E-24. The sniper should dust off the scope and keep it free of dirt. He should dust the lenses with a lens cleaning brush and only clean them with lens cleaning solvent and lens tissue.

NOTE: The Barrett is easy to maintain, but because of the size of its components the sniper must pay attention to what he is doing, or he may damage the weapon, injure himself, or hurt others around him if not careful.

Appendix F

Foreign/Nonstandard Sniper Weapon Systems Data

Several countries have developed SWSs comparable to the U.S. systems. The designs and capabilities of these weapon systems are similar. This appendix describes the characteristics of sniper weapon systems that could be encountered on deployments. This is not an all-inclusive list, and not all weapons are current issue. The country listed is either the last country of issue or the manufacturer.

AUSTRIA

F-1. The following systems are currently in use: Steyr Model SSG 69 and SSG-PII rifles with Kahles ZF69, ZF84, or RZFM86 telescopes.

F-2. The Austrian Scharf Schutzen Gewehr (Sharp Shooter's Rifle) 69 (SSG-69) is the current sniper weapon of the Austrian Army and several foreign military forces. It is available in either 7.62- x 51-mm NATO or the .243 Winchester calibers. Recognizable features include a synthetic stock (green or black) that is adjustable for length of pull by a simple spacer system; hammer-forged, medium-heavy barrel; two-stage trigger, adjustable for weight of pull (a set trigger system is frequently seen); and a machined, longitudinal rib on top of the receiver that accepts several types of optical mounts. The mounting rings have a quick-release lever system that allows removal and reattachment of the optics with no loss of zero. The typical sighting system consists of the Kahles ZF69 6- x 42-mm telescope; iron sights are permanently affixed to the rifle for emergency use. The SSG-PII (Politzei II) has a heavy barrel and does not have iron sights. The telescope comes equipped with a bullet drop compensator graduated to 800 meters, and a reticle that consists of a post with broken crosshairs. The Steyr SSG-69 has a well-deserved reputation for accuracy. The Kahles ZF-series of telescopes are zeroed with the same procedure used for Soviet telescopes.

Steyr SSG-69 Characteristics

System of operation: bolt-action.
Caliber: 7.62- x 51-mm NATO.
Overall length: 44.5 inches.
Barrel length: 25.6 inches.
Rifling: 4-groove, 1/12-inch right-hand twist.
Weight: 10.3 pounds.

Magazine capacity: 5- or 10-round detachable magazine.
Telescope: Kahles ZF69 6 x 42 mm; BDC: 100 to 800 m.
Front: hooded post.
Rear: notch.

F-3. Ammunition requirement: The ZF69 is designed for the NATO ball ammunition: 147/150 gn FMJBT @ 2,800 fps. Some models of this telescope were designed for export to the United States and the BDC is calibrated for Federal's 308M load (168 HPBT @ 2,600 fps). The Kahles ZF84 telescope is available with the following ballistic cams: .223/62 gn; .308/143 gn; .308/146 gn; .308/168 gn; .308/173 gn; .308/185 gn; and .308/190 gn.

BELGIUM

F-4. This system is currently in use: Fabrique Nationale (FN) Model 30-11.

F-5. The FN Model 30-11 is the current sniper rifle of the Belgian Army. It is built on a Mauser bolt-action with a heavy barrel and a stock with an adjustable length of pull. The sighting system consists of the FN 4x, 28-mm telescope and aperture sights with 1/6 MOA adjustment capability. Accessories include the bipod of the MAG machine gun, butt-spacer plates, sling, and carrying case.

FN Model 30-11 Characteristics

System of operation: bolt-action.	Telescope: 4x with post reticle, range-finding stadia, and BDC: 100 to 600 m.
Caliber: 7.62- x 51-mm NATO.	Front: hooded aperture.
Overall length: 45.2 inches.	Rear: Anschutz match-aperture micrometer adjustable for windage/elevation, and fitted to mount on the rifle's scope base with a quick-detachable mount.
Barrel length: 20.0 inches.	
Rifling: 4-groove, 1/12-inch right-hand twist.	
Weight: 15.5 pounds.	
Magazine capacity: 10-round detachable magazine.	Ammunition requirement: 7.62- x 51-mm NATO ball (147/150 gn FMJBT @ 2,800 fps).

CANADA

F-6. This system is currently in use: Parker Hale Model C3.

F-7. The Parker Hale Model C3 is a modified target rifle (commercial Model 82 rifle, Model 1200 TX target rifle) built on the Mauser action. It was adopted in 1975. The receiver is fitted with two male dovetail blocks to accept either the Parker Hale 5E vernier rearsight or the Kahles 6- x 42-mm telescope. The stock has a spacer system to adjust the length of pull.

Parker Hale Model C3 Characteristics

System of operation: bolt-action.	Telescope: Kahles ZF69 6 x 42 mm; BDC: 100 to 800 m.
Caliber: 7.62- x 51-mm NATO.	
Overall length: 48.0 inches.	Front: detachable hooded post.
Barrel length: 26.0 inches.	Rear: detachable aperture.
Weight: 12.8 pounds.	Ammunition requirement: 7.62- x 51-mm NATO ball (147/150 gn FMJBT @ 2,800 fps).
Magazine capacity: 4-round internal magazine.	

CZECH REPUBLIC AND SLOVAKIA

F-8. This system is currently in use: Model 54.

F-9. The current SWS is the VZ 54 sniper rifle ("vzor" is the Czech word for "model"; therefore, "VZ 54" is the same as "Model 54"). It is a manually operated, bolt-action, 10-round box, magazine-fed, 7.62- x 54-mm rimmed weapon. It is built with a free-floating barrel. This weapon is similar to the Soviet M1891/30 sniping rifle, but shorter and lighter. The rifle is 45.2 inches long and weighs 9.0 pounds with the telescope. It has a muzzle velocity of 2,659 fps with a maximum effective range of 1,000 meters.

FINLAND

F-10. This system is currently in use: Vaime Silenced Sniper Rifle Mark 2 (SSR Mk2).

F-11. The Finnish armed forces are using a 7.62- x 51-mm NATO sniper rifle that is equipped with an integral barrel and silencer assembly. The SSR Mk2 has a fixed, self-cleaning, and noncorrosive silencer. It has a nonreflective plastic stock and an adjustable bipod. Through the use of adapters, any telescopic or electro-optical sight may be mounted. The weapon is not equipped with metallic sights. With subsonic ammunition, the SSR Mk2 has a maximum effective range of 200 meters.

SSR Mk2 Characteristics

System of operation: bolt-action.
Caliber: 7.62- x-51-mm NATO.
Overall length: 46.5 inches.
Barrel length: 18.3 inches.
Rifling: not known.
Weight: 11 pounds.

Magazine capacity: 10-round internal magazine.
Telescope: various.
Front: none.
Rear: none.
Ammunition requirements: subsonic (185 gn FMJBT @ 1,050 fps).

FRANCE

F-12. These systems are currently in use: MAS-GIAT FR-F1 and FR-F2.

F-13. The FR-F1 sniping rifle, known as the Tireur d'Elite (sniper), was adopted in 1966. It is based on the MAS 1936 bolt-action rifle. The length of pull may be adjusted with the removable butt-spacer plates. This weapon's sighting system consists of the Model 53 bis 3.8x telescopic sight and integral metallic sights with luminous spots for night firing. Standard equipment features a permanently affixed bipod whose legs may be folded forward into recesses in the fore-end of the weapon. The barrel has an integral muzzle brake or flash suppressor. This weapon has a muzzle velocity of 2,794 fps and a maximum effective range of 800 meters.

MAS-GIAT FR-F1 Characteristics

System of operation: bolt-action.
Caliber: 7.62- x 51-mm NATO or 7.5- x 54-mm French.
Overall length: 44.8 inches.
Barrel length: 22.8 inches.
Rifling: not known.
Weight: 11.9 pounds.

Magazine capacity: 10-round detachable box magazine.
Telescope: Model 53, 3.8x.
Front: hooded post.
Rear: notch.
Ammunition requirement: not known.

F-14. The FR-F2 sniping rifle is an updated version of the F1. Dimensions and operating characteristics remain unchanged; however, functional improvements have been made. A heavy-duty bipod has been mounted more toward the butt-end of the rifle, adding ease of adjustment for the firer. Also, the major change is the addition of a thick, plastic thermal sleeve around and along the length of the barrel. This addition eliminates or reduces barrel mirage and heat signature.

MAS-GIAT FR-F2 Characteristics

System of operation: bolt-action.
Caliber: 7.62- x 51-mm NATO.
Overall length: 47.2 inches.
Barrel length: 22.9 inches.
Rifling: 3-groove, 1/11.6-inch right-hand twist.
Weight: 13.6 pounds.

Magazine capacity: 10-round detachable magazine.
Telescope: 6- x 42-mm or 1.5–6- x 42-mm Schmidt and Bender; BDC: 100 to 600 m.
Front: post.
Rear: notch.
Ammunition requirement: 150 gn FMJBT @ 2,690 fps.

GERMANY

F-15. These systems are currently in use: Mauser Model SP66, Walther WA 2000, and the Heckler and Koch PSG-1.

F-16. The Mauser Model SP66 is used by the Germans and also by about 12 other countries. This weapon is a heavy-barrelled, bolt-action rifle built upon a Mauser short-action. It has a completely adjustable thumbhole-type stock. The muzzle of the weapon is equipped with a flash suppressor and muzzle brake.

Mauser SP66 Characteristics

System of operation: bolt-action.
Caliber: 7.62- x 51-mm NATO.
Overall length: not known.
Barrel length: 26.8 inches.
Rifling: not known.
Weight: not known.

Magazine capacity: 3-round internal magazine.
Telescope: Zeiss-Diavari ZA 1.5–6x.
Front: detachable hooded post.
Rear: detachable aperture.
Ammunition requirement: not known.

F-17. The Walther WA 2000 is built specifically for sniping. The entire weapon is built around the 25.6-inch barrel; it is a semiautomatic gas-operated bull-pup design that is 35.6 inches long. This unique weapon is chambered for .300 Winchester Magnum, but it can be equipped to accommodate calibers 7.62- x 51-mm NATO or 7.5- x 55-mm Swiss. The weapon's trigger is a single- or two-staged type. It can be fitted with various optics, but is typically found with a Schmidt & Bender 2.5–10- x 56-mm telescope. It has range settings from 100 to 600 meters and can be dismounted and mounted without loss of zero.

Walther WA 2000 Characteristics

System of operation: semiautomatic.
Calibers: .300 Winchester Magnum, 7.62- x 51-mm NATO, 7.5- x 55-mm Swiss.
Overall length: 35.6 inches.
Barrel length: 25.6 inches.
Rifling: not known.
Weight: 18.3 pounds.

Magazine capacity: 3-round detachable magazine.
Telescope: Schmidt and Bender 2.5–10 x 56 mm, BDC: 100 to 600 m.
Front: none.
Rear: none.
Ammunition requirement: not known.

F-18. The Heckler and Koch Prazisions Schutzen Gewehr (Precision Marksman's Shooting Rifle) PSG-1 is an extremely accurate version of the G-3. It is a gas-operated, magazine-fed, semiautomatic weapon with a fully adjustable, pistol-grip-style stock. Heckler & Koch claims that this weapon will shoot as accurately as the inherent accuracy of the ammunition. The 6- x 42-mm Hensoldt has light emitting diode (LED)-enhanced, illuminated crosshairs, elevation adjustments from 100 to 600 meters, and point-blank settings from 10 to 75 meters. Sighting requires loosening two small screws located in the center of the windage and elevation knobs. Once the screws are loosened, the adjustment can be made to center the shot group to correspond with one of the range settings on the knobs. The adjustments for both elevation and windage move the impact of the bullet one centimeter (0.4 inches) at 100 meters.

Heckler and Koch PSG-1 Characteristics

System of operation: semiautomatic.
Caliber: 7.62- x 51-mm NATO.
Overall length: 47.5 inches.
Barrel length: 25.6 inches.
Rifling: polygonal, 1/12-inch right hand twist.
Weight: 17.8 pounds.

Magazine capacity: 5- and 20-round detachable magazine.
Telescope: 6- x 42-mm Hensoldt with illuminated reticle, BDC: 100 to 600 m.
Front: none.
Rear: none.
Ammunition requirement: Lapua 7.62- x 51-mm NATO Match: 185 FMJBT D46/D47 @ 2,493 fps.

ISRAEL

F-19. These systems are currently in use: Galil and M21 Sniping Rifles.

F-20. The Israelis copied the basic design, operational characteristics, and configuration of the Soviet AK-47 assault rifle to develop an improved weapon to meet the demands of the Israeli Army. The Galil sniping rifle is a further evolution of this basic design. Like most service rifles modified for sniper use, the weapon is equipped with a heavier barrel fitted with a flash suppressor; it can be equipped with a silencer and fired with subsonic ammunition. The weapon features a pistol-grip-style stock, a fully adjustable cheekpiece, a rubber recoil pad, a two-stage trigger, and an adjustable bipod mounted to the rear of the fore-end of the rifle. Its sighting system consists of a side-mounted 6- x 40-mm telescope and fixed metallic sights. When firing FN Match ammunition, the weapon has a muzzle velocity of 2,672 fps; when firing M118 special ball ammunition, it has a muzzle velocity of 2,557 fps. The specifications on the M21 can be found in the U.S. section.

Galil Sniper Rifle Characteristics

System of operation: semiautomatic.
Caliber: 7.62- x 51-mm NATO.
Overall length: 43.9 inches.
Barrel length: 20 inches.
Rifling: 4-groove, 1/12-inch right-hand twist.
Weight: 18.3 pounds.

Magazine capacity: 5- or 25-round detachable magazine.
Telescope: 6- x 40-mm Nimrod, BDC: 100 to 1,000 m.
Front: hooded post with tritium night sight.
Rear: aperture with flip-up tritium night sight.
Ammunition requirement: M118 (173 gn FMJBT @ 2,610 fps).

ITALY

F-21. This system is currently in use: Beretta Sniper Rifle.

F-22. The Beretta rifle is the Italian sniper rifle. This rifle is a manually operated, bolt-action, 5-round box, magazine-fed weapon that fires the 7.62- x 51-mm NATO. Its 45.9-inch length consists of a 23-inch heavy, free-floating barrel, a wooden thumbhole-type stock with a rubber recoil pad, and an adjustable cheekpiece. Target-quality, metallic sights consist of a hooded front sight and a fully adjustable, V-notch rear sight. The optical sight consists of a Zeiss-Diavari ZA 1.5–6x variable telescope. The weapon weighs 15.8 pounds with a bipod, and 13.75 pounds without a bipod. The NATO-standard telescope mount allows almost any electro-optical or optical sight to be mounted to the weapon.

PEOPLE'S REPUBLIC OF CHINA

F-23. This system is currently in use: Norinco Type 79.

F-24. The standard sniper rifle of the People's Republic of China is the Norinco Type 79, which was adopted in 1980. It is a virtual copy of the Soviet SVD. In many instances, these rifles are nothing more than refinished and restamped Soviet SVDs that were once sold to the PRC. They have been imported into the U.S. under the designation of NDM-86. The specifications can be found under the Soviet SVD.

ROMANIA

F-25. This system is currently in use: Model FPK.

F-26. The FPK was adopted in 1970. This sniper rifle fires the Mosin/Nagant M1891 cartridge, which has a case length that is 15 mm longer than the 7.62- x 39-mm Warsaw Pact cartridge. Since the bolt of the AKM travels 30 mm (1.18 inches) farther to the rear than is necessary to accommodate the 7.62- x 39-mm cartridge, the Romanian designers were able to modify the standard AKM-type receiver mechanism to fire the more powerful and longer-ranged 7.62- x 54-mm rimmed cartridge. First, they altered the bolt face to take the larger-rimmed base of the M1891 cartridge, added a new barrel, and lengthened the RPK-type gas piston system. The gas system of the Soviet SVD (Dragunov) sniping rifle is more like that of the obsolete Tokarev rifle. Second, the Romanians developed their own 10-shot magazine, and they fabricated a skeleton stock from laminated wood (plywood). This buttstock, with its molded cheek rest, is probably slightly better than the one used on the Dragunov. Third, the Romanians have riveted two steel reinforcing plates to the rear of the receiver to help absorb the increased recoil forces of the more powerful M1891 cartridge. Finally, they have attached a muzzle brake of their own design. The standard AKM wire cutter bayonet will attach to this sniper rifle. The telescopic sight has English language markings.

FPK Characteristics

System of operation: semiautomatic.
Caliber: 7.62- x 54-mm rimmed.
Overall length: 45.4 inches.
Barrel length: 26.7 inches.
Rifling: not known.
Weight: 10.6 lbs.
Magazine capacity: 10-round detachable box magazine.

Telescope: LSP (Romanian copy of the Soviet PSO-1); BDC: 100 to 1,000 m with 1,100, 1,200, and 1,300 m reference points.
Front: hooded post.
Rear: sliding U-shaped notch.
Ammunition requirements: see Soviet SVD comments.

SPAIN

F-27. This system is currently in use: Model C-75.

F-28. The 7.62- x 51-mm NATO C-75 Special Forces rifle is the current sniper rifle of Spain. This bolt-action weapon is built upon the Mauser 98 action. It is equipped with iron sights and has telescope mounts machined into the receiver to allow for the mounting of most electro-optic or optic sights. The weapon weighs 8.14 pounds.

SWITZERLAND

F-29. This system is currently in use: SIG Model 510-4.

F-30. The Swiss use the 7.62- x 51-mm NATO SIG Model 510-4 rifle with a telescopic sight. The 510-4 is a delayed, blow-back-operated, 20-round, magazine-fed, semiautomatic or fully automatic weapon. With bipod, telescope, and empty 20-round magazine, the weapon weighs 12.3 pounds. It is 39.9 inches long with a 19.8-inch barrel and has a muzzle velocity of 2,591 fps.

UNITED KINGDOM

F-31. These systems are currently in use: Lee Enfield Model L42A1, Parker-Hale models 82 and 85, and the Accuracy International L96A1. The Lee Enfield No. 4 Mark 1 (T) is obsolete but still found in use around the world.

F-32. The L42A1 is the current standard sniper rifle. It is a conversion of the Lee Enfield No. 4 Mark 1 (T) .303, and was adopted in 1970. It has a heavy 7.62- x 51-mm NATO barrel, and the fore-end is cut back. The original No. 32 telescope was renovated, regraduated, and redesignated the "Telescope Straight Sighting L1A1," which is marked on the tube along with the part number, O.S. 2429 G.A. The original No. 32 markings are usually still visible, cancelled out, and painted over. New range graduations are read in meters instead of yards. Receivers from No. 4 Mark 1 (T) or Mark 1* (T) are used for this rifle. The magazine of the L42A1 is designed for 7.62-mm NATO cartridges and has a capacity of 10 rounds. The buttstock has the same type "screw on" wooden cheek piece as used with the No. 4 Mark 1 (T). The left side of the receiver has a telescope bracket for the telescope No. 32 Mark 3. A leaf-type rear sight and a protected blade-type front sight are also used.

Lee Enfield L42A1 Characteristics

System of operation: bolt-action.
Caliber: 7.62- x 51-mm NATO.
Overall length: 46.5 inches.
Barrel length: 27.5 inches.
Rifling: 4-groove, 1/12-inch right-hand twist.
Weight: 12.5 pounds.

Magazine capacity: 10-round detachable magazine.
Telescope: L1A1, 3x, BDC: 0 to 1,000 m.
Front: blade, with protecting ears.
Rear: aperture.
Ammunition requirement: NATO ball, 147/150 gn FMJBT @ 2,800 fps.

F-33. The Parker-Hale Model 82 sniper rifle is a bolt-action 7.62- x 51-mm NATO rifle built upon a Mauser 98 action. It is a militarized version of the Model 1200 TX target rifle. It is equipped with metallic target sights and the Pecar V2S 4 to 10x variable telescope. An optional, adjustable bipod is also available.

Parker-Hale Model 82 Characteristics

System of operation: bolt-action.
Caliber: 7.62- x 51-mm NATO.
Overall length: 48.0 inches.
Barrel length: 26.0 inches.
Rifling: not known.
Weight: 12.8 pounds.

Magazine capacity: 4-round internal magazine.
Telescope: Pecar V2S 4–10x.
Front: detachable hooded post.
Rear: detachable aperture.
Ammunition requirement: 7.62- x 51-mm NATO ball (147/150 gn FMJBT @ 2,800 fps).

F-34. The Model 85 sniper rifle is a bolt-action 7.62- x 51-mm rifle designed for extended use under adverse conditions. It uses a McMillan fiberglass stock that is adjustable for length of pull. The telescope is mounted on a quick-detachable mount that can be removed in emergencies to reveal a flip-up rear aperture sight that is graduated from 100 to 900 meters.

Parker-Hale Model 85 Characteristics

System of operation: bolt-action.
Caliber: 7.62- x 51-mm NATO.
Overall length: 47.5 inches.
Barrel length: 24.8 inches.
Rifling: 4-groove, 1/12-inch right-hand twist.
Weight: 12.5 pounds.
Magazine capacity: 10-round detachable magazine.

Telescope: Swarovski ZFM 6 x 42 mm (BDC: 100 to 800 m) or ZFM 10 x 42 mm (BDC: 100 to 1,000 m).
Front: protected blade.
Rear: folding aperture.
Ammunition requirement: NATO ball, 147/150 gn FMJBT @ 2,800 fps.

F-35. The L96A1 sniper rifle is built by Accuracy International using a unique bedding system designed by Malcolm Cooper. It features an aluminum frame with a high-impact plastic, thumbhole-type stock; a free-floated barrel; and a lightweight-alloy, fully adjustable bipod. The rifle is equipped with metallic sights that can deliver accurate fire out to 700 meters and can use the L1A1 telescope. The reported accuracy of this weapon is 0.75 MOA at 1,000 meters. One interesting feature of the stock design is a spring-loaded monopod concealed in the butt. Fully adjustable for elevation, the monopod serves the same purpose as the sand sock that the U.S. Army uses.

Accuracy International Model PM/L96A1 Characteristics

System of operation: bolt-action.

Caliber: 7.62- x 51-mm NATO, .243 Winchester, 7 mm Remington Magnum, 300 WM.

Overall length: 47.0 inches.

Barrel length: 26 inches.

Rifling: 1/12-inch right-hand twist.

Weight: 15 pounds.

Magazine capacity: 10-round detachable magazine.

Telescope: 6- x 42-mm or 12- x 42-mm Schmidt and Bender.

Front: none.

Rear: none.

Ammunition requirement: not known.

F-36. The Lee Enfield Rifle No. 4 Mark 1 (T) and No. 4 Mark 1* (T) are sniper versions of the No. 4. They are fitted with scope mounts on the left side of the receiver and have a wooden cheek rest screwed to the butt. The No. 32 telescope is used on these weapons.

Lee Enfield No. 4 Mark 1 (T) Characteristics

System of operation: bolt-action.

Caliber: .303 British.

Overall length: 44.5 inches.

Barrel length: 25.2 inches.

Rifling: not known.

Weight: 11.5 pounds.

Magazine capacity: 10-round detachable magazine.

Telescope: No. 32, 3x, BDC: 100 to 1,000 yards.

Front: blade with protecting ears.

Rear: vertical leaf with aperture battle sight or L-type.

Ammunition requirement: .303 ball with a muzzle velocity (at date of adoption) of 2,440 fps.

FORMER UNION OF SOVIET SOCIALIST REPUBLICS

F-37. This system is currently in use: SVD (Dragunov).

F-38. The self-loading rifle, SVD (Dragunov) is a purpose-designed system that replaced the M1891/30 sniper rifle in 1963. The bolt operation of the SVD is similar to that of the AK/AKM. The principal difference is that the SVD has a short stroke piston system. It is not attached to the bolt carrier like that of the AK/AKM, and delivers its impulse to the carrier, which then moves to the rear. The remainder of the operating sequence is quite similar to the Kalashnikov-series assault rifle. The rifle has a somewhat unusual stock in that a large section has been cut out of it immediately to the rear of the pistol grip. This lightens the weight of the rifle considerably. It has a prong-type flash suppressor similar to those used on current U.S. small arms. It is equipped with metallic sights that are graduated to 2,000 meters and the PSO-1 4x telescopic sight with a battery-powered, illuminated reticle. The PSO-1 also incorporates a metascope that, when activated, is capable of detecting an active, infrared source. The PSO-1 is designed for the ballistic trajectory of the LPS ball round. The windage knob provides 2 MOA per click and 4 MOA per numeral. The reticle pattern has 10 vertical lines to the left and right of the aiming chevron. These lines are spaced 4 MOA from each other, which provide 40 MOA to the left and right of the aiming chevron.

SVD Characteristics	
System of operation: semiautomatic.	Magazine capacity: 10-round detachable magazine.
Caliber: 7.62- x 54-mm rimmed.	Telescope: 4x PSO-1, BDC: 0 to 1,300 m.
Overall length: 47.9 inches.	Front: hooded post.
Barrel length: 24.5 inches.	Rear: tangent with notch.
Rifling: 4 grooves, 1/10-inch right-hand twist.	Ammunition requirement: LPS ball (149 gn FMJBT @ 2,800 fps).
Weight: 9.7 pounds.	

FORMER WARSAW PACT AMMUNITION

F-39. The standard M1908 Russian L ball cartridge features a 149 grain lead-core spitzer bullet with a gilding metal jacket and a conical hollow base. The L ball gives about 2,800 fps from the M1891/30 rifles. It can be identified with a plain, unpainted, copper-colored bullet.

F-40. The LPS ball cartridge is a 149 grain boat tail with a gilding metal-clad steel jacket and mild steel core. The LPS cartridge can be identified by a white or silver bullet tip, distinguishing it from the lead-core L ball. Velocity is around 2,820 fps.

F-41. The M1930 heavy ball sniper load is known as the Type D and is sometimes identified by a yellow bullet tip. It features a 182 grain full metal jacket bullet with a hollow-base boat tail and develops 2,680 fps from the M1891/30 or the SVD.

F-42. The general rule for identifying Soviet/Warsaw Pact ammunition is as follows: when the head of the cartridge case is oriented so that both numbers can be read, the factory number appears at 12 o'clock and the date of manufacture appears at 6 o'clock.

MOSIN-NAGANT BOLT-ACTION SNIPER RIFLE MODEL M1891/30

F-43. The M1891 was adopted as the Russian Army service rifle in 1891. It has a blade front sight with a leaf rear sight graduated in arshins (paces) from 100 to 3,200 (2,496 yards). In the 1930s, the improved M1891/30 was fielded. The M1891/30 has a hooded front sight and a tangent rear sight graduated from 100 to 2,000 meters. The M1891/30 sniper rifle was adopted shortly thereafter, with its only modification being the addition of a telescopic sight. Details on the telescope are found in Appendix E.

M1891/30 Characteristics

System of operation: bolt-action.
Caliber: 7.62- x 54-mm rimmed.
Overall length: 48.5 inches.
Barrel length: 28.7 inches.
Rifling: 4-groove, 1/10-inch right-hand twist.
Weight: 11.3 pounds.

Magazine capacity: 5-round semifixed magazine.
Telescope: PU 3.5x or PE 4x.
Front: hooded post.
Rear: tangent rear, graduated from 100 to 2,000 m.
Ammunition requirement: L or LPS ball (149 gn FMJ @ 2,800 fps).

UNITED STATES OF AMERICA

F-44. These systems are currently in use: M24 and M21 (used by the Army). The USMC has adopted a product-improved version of the Remington 700 that is currently known as the M40A1. Special application sniper rifles, such as the Barrett Model 82 and the RAI Model 500, are used on an organized but limited basis. Numerous nonstandard sniper rifles are used by different U.S. Government and DOD agencies. Also, obsolete sniper rifles are still being used abroad. These include the M1903A4, M1C, M1D, and the M21.

M21 SNIPER SYSTEM

F-45. In September 1968, the Army Materiel Command was directed to produce 1,800 National Match M-14s for immediate shipment to Vietnam. From 1968 until 1975, when the XM-21 was adopted, several NM M-14 variants with different telescopes were shipped to Vietnam for use. The first XM-21s used the WW II-era M84 telescope. James Leatherwood, the designer of the ART-series, provided most of the telescopes, although others were used. The M21 is carefully assembled to National Match standards with selected components. The stock was originally an epoxy-impregnated walnut or birch stock. The rifle has NM iron sights. The elevation and windage adjustments provide 1/2 MOA corrections. The scope mount is mounted to the side of the receiver with a large knurled knob. Later mounts provided two points of attachment with an additional knob threaded into a modified clip guide. The M21 was type-classified with the ART I. The ART II was later used on a limited basis, and the M3A Ultra has been used to upgrade the M21 system.

M21 Characteristics

System of operation: semiautomatic.
Caliber: 7.62- x 51-mm NATO.
Overall length: 44.3 inches.
Barrel length: 22 inches.
Rifling: 4-groove, 1/12-inch right-hand twist.
Weight: 14.4 pounds.

Magazine capacity: 20-round detachable magazine.
Telescope: ART I or ART II, BDC: 300 to 900 m.
Front: protected post.
Rear: hooded aperture.
Ammunition requirement: M118 Match or Special Ball (173 gn FMJBT @ 2,610 fps).

USMC M40A1

F-46. The M40A1 is the current USMC sniping rifle that is the culmination of 20 years of use of the Remington Model 700 since the Vietnam War. It is built by match armorers to exacting standards using selected components. It uses Remington M700 and 40x receivers mated to a heavy McMillan/Wiseman stainless steel match barrel. The stock is made by McMillan. The Unertl 10x USMC sniper scope has the mil-dot reticle and a BDC designed to range from 100 to 1,000 yards.

M40A1 Characteristics

System of operation: bolt-action.
Caliber: 7.62- x 51-mm NATO.
Overall length: 44 inches.
Barrel length: 24 inches.
Rifling: 6-groove, 1/12-inch right-hand twist.
Weight: 14.4 pounds.

Magazine capacity: 5-round internal magazine.
Telescope: Unertl 10x, BDC: 100 to 1,000 yards.
Front: none.
Rear: none.
Ammunition requirement: M118 Match or Special Ball (173 gn FMJBT @ 2,610 fps), or Federal Match with the 180 gn Sierra MatchKing bullet.

BARRETT MODEL 82A1

F-47. The Barrett Model 82A1 sniping rifle is a recoil-operated, 11-round detachable box, magazine-fed, semiautomatic chambered for the caliber .50 Browning cartridge. Its 36.9-inch fluted barrel is equipped with a six-port muzzle brake that reduces recoil by 30 percent. It has an adjustable bipod and can also be mounted on the M82 tripod or any mount compatible with the M60 machine gun. This weapon has a pistol-grip-style stock, is 65.9 inches long, and weighs 32.9 pounds. The sighting system consists of a telescope, but no metallic sights are provided. The telescope mount may accommodate any telescope with 1-inch rings. Muzzle velocity of the Model 82A1 is 2,849 fps.

IVER JOHNSON MODEL 500

F-48. The Iver Johnson Model 500 is the old version of the Research Armaments Industry (RAI) Model 500/Daisy Model 500. The Model 500 long-range rifle is a bolt-action, single-shot weapon, which is chambered for the caliber .50 Browning cartridge. It has a 33-inch heavy, fluted, free-floating barrel. With its bipod, fully adjustable stock, cheek piece, and telescope, it weighs a total of 29.92 pounds. The weapon is equipped with a harmonic balancer that dampens barrel vibrations, a telescope with a ranging scope base, a muzzle velocity of 2,912 fps, and a muzzle brake with flash suppressor. The USMC and United States Navy (USN) have used this weapon in the past.

U.S. SNIPER RIFLES

F-49. These systems are currently in use: Remington Models 40XB, 40XC, and 700 rifles.

F-50. These variations of the Remington M700 bolt-action rifle are widely used. The M700 is the standard rifle. The M700 and its variants have tubular/round actions which are preferred by many competitors due to its ease of trueing and bedding. It is most frequently seen in the heavy-barreled "Varmint Special" version. The 40XB is a single-shot competition rifle and extremely accurate. The 40XB has a solid magazine well that adds to the action's rigidity. The 40X or 40XC is similar to the XB except they have a magazine well, stripper clip guide, and are designed for use in high-powered rifle competition. The M24 SWS is built on a Remington M700 action marked M24 M700. It is built to the same exacting standards as the 40XBs. The original M24 came with a Rock 5R barrel. The new M24s from Remington come with a Remington hammer-forged barrel. Most Remington .308 rifles (M700, 40XB, and 40XC) come with a short action for reduced action size, increased action rigidity, and reduced bolt-cycling distance. The M24 was adopted with a long action so that it could be converted to the .300 Winchester Magnum cartridge at a later date. This change may be accomplished by replacing the barrel and bolt. In this magnum chambering, the M24 will be designated the Medium Sniper Rifle (MSR) and be effective out to 1,200 meters. The BDC on the Leupold and Stevens M3A will be replaced to match the different ballistic trajectory.

Remington Models 40X/700 Characteristics

System of operation: bolt-action.

Caliber: 7.62- x 51-mm NATO (.308 Winchester), .300 Winchester Magnum, and others.

Overall length: approximately 42 inches (dependent on barrel length).

Barrel length: 22 to 26 inches.

Rifling: 4- to 6-groove, 1/10- to 1/12-inch right-hand twist.

Weight: 10 to 15 pounds.

Magazine capacity: 5-round standard calibers, 3-round magnum calibers, internal magazine; 40XB is single shot.

Telescopes: Leupold and Stevens Ultra/Mark IV M1A, M3A; Unertl 10x USMC; Bausch and Lomb 10- x 40-mm.

Front: none.

Rear: none.

Ammunition requirement: varied.

WINCHESTER MODEL 70

F-51. The Model 70 in .308 or .300 Winchester Magnum, when properly built, is also a very effective and accurate rifle, as proven by the multiple national and international competitors that use them. Winchester now makes a true short action, in caliber .308 as a varmint rifle that can be an alternative to the M700 Remington. The Winchester Model 70 has a square-bottomed action.

MCMILLAN SYSTEMS

F-52. The M-86SR (.308 Win), M-86LR (.300 Win Mag), and M-89 (.308 suppressed) are bolt-action rifles built on McMillan actions. The M-88ELR, M-87ELR, and M-87R are caliber .50 bolt rifles. The McMillan M-40 is a Remington short-action barrel with a McMillan .308 match barrel. A variety of optics are available: Leupold and Stevens Ultra/Mark IV M1A, M3A, 3.5–10x Law Enforcement; the Bausch and Lomb 10- x 40-mm tactical; and the Phrobis tactical rifle telescopes.

BARRETT FIREARMS

F-53. These firearms consist of the M82, M82A1 light semiautomatic caliber .50 rifles, and the M90 bolt-action caliber .50 rifle. Appendix E provides additional information.

OTHER SYSTEMS

F-54. *Robar Systems.* Accurized Remington Model 700 rifle.

F-55. *Iver Johnson Convertible Long-Range Rifle System.* The characteristics of this bolt-action rifle are listed below.

Iver Johnson Convertible Long-Range Rifle Characteristics

System of operation: bolt-action.	Magazine capacity: 4-round (7.62) or 5-round (8.58) detachable magazine.
Caliber: 7.62- x 51-mm NATO, 8.58 x 71 mm (.338/.416).	Telescope: varied.
Overall length: 46.5 inches.	Front: none.
Barrel length: 24 inches.	Rear: none.
Rifling: 4-groove; 1/12 (7.62)-, 1/10 (8.58)-inch right-hand twist.	Ammunition requirement: 8.58 x 71 mm; 250 gn HPBT @ 3,000 fps.
Weight: 15 pounds.	

OBSOLETE U.S. SNIPER RIFLES

F-56. The M1903A4 Springfield was adopted in December 1942 as a sniper rifle during WW II. The only modification to the standard service rifle was the addition of a pistol grip and optical sight. There were numerous telescopic sights used, but the most common were the M84 and the Weaver Model 330C (marked M73B1 for the contract). There are a few 1903s that were meticulously assembled with selected parts for sniper use, but as a general rule, the majority were standard service rifles. The low magnification of the telescopes (2.2x for the M84) made long-range target interdiction difficult. The M84 scope is discussed in Appendix G. The Model 1942 is a USMC modification of the 1903A1, fitted with an 8x Unertl scope. These rifles were manufactured by Remington, Springfield Armory, and the L.C. Smith Corona Typewriter Company.

M1903A4 Characteristics

System of operation: bolt-action.

Caliber: .30 M1/M2 ball (7.62 x 63 mm/30-06).

Overall length: 43.5 inches.

Barrel length: 24 inches.

Rifling: 4-groove (and 2-groove), 1/10-inch right-hand twist.

Weight: 9.4 pounds.

Magazine capacity: 5-round internal magazine.

Telescope: M84, M73B1 Weaver (Model 330C), or the M73 Lyman Alaska; BDC: 0 to 900 yards.

Front: none.

Rear: none.

Ammunition requirement: Caliber .30 M1/M2 ball (150 FMJ flat base @ 2,800 fps.

GARAND M1C AND M1D

F-57. In 1939, the Springfield Armory and Winchester began production of the M1. The M1 was the first self-loading rifle that withstood battlefield use. The M1C and M1D were developed for designated marksman use. The M1D was fitted with a steel collar around the barrel in front of the receiver, which was tapped for a side-mounted scope mount, because the weapon loads through the top of the receiver. An M84 2.2x scope was used. A specially fabricated leather extension was affixed to the left side of the stock to provide a solid stock weld to accommodate the side-mounted telescope. This piece allowed the sniper to rest his cheek and fire left-eyed. Although the rifle can be fired right-eyed, it was designed to be fired left-eyed. It is a fallacy to this day that the leather stock extension is a cheek piece; it is not. It was and is a rest for use with the side-mounted scope. The majority of the M1Ds were also fitted with a prong-flash hider. The M1C is identical to the M1D except in one respect: the M1C has a side mount that was tapped into the left side of the receiver directly instead of using a collar around the barrel. Like the M1903A4, nothing was done to the majority of the rifles to accurize them. Eventually, hand-assembled M1Ds and M1Cs were made and used.

M1C/D Characteristics

System of operation: semiautomatic.

Caliber: .30 Caliber M1/M2 ball (7.62 x 63 mm/ 30-06).

Overall length: 43.6 inches.

Barrel length: 24 inches.

Rifling: 4-groove, 1/10-inch right-hand twist.

Weight: 11.8 pounds.

Magazine capacity: 8-round en-bloc metallic clip.

Telescope: M84, 2.2x, BDC: 0 to 900 yards.

Front: protected post.

Rear: aperture.

Ammunition requirement: M1/M2 ball (150 gn FMJ flat base bullet @ 2,800 fps).

FORMER YUGOSLAVIA

F-58. This system is currently in use: Model M76.

F-59. The Yugoslav armed forces use the M76 semiautomatic sniping rifle. It is believed to be based upon the FAZ family of automatic weapons. It features permanently affixed metallic sights, a pistol-grip-style wood stock, and a 4x telescopic sight. The telescopic sight is graduated in 100-meter increments from 100 to 1,000 meters, and the optical sight mount allows the mounting of passive nightsights. It has a muzzle velocity of 2,361 fps.

M76 Characteristics

System of operation: semiautomatic.

Caliber: 7.92 x 57 mm (8-mm Mauser), 7.62- x 54-mm R, 7.62- x 51-mm NATO.

Overall length: 44.7 inches.

Barrel length: 21.6 inches.

Rifling: not known.

Weight: 11.2 pounds.

Magazine capacity: 10-round detachable.

Telescope: 4x, BDC: 100 to 1,000 m.

Front: hooded post.

Rear: tangent.

Ammunition requirement: 7.92 x 57 mm (2,361 fps); 7.62- x 51-mm NATO (2,657 fps).

Appendix G

Sniper Rifle Telescopes

A scope mounted on the rifle allows the sniper to detect and engage targets more effectively. Another advantage of the scope is its ability to magnify the target. As previously stated, a scope does not make a soldier a better sniper, it only helps him see better. This appendix explains the characteristics and types of scopes.

CHARACTERISTICS OF RIFLE TELESCOPES

G-1. The telescope is an optical instrument that the sniper uses to improve his ability to see his target clearly in most situations. It also helps him to quickly identify or recognize the target and enables him to engage with a higher rate of success. The following characteristics apply to most types of scopes.

TELESCOPE MAGNIFICATION

G-2. The average unaided human eye can distinguish 1-inch detail at 100 meters. Magnification, combined with quality lens manufacture and design, permits resolution of this 1 inch divided by the optical magnification. The general rule is 1x magnification per 100 meters. The magnification (power) of a telescope should correspond to the maximum effective range of the weapon system being used. This amount of power will enable the operator to identify precise corrections. For example, a 5x telescope is adequate out to 500 meters; a 10x is good out to 1,000 meters. The best all-around magnification determined for field-type sniping is the 10x because it permits the operator to identify precise corrections out to 1,000 meters. The field of view of a 10x at close range, while small, is still enough to see large and small targets. Higher-powered telescopes have very limited fields of view, making close range and snap target engagements difficult. Substandard high-powered telescopes may be hard to focus and have parallax problems. Some marksmen still prefer lower-powered telescopes. Recent advances in the construction of variable-powered rifle telescopes have negated the problems that once plagued them. Advantages of the variable power scopes in both urban and clip-on style night vision devices make them very desirable in the long run. A number of scope manufacturers now make reliable variable powers in the 2.5 to 10 or 14 range. This allows the sniper to power down so as not to overpower the phosphorus matrix of the NVD or to gain a wider field of view in close in sniping.

PARALLAX

G-3. Parallax results when the target is not focused on the same focal plane as the reticle. When parallax is present, the target will move in relation to the reticle when the sniper moves his head (changes his spot weld) while looking through the telescope. It is more apparent in high-powered telescopes. With parallax, the error will affect the strike of the bullet by the amount seen in the scope. If the crosshairs move from one side of the target to the other, then the potential error is from one side of the target to the other. Therefore, the sniper should zero his system, for elevation, at the greatest distance possible. For a 1,000-meter system, the sniper should confirm his zero at 500 meters for elevation. The initial zero of the weapon system for elevation and windage should be at 100 or 200 meters. This will keep the shooter on paper. The 100-meter range will negate most wind effect; however, the shooter is capable of computing the wind effect and zero with the bullet strike at that point on the target. The M1A and M3A Ultra/Mark IV by Leupold and Stevens have a focus/parallax knob on the left side of the telescope. With the M1A and M3A, it is imperative that the sniper adjusts his focus/parallax when he zeros his system for each range and that he records this data in his shooter's log. If there is a zero-shift while adjusting parallax from range to range, then the scope is defective and requires replacement. The reticle must be focused for the eye prior to focusing on the target. If the reticle is perfectly focused, then the target will be in focus and the scope will be parallax-free. If the target is focused and the scope is not parallax-free, then the shooter may wish to refocus the reticle and recheck parallax. Once both reticle and target are focused on the same plane, the scope will be parallax-free at that range only. The snipers will be required to then focus the target for each range to obtain a parallax-free scope. This information is then recorded in the shooter's log for use when firing over unknown distances and will become part of the sniper and observer dialogue. As an example, "Range 650+1, windage left 1 click spin drift, parallax second ball."

ADJUSTABLE OBJECTIVE LENS

G-4. Adjustable objective lenses for focusing at different magnifications and ranges are becoming quite common. Some target telescopes (such as the M1A and M3A) have a third turret knob on the side of the telescope that will focus the objective lens. Doing so will focus the target and reticle on the same plane, eliminating parallax. Unfortunately, many telescopes have neither and must be dealt with on an individual basis. The best way to deal with the problem is to eliminate shadow. Once shadow is eliminated, the sniper must ensure that the reticle moves the same distance left and right on the target as well as up and down. Doing so will assist in attaining the same aim point even with parallax. If shadow is present, with parallax error, then the strike of the bullet will be opposite of the shadow. The shadow indicates that the sniper is looking down that side of the scope's exit pupil and the crosshairs will appear to have moved to that side as well. The sniper will then compensate by moving the weapon right to get the crosshair back on target, causing the strike of the bullet to be opposite of the shadow.

VARIABLE-POWERED TELESCOPES

G-5. Older variable-powered telescopes often shifted the POI if the magnification was changed from its original setting when sighting the system. Modern, high-quality, variable-powered telescopes do not have this problem. This type of movement has been tested on a number of quality variable-powered telescopes. After zeroing, the scopes showed no variation in the POA versus the POI at any range or any power. Of course, it is prudent to test the system during live-fire exercises to establish the optic's reliability.

TELESCOPE ADJUSTMENTS

G-6. One telescope will not automatically work in the same manner for every sniper. Each sniper's vision is different and requires different adjustments. The following factors vary with each use.

FOCUSING

G-7. Focusing the telescope to the sniper assigned the weapon is important. He can adjust the ocular lens of most telescopes to obtain a crisp, clear picture of the reticle. To do this, the sniper should look at a distant object for several seconds without using the telescope. Then he should shift his vision quickly, looking through the telescope at a plain background. The reticle pattern should be sharp and clear before his eye refocuses. If he needs to make an adjustment to match his eyes, he should hold the eyepiece lock ring and loosen the eyepiece by turning it two turns clockwise to compensate for nearsightedness and counterclockwise to compensate for farsightedness. Then, with a quick glance he should recheck the image. If the focus is worse, he then turns it four revolutions in the opposite direction. It will normally take two full revolutions to see a noticeable difference in the focus. Once the reticle appears focused, the sniper leaves the sights alone and allows the eye to rest for 5 to 10 minutes and then rechecks the reticle. If he force-focuses the reticle, his eye will tire and he will see two reticles after shooting for a period of time. After determining the precise focus for his eye, the sniper should make sure to retighten the lock ring securely against the eyepiece to hold it in position.

> **CAUTION**
> Never look at the sun through the telescope. Concentration of strong solar rays can cause serious eye damage.

EYE RELIEF

G-8. Proper eye relief is established very simply. First, the sniper loosens the scope rings' Allen screws so that the telescope is free to move. He gets into the shooting position that will be used most frequently and slides the scope forward or back until a full, crisp picture is obtained. There should be no shading in the view. This view will be anywhere from 2 to 4 inches from his eye depending on the telescope. He rotates the telescope until the reticle crosshairs are perfectly vertical and horizontal, then he tightens the rings' screws.

G-9. The M24 has a one-piece telescope base that has two sets of machined grooves that allow the telescope to be mounted either forward or back to adjust for personal comfort. If that range of adjustment is not sufficient, the telescope can be adjusted after the mounting ring lock screws are loosened.

UNITED STATES TELESCOPES

G-10. The sniper team carries the telescope on all missions. The observer uses the telescope to determine wind speed and direction. The sniper uses this information to make quick and accurate adjustments for wind conditions. The team also uses the telescope for quicker and easier target identification during troop movement. The following discussion applies to the U.S. rifle telescopes currently in use.

M84 TELESCOPIC SIGHT

G-11. The M84 telescopic sight has a magnification of 2.2x. It has a field of view of 27 feet at 100 yards. The maximum field of view is obtained with an eye relief of 3 1/2 to 5 inches. The reticle consists of a vertical post and a horizontal crosshair. The post is 3 MOA in width. The sight is sealed with rubber seals and may be submerged without damage (not recommended due to age). The windage knob has 60 MOA of adjustment, 30 MOA from center left or right. However, there are a total of 100 MOA adjustments available to zero the telescope for misalignment. To adjust the strike of the bullet vertically, the sniper turns the knob to the higher numbers to raise the POI, and to lower numbers to lower the POI. A complete turn of the elevation knob provides 40 MOA of adjustment. One click of the elevation or windage knob equals 1 MOA. The elevation scale starts at 0 yards and goes up to 900 yards with graduations every 50 yards. There is a numbered graduation every 100 yards.

G-12. To zero the scope, the sniper shoots at a target at 100 or 200 yards. He adjusts the elevation and windage until the POA and POI are the same. He turns out the setscrews on both the elevation and windage knobs to "zero" them. The sniper then lifts and rotates the windage dial until the windage (deflection) is on the zero marking for the no-wind zero. He lifts and rotates the range (elevation) knob to the distance used for the zeroing procedure. He can mount this telescope on both the M1C and M1D, the M1 Marine sniper, and the 1903A4 Springfield using a Redfield scope mount.

ADJUSTABLE RANGING TELESCOPE I (ART I)

G-13. The ART I automatically compensates for trajectory when a target of the proper size is adjusted between the stadia lines. It is a 3–9x variable that compensates for targets from 300 to 900 meters. It has a one-piece ballistic cam/power ring. The ballistic cam is set for the ballistic trajectory of the M118 Match or Special Ball ammunition (173 grain FMJBT @ 2610 fps). Each click or tick mark on the adjustment screws is worth 1/2 MOA in value. The ART I is zeroed at 300 meters. The sniper sets the power ring to 3 (3x/300m) and removes the adjustment turret caps. He fires the rifle and adjusts the elevation and windage adjustment screws until the POI is the same as the POA. Then he screws the turret caps back on to maximize the waterproofing of the telescope.

G-14. The reticle has four stadia lines on it (Figure G-1). The two horizontal stadia lines are on the vertical crosshair, are 30 inches apart at the designated distance, and are used for ranging. The vertical crosshair and horizontal stadia lines are used to range targets from the beltline to the top of the head. The sniper adjusts the power/cam until the stadia lines are bracketing the target's beltline and top of head. The numeral on the power ring is the target distance. For example, if the power ring reads 5, the target is at 500 m, and the scope is at 5x magnification. The ballistic cam has automatically adjusted the telescope for the trajectory of the round by changing the telescope's POA. The sniper aims center mass on the target to obtain a hit in a no-wind situation. The two vertical stadia lines are on the horizontal crosshair, are 60 inches apart at the designated distance, and are used for wind hold-offs and leads. If necessary, he holds off for environmental effects or target movement.

NOTE: It is imperative to keep the scope base clean. The cam slides along the mount and pushes the telescope off from the bearing surface. Debris can interfere with the precise camming and ranging functions.

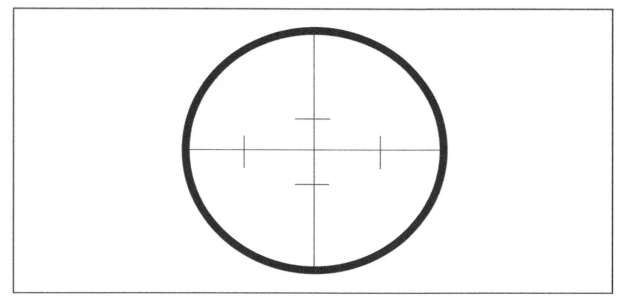

Figure G-1. ART I Telescope Reticle

ADJUSTABLE RANGING TELESCOPE II (ART II)

G-15. The ART II is similar in operation and design to the ART I, with two major modifications. The ballistic cam and the power ring are now separate and can be moved independently of each other. This modification was made so that after ranging a target, the ballistic cam can be locked to permit the sniper to increase the magnification for greater definition. The problem with this system is that it seldom works correctly. The two rings are locked together in poker-chip-tooth fashion, and even when locked together, they can move independently. When unlocked, it is very difficult to move one without the other moving, creating a change in the camming action, and ultimately, causing misses. It is best to lock them together and keep them together. The mount is similar to the ART I mount, and the bearing surface must be kept clean. The ART II mount has two mounting screws, one of which is threaded into a modified clip guide. The reticle is the second major modification. The reticle pattern is a standard crosshair, with thick outer bars on the left, right, and bottom crosshairs (Figure G-2). The horizontal crosshair has two dots, one on each side of the crosshair intersection. Each dot is 30 inches from the center and a total of 60 inches apart. The heavy bars are 1 meter in height or thickness at the range indicated. To determine the range to a target, the sniper adjusts the power ring and cam together until the target is of equal height to the bar. The correct placement of the bar is from the crotch to the top of head (1 meter). He aims center mass for a no-wind hit. He can read the cam to determine the range.

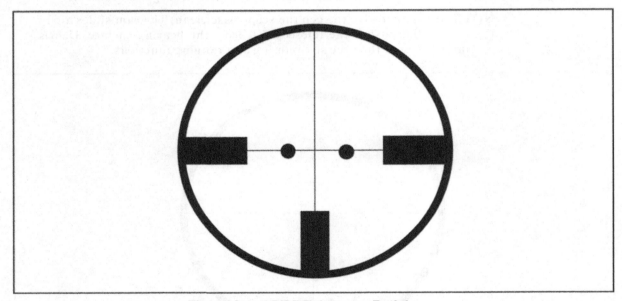

Figure G-2. ART II Telescope Reticle

LEUPOLD AND STEVENS M1A AND M3A ULTRA/MARK 4 10X OR 16X

G-16. The M1A comes in either 10x or 16x. It has three large, oversized target knobs. The left knob (as seen from the sniper) is for focus/parallax adjustment. The top knob is for elevation adjustment. The right knob is for windage adjustment. Table G-1, page G-7, explains the scope adjustments. The M3A is only available in 10x. It has the same knob arrangement as the M1A, but the knobs are smaller, and they have different click values. All Leupold and Stevens sniper telescopes use the mil-dot reticle (Figure G-3, page G-7).

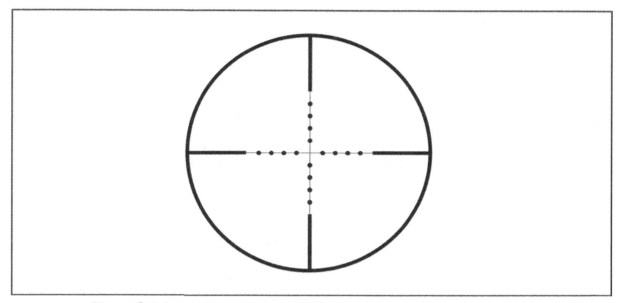

Figure G-3. Leupold and Stevens Telescope With a Mil-Dot Reticle

G-17. The M3A has a ballistic collar to compensate for the trajectory of the specified cartridge. The collar is calibrated for bullet drop compensation from 100 meters to 1,000 meters. The following are available:

- 7.62 mm NATO M118 (173 FMJBT @ 2,610 fps).
- .300 Winchester Magnum (190 HPBT @ 2900 fps).

NOTE: This collar is erroneously marked as 220 grain.

- .30-06 Springfield (180 HPBT @ 2,700 fps).
- 5.56 mm M193/.223 Remington (55 FMJBT @ 3,200 fps).

Table G-1. Adjustments of Leupold and Stevens Telescopes

Model	Elevation	Windage	Complete Revolution
M1A	1/4 MOA	1/4 MOA	15 MOAs
M3	1 MOA	1 MOA	100–1,000 M
M3A	1 MOA	1/2 MOA	100–1,000 M

NOTE: If the scope exhibits a zero shift after focusing the scope for the range, then the sniper should send the scope in for maintenance.

BAUSCH AND LOMB TACTICAL RIFLE TELESCOPE

G-18. The Bausch and Lomb is a 10- x 40-mm fixed-magnification telescope with 1/4 MOA adjustments. It has two large, target-type knobs. The upper knob is for elevation, and the knob on the right is for windage. The eyepiece houses the range-focus adjustment ring that is calibrated from 50 yards to infinity. It has the same mil-dot reticle pattern as the Leupold and Stevens series and USMC Unertl telescope. Each revolution of the adjustment knobs provides 12 MOA. This scope is no longer manufactured and only a few are in the system.

UNERTL USMC SNIPER TELESCOPE

G-19. John Unertl was a USMC sniper during World War I and later became the manufacturer of some of the finest U.S.-made optics. This telescope was designed and built by the John Unertl Company. It is a fixed 10x, steel-tubed, mil-dot telescope with a BDC for the M118 ammunition. The lens is coated with a high-efficiency, low-reflection (HELR) film that transmits up to 91 percent of the ambient light. This telescope has 1/2 MOA adjustments, fine adjustments for zeroing, and 1 MOA adjustments for elevation under normal use. For windage, the adjustments are in .5 MOA. It has a fine-tune elevation capability that permits +/− 3 MOA, in .5 MOA adjustments, to adjust for differences in sniper's zero, temperatures, ammunition lots, and ammunition. The windage adjustment has 60 MOA of main adjustment with +/− 4 MOA fine adjustment. This telescope also has a parallax-adjusting capability. The reticle is identical to that of the Leupold and Stevens series and the Bausch and Lomb Tactical scope. Care must be taken as most of these scopes are not waterproof and can fog badly under high-humidity use.

SOVIET TELESCOPES

G-20. The Soviet telescopes are made on machinery purchased from Carl Zeiss of Germany during the 1930s. Their optical quality is therefore good to excellent. Their operation is rather simple. Only the PE series has the capability of individually focusing to the user. The top turret is for elevation adjustment and has a ballistic cam that is calibrated for the 7.62- x 54-mm Rimmed L ball ammunition (150 gn FMJ flat base @ 2,800 fps). The turret on the left is for windage adjustments. Table G-2, page G-9, lists various models and their characteristics.

G-21. The zeroing procedures are identical for all Soviet telescopes. The sniper should zero at 100 meters. To do so, he loosens the small screws on the turrets that hold the top plate to the cam that is engraved with the tick marks and numerals. Several turns are all that is necessary. He should not remove these screws completely; they are not captive and are easily lost. Using a small screwdriver, he gently pries the top plate and cam apart so that the top plate can move independently of the cam. Firing three-shot groups, he adjusts the elevation and windage knobs until the POA and the POI are the same. When making adjustments, the sniper should **move the reticle to the shot group**. This adjustment is the major difference from zeroing these telescopes when compared to zeroing modern, U.S.-style telescopes where the shot group is moved to the reticle (POA). When the rifle and telescope system is zeroed, the sniper should "zero out" the cams. He should turn the elevation cam until the "1," which represents 100 meters, is aligned with the reference tick mark. He makes sure the top plate does not rotate when the cam is moved. The windage cam is also centered on its "0" marking. The sniper then pushes down on the top plates until they mate with the cams. He carefully tightens the small metal screws. The telescope is now zeroed.

Table G-2. M1891/30 Sniper Telescopes

Model	Magnification	BDC (out to)	Tube Diameter
PE	4x	1,400 m	1 inch
PU	3.5x	1,300 m	30 mm
PV	3.5x	1,300 m	30 mm

SOVIET MANUAL DESCRIPTION

G-22. Telescopes can be any of various tubular optical instruments. Soviet technical manuals describe the telescope in two parts: a telescope tube and a mount.

Telescope Tube

G-23. On the top of the tube is an elevation range knob, consisting of a screw and a drum, marked with numbers from 1 to 14 on the PE scope and from 1 to 13 on the PU scope. Each graduation is equivalent to 100 meters in distance.

G-24. At the left rear side of the scope is a windage knob. The components of the windage knob are the same as that of the elevation-range knob. The sniper uses the windage knob to compensate for the effects of wind on the trajectory of the bullet. The windage knob has 10 graduations; the middle one is marked with the number 0.

G-25. To move the strike of the bullet to the right, the sniper turns the windage knob to the direction of the mark "+," and conversely, turns the knob to the direction of the mark "−" to move the strike of the bullet to the left. Each click of windage corresponds to 1 mil.

G-26. The telescope tube contains a system of optical glasses including convex lenses, prisms, and an eyepiece. The reticle is a cross-wire type. When aiming the rifle at the objective, the sniper places the vertical line of the reticle right on the objective. He uses the horizontal line to adjust the aim. The two knobs provide horizontal and vertical movement of the reticle.

G-27. The telescope tube PE has adjusting devices. When taking aim, the sniper adjusts the knobs on the tube to fit with the observer's eye.

G-28. The telescope tube PU has no adjusting (focusing) devices. Therefore, when aiming, the observer looks through the telescope and moves his head until the sighted object is in focus.

G-29. When using a telescope to aim at the objective, the sniper places the eye at the center of the eyepiece, thus forming a sight alignment toward the objective. If aiming inaccurately, the sniper will see a small, black, crescent-shaped spot in the telescope.

Mount

G-30. The mount for PE consists of a base and a body. The sniper fixes the base to the receiver of the rifle with six screws. He uses the body of the mount, after it is fastened to the base, to fix the telescope to the rifle.

G-31. The mount for PU also includes a base and a body. The sniper connects the base, after it is screwed to the receiver of the rifle, with the body of the mount by guide lugs and screws. The body of the mount may be moved up and down on the base using the two screws on the upper side and the rear lower side of the base. The sniper uses the body to fix the telescope to the rifle.

G-32. The sniper then loosens three screws to rotate the sighting telescope, but only loosens the screws when firing for adjustment at the repair station of the regiment.

PSO-1

G-33. The PSO-1 scope will be found mounted on the Soviet SVD and the Romanian FPK. The PSO-1 is 4x, and has an illuminated reticle powered by a small battery. The battery housing is located at the bottom rear of the telescopic sight mount. To change batteries, the sniper presses in and rotates the battery housing counterclockwise. He removes the old battery and replaces it with the same type. He can replace the reticle lamp by unscrewing its housing and removing the bulb (the RPG-7 sight uses the same bulb). The reticle light is turned on or off by its switch. The lens cap should always be in place except when the telescope is in use. Two covers are issued with each rifle: one is for the telescopic sight alone and the other covers the sight and breech when the PSO-1 is mounted. A belt pouch is provided for carrying the telescope when dismounted from the rifle, four magazines, a cleaning kit, and an extra battery and lamp.

G-34. If the sniper needs to use open sights, he sets the rear sight by pressing in the locks on the rear sight slide, then moves the slide along the rear sight leaf. He then aligns the front edge of the slide with the numeral that corresponds in hundreds of meters. He can use the same sight picture as for firing a pistol.

G-35. If the sniper uses the PSO-1, he rotates the elevation knob until the index aligns with the figure that corresponds to the range in hundreds of meters. He can closely determine the range by using the range finder located in the lower left of the telescopic reticle. This range finder is graduated to the height of a man (5 feet 7 inches) from 200 to 1,000 meters. The sniper looks through the telescope and places the horizontal line at the bottom of the target. He moves the telescope until the upper (curved) line just touches the top of the target's head. The number indicates the range in hundreds of meters. If the target falls between numbers, he must estimate the remaining distance. When the range is determined and set into the elevation knob, he uses the point of the top chevron on the reticle as an aiming point. He uses the three lower chevrons for firing at 1,100, 1,200, and 1,300 meters with the elevation knob set at 10.

G-36. The sniper uses the horizontal scale extending out from the sides of the top chevron for hasty wind and lead corrections; each tick mark is worth 1 Soviet mil (6,000 Soviet mils per 360 degrees). The horizontal scale is numbered every 5 and 10 mils. Rotating the windage knob makes deliberate changes. The windage knob is graduated every 1/2 Soviet mil. The windage knob scale has two clicks per graduation, each click representing 1/2 mil (.5 mil), each graduation one mil. At 1,000 meters, each click moves the impact of the round .5 meters (20 inches), each graduation moves the impact 1 meter (40 inches). The numbers on the windage knob are colored. Right windage corrections are black and are obtained when the knob is rotated clockwise. Left windage corrections are red and are obtained when the knob is rotated counterclockwise.

G-37. When the sniper must fire in dim light, he illuminates the reticle by turning on the switch in the telescopic sight mount. If active infrared light sources are believed to be used by the enemy, he sets the range drum at four and switches the infrared detector into place. He then scans the area to the front; if any active infrared light sources are in use, they will appear as orange-red blobs in the telescope. He aligns the point of the reticle on the light and fires. The sniper should turn off the reticle when not in use to conserve the battery and swing the infrared detector out of the way so that it will be activated by light during the day. Several hours of direct sunlight are required to activate the infrared detector.

G-38. If the sniper is unable to obtain the correct dry cell batteries, he can easily assemble a suitable expedient. The Soviet dry cell is 5.0 volts. The following are required:

- Two 1.25 volt/625 camera batteries (lithium).
- One 3.0 volt/DL2025 camera battery (lithium).
- One plastic bushing Outside Diameter–0.85," Inside Diameter–0.60," Length–0.73."

G-39. The sniper should place the batteries' positive "+" side into the battery compartment first. He places the large, flat DL2025 in first, then the bushing, then the two 625 batteries, and replaces the battery compartment cap.

Appendix H

Ballistics Chart

SIERRA BALLISTICS III

Data for: 7.62mm M118 **Bullet:** 173 grains **BCs:** .515(H), .503(M), .491(L)
Company: Sierra **Temperature:** 59 **Pressure:** 29.53 **Humidity:** 78%
Zero: 100 meters **Crosswind:** 10.00 mph **Tail Wind:** + 0.00 mph
Elevation Angle: 0 degrees **Altitude:** 0 feet **Sight Height:** 1.7 inches

Range (meters)	Velocity (fps)	Energy (ft-lb)	Bullet Path (inches)	Drop (inches)	Drift (inches)	Time of Flight (sec)
0	2,610.0	2,616	− 1.7	+ 0.0	+ 0.0	0.000000
25	2,561.9	2,521	− 0.7	− 0.2	− 0.1	0.031719
50	2,514.3	2,428	− 0.0	− 0.7	− 0.2	0.064037
75	2,467.2	2,338	+ 0.2	− 1.7	− 0.5	0.096969
100	2,420.6	2,250	+ 0.0	− 3.1	− 0.9	0.130533
125	2,374.4	2,165	− 0.6	− 4.9	− 1.3	0.164745
150	2,328.8	2,083	− 1.7	− 7.2	− 1.9	0.199625
175	2,283.7	2,003	− 3.3	− 9.9	− 2.7	0.235191
200	2,239.1	1,925	− 5.4	− 13.2	− 3.5	0.271464
225	2,194.9	1,850	− 8.0	− 17.0	− 4.5	0.308463
250	2,151.2	1,777	− 11.1	− 21.3	− 5.6	0.346210
275	2,108.0	1,707	− 14.8	− 26.2	− 6.9	0.384726
300	2,065.3	1,638	− 19.1	− 31.6	− 8.3	0.424036
325	2,023.2	1,572	− 24.0	− 37.7	− 9.8	0.464162
350	1,981.5	1,508	− 29.5	− 44.4	− 11.5	0.505128
375	1,940.4	1,446	− 35.6	− 51.8	− 13.3	0.546959
400	1,899.8	1,386	− 42.5	− 59.8	− 15.3	0.589681
425	1,859.7	1,328	− 50.1	− 68.6	− 17.4	0.633319
450	1,820.3	1,273	− 58.4	− 78.1	− 19.8	0.677901
475	1,781.2	1,219	− 67.5	− 88.4	− 22.2	0.723454
500	1,742.0	1,165	− 77.5	− 99.5	− 24.9	0.770022
525	1,703.4	1,114	− 88.2	− 111.5	− 27.8	0.817641
550	1,665.5	1,065	− 99.9	− 124.3	− 30.8	0.866341
575	1,628.4	1,018	− 112.5	− 138.1	− 34.0	0.916152
600	1,591.9	973	− 126.1	− 152.9	− 37.5	0.967102
625	1,556.3	930	− 140.7	− 168.7	− 41.1	1.019220
650	1,521.4	889	− 156.4	− 185.6	− 45.0	1.072533
675	1,487.3	850	− 173.2	− 203.6	− 49.0	1.127070

Data for: 7.62mm M118 (Continued)						
Range (meters)	Velocity (fps)	Energy (ft-lb)	Bullet Path (inches)	Drop (inches)	Drift (inches)	Time of Flight (sec)
700	1,454.1	812	−191.2	−222.8	−53.3	1.182854
725	1,421.8	776	−210.4	−243.2	−57.8	1.239911
750	1,390.4	743	−231.0	−264.9	−62.6	1.298263
775	1,360.0	710	−252.8	−287.9	−67.5	1.357928
800	1,330.6	680	−276.1	−312.4	−72.7	1.418923
825	1,302.2	651	−300.8	−338.3	−78.2	1.481260
850	1,274.9	624	−327.0	−365.7	−83.9	1.544946
875	1,248.8	599	−354.9	−394.8	−89.8	1.609982
900	1,223.9	575	−384.4	−425.5	−95.9	1.676366
925	1,200.1	553	−415.7	−458.0	−102.3	1.744087
950	1,177.6	533	−448.8	−492.2	−108.9	1.813130
975	1,156.3	514	−483.8	−528.4	−115.8	1.883472
1,000	1,136.3	496	−520.7	−566.5	−122.9	1.955087

Range	Path	Drift
0	−1.7	+0.0
100	+0.0	−0.9
200	−5.4	−3.5
300	−19.1	−8.3
400	−42.5	−15.3
500	−77.5	−24.9

Environmental Conditions

Actual barometric pressure at firing site..................................29.53 inches
Actual speed of sound at firing site ....1,121 fps
Effective ballistic coefficient at firing site................................0.514

Animal Lead Calculations

Average lead for a running deer at 100 meters is...................3 feet
Average lead for a running elk at 100 meters is ....5 feet
Average lead for a running antelope at 100 meters is ....8 feet

SIERRA BALLISTICS III

Data for: 7.62mm M852   **Bullet:** 168 grains Match King   **BCs:** .462(H), .447(M), .424(L)
Company: Sierra   **Temperature:** 59   **Pressure:** 29.53   **Humidity:** 78%
Zero: 600 meters   **Crosswind:** 10.00 mph   **Tail Wind:** + 0.00 mph
Elevation Angle: 0 degrees   **Altitude:** 0 feet   **Sight Height:** 1.5 inches

Range (meters)	Velocity (fps)	Energy (ft-lb)	Bullet Path (inches)	Drop (inches)	Drift (inches)	Time of Flight (sec)
0	2,600.0	2,521	−1.50	0.00	0.00	0.000000
25	2,546.3	2,418	5.20	0.19	−0.06	0.031865
50	2,493.3	2,319	11.49	0.77	−0.23	0.064404
75	2,440.6	2,222	17.37	1.78	−0.53	0.097641
100	2,388.5	2,128	22.81	3.22	−0.95	0.131599
125	2,337.2	2,037	27.80	5.12	−1.51	0.166299
150	2,286.5	1,950	32.31	7.49	−2.20	0.201764
175	2,236.5	1,866	36.32	10.35	−3.03	0.238020
200	2,187.1	1,784	39.82	13.74	−4.00	0.275089
225	2,138.5	1,706	42.77	17.67	−5.12	0.312998
250	2,090.0	1,629	45.16	22.16	−6.39	0.351774
275	2,040.1	1,552	46.95	27.25	−7.82	0.391476
300	1,991.0	1,478	48.12	32.97	−9.43	0.432153
325	1,942.5	1,407	48.63	39.34	−11.22	0.473840
350	1,894.8	1,339	48.45	46.39	−13.18	0.516572
375	1,847.8	1,273	47.56	54.17	−15.34	0.560386
400	1,801.5	1,210	45.90	62.71	−17.70	0.605319
425	1,755.9	1,150	43.44	72.05	−20.26	0.651414
450	1,711.0	1,092	40.14	82.23	−23.03	0.698711
475	1,667.0	1,036	35.96	93.30	−26.02	0.747254
500	1,624.2	984	30.84	105.30	−29.24	0.797077
525	1,581.7	933	24.73	118.29	−32.69	0.848218
550	1,539.3	884	17.59	132.31	−36.38	0.900757
575	1,498.1	837	9.35	147.43	−40.33	0.954745
600	1,458.0	793	−0.04	163.70	−44.54	1.010219
625	1,419.0	751	−10.65	181.20	−49.02	1.067219
650	1,381.1	711	−22.56	199.98	−53.78	1.125786
675	1,345.0	675	−35.82	220.13	−58.81	1.185945

| Data for: 7.62mm M852 (Continued) |||||||
Range (meters)	Velocity (fps)	Energy (ft-lb)	Bullet Path (inches)	Drop (inches)	Drift (inches)	Time of Flight (sec)
700	1,310.7	641	−50.52	241.71	−64.13	1.247702
725	1,278.0	609	−66.73	264.80	−69.73	1.311059
750	1,246.9	580	−84.53	289.48	−75.61	1.376018
775	1,217.2	553	−104.00	315.83	−81.77	1.442582
800	1,189.0	527	−125.22	343.93	−88.22	1.510752
825	1,163.0	504	−148.28	373.87	−94.94	1.580501
850	1,139.0	484	−173.25	405.73	−101.93	1.651770
875	1,116.8	465	−200.23	439.60	−109.18	1.724503
900	1,096.1	448	−229.30	475.54	−116.68	1.798649
925	1,076.9	433	−260.52	513.65	−124.42	1.874162
950	1,058.9	418	−293.98	554.00	−132.39	1.950999
975	1,042.0	405	−329.77	596.66	−140.58	2.029121
1,000	1,026.0	393	−367.95	641.72	−149.00	2.108491

Range	Path	Drift	3-mph Target Lead	3-mph Mil Dot Lead
0	−1.50	0.00	0"	0.00
100	+22.81	−0.95	6"	Light 1.50
200	+39.82	−4.00	12"	1.50
300	+48.12	−9.43	19"	Heavy 1.50
400	+45.90	−17.70	26"	Light 1.75
500	+30.84	−29.24	35"	1.75
600	−0.04	−44.54	44"	Heavy 1.75

SIERRA BALLISTICS III

Data for: 7.62mm M118LR **Bullet:** 175 grains Match King **BCs:** .505(H), .496(M), .485(L)
Company: Sierra **Temperature:** 59 **Pressure:** 29.53 **Humidity:** 78%
Zero: 600 meters **Crosswind:** 10.00 mph **Tail Wind:** + 0.00 mph
Elevation Angle: 0 degrees **Altitude:** 0 feet **Sight Height:** 1.5 inches

Range (meters)	Velocity (fps)	Energy (ft-lb)	Bullet Path (inches)	Drop (inches)	Drift (inches)	Time of Flight (sec)
0	2,600.0	2,626	−1.50	0.00	0.00	0.000000
25	2,551.6	2,529	0.20	0.19	−0.05	0.031833
50	2,503.8	2,435	1.50	0.77	−0.21	0.064271
75	2,456.1	2,344	2.38	1.77	−0.47	0.097334
100	2,409.0	2,255	2.84	3.20	−0.85	0.131040
125	2,362.5	2,168	2.84	5.08	−1.35	0.165408
150	2,316.5	2,085	2.39	7.42	−1.97	0.200455
175	2,271.0	2,004	1.44	10.25	−2.71	0.236201
200	2,226.1	1,925	0.00	13.58	−3.57	0.272665
225	2,181.8	1,849	−1.97	17.43	−4.57	0.309867
250	2,138.0	1,776	−4.48	21.83	−5.70	0.347829
275	2,094.7	1,705	−7.57	26.79	−6.96	0.386572
300	2,051.9	1,636	−11.24	32.35	−8.37	0.426119
325	2,009.7	1,569	−15.53	38.52	−9.92	0.466494
350	1,968.0	1,505	−20.46	45.34	−11.63	0.507720
375	1,926.8	1,442	−26.06	52.83	−13.49	0.549823
400	1,886.2	1,382	−32.37	61.02	−15.50	0.592830
425	1,846.1	1,324	−39.40	69.93	−17.68	0.636767
450	1,806.5	1,268	−47.19	79.61	−20.03	0.681663
475	1,766.7	1,213	−55.78	90.08	−22.56	0.727555
500	1,727.3	1,159	−65.20	101.39	−25.27	0.774488
525	1,688.5	1,108	−75.49	113.56	−28.16	0.822498
550	1,650.5	1,058	−86.69	126.65	−31.26	0.871613
575	1,613.4	1,011	−98.84	140.68	−34.55	0.921859
600	1,577.2	966	−112.00	155.72	−38.04	0.973260
625	1,541.8	924	−126.19	171.80	−41.74	1.025842
650	1,507.2	883	−141.48	188.97	−45.66	1.079631
675	1,473.4	843	−157.91	207.28	−49.79	1.134655

Data for: 7.62mm M118LR (Continued)						
Range (meters)	Velocity (fps)	Energy (ft-lb)	Bullet Path (inches)	Drop (inches)	Drift (inches)	Time of Flight (sec)
700	1,440.5	806	− 175.53	226.79	− 54.14	1.190941
725	1,408.2	770	− 194.41	247.56	− 58.73	1.248516
750	1,376.9	737	− 214.60	269.63	− 63.54	1.307409
775	1,346.8	705	− 236.15	293.07	− 68.59	1.367634
800	1,318.0	675	− 259.14	317.94	− 73.87	1.429192
825	1,290.3	647	− 283.62	344.31	− 79.38	1.492086
850	1,263.8	620	− 309.67	372.23	− 85.14	1.556318
875	1,238.3	596	− 337.33	401.79	− 91.13	1.621889
900	1,213.8	572	− 366.70	433.03	− 97.35	1.688801
925	1,190.3	550	− 397.82	466.05	− 103.81	1.757055
950	1,168.3	530	− 430.79	500.89	− 110.50	1.826631
975	1,147.8	512	− 465.65	537.64	− 117.42	1.897487
1,000	1,128.7	495	− 502.49	576.36	− 124.56	1.969582

Range	Path	Drift	3-mph Target Lead	3-mph Mil Dot Lead	
0	− 1.50	0.00	0"		0.00
100	+ 2.84	0.85	6"	Light	1.50
200	0.00	− 3.57	12"		1.50
300	− 11.24	− 8.37	19"	Heavy	1.50
400	− 32.37	− 15.50	26"	Light	1.75
500	− 65.20	− 25.27	34"		1.75
600	− 112.00	− 38.04	43"	Heavy	1.75

SIERRA BALLISTICS III

Data for: 5.56mm **Bullet:** 77 grains SPR **BCs:** .372(H), .372(M), .372(L)
Company: Sierra **Temperature:** 59 **Pressure:** 29.53 **Humidity:** 78%
Zero: 200 meters **Crosswind:** 10.00 mph **Tail Wind:** + 0.00 mph
Elevation Angle: 0 degrees **Altitude:** 0 feet **Sight Height:** 2 inches

Range (meters)	Velocity (fps)	Energy (ft-lb)	Bullet Path (inches)	Drop (inches)	Drift (inches)	Time of Flight (sec)
0	2,600.0	1,156	− 2.00	0.00	+ 0.00	0.000000
25	2,535.6	1,099	− 0.18	0.19	− 0.07	0.031929
50	2,471.9	1,045	1.24	0.78	− 0.28	0.064674
75	2,409.0	992	2.24	1.79	− 0.64	0.098268
100	2,347.1	942	2.78	3.26	− 1.15	0.132744
125	2,286.1	893	2.86	5.19	− 1.83	0.168100
150	2,226.1	847	2.44	7.62	− 2.67	0.204500
175	2,167.1	803	1.50	10.58	− 3.69	0.241800
200	2,109.0	760	0.00	14.08	− 4.89	0.280100
225	2,051.9	720	− 2.08	18.18	− 6.27	0.319600
250	1,995.7	681	− 4.78	22.88	− 7.85	0.360100
275	1,940.4	644	− 8.13	28.24	− 9.63	0.401700
300	1,886.1	608	− 12.16	34.29	− 11.62	0.444600
325	1,832.7	574	− 16.93	41.07	− 13.83	0.488700
350	1,780.3	542	− 22.47	48.62	− 16.26	0.534000
375	1,728.8	511	− 28.83	56.99	− 18.93	0.580800
400	1,678.2	481	− 36.06	66.22	− 21.85	0.628900
425	1,629.2	454	− 44.21	76.38	− 25.02	0.678500
450	1,581.6	428	− 53.33	87.52	− 28.46	0.729500
475	1,535.5	403	− 63.50	99.69	− 32.17	0.782200
500	1,490.8	380	− 74.76	112.97	− 36.15	0.836300
525	1,447.4	358	− 87.19	127.41	− 40.42	0.892200
550	1,405.3	338	− 100.87	143.09	− 44.99	0.949600
575	1,364.8	318	− 115.85	160.09	− 49.86	1.008800
600	1,326.4	301	− 132.23	178.48	− 55.03	1.069800
625	1,290.1	285	− 150.09	198.35	− 60.51	1.132470
650	1,255.7	270	− 169.50	219.77	− 66.30	1.196890
675	1,223.0	256	− 190.56	242.84	− 72.39	1.263070

Data for: 5.56mm (Continued)						
Range (meters)	Velocity (fps)	Energy (ft-lb)	Bullet Path (inches)	Drop (inches)	Drift (inches)	Time of Flight (sec)
700	1,192.0	243	− 213.36	267.65	− 78.79	1.33099
725	1,163.5	231	− 237.98	294.29	− 85.50	1.40064
750	1,137.4	221	− 264.52	322.84	− 92.50	1.47194
775	1,113.4	212	− 293.07	353.40	− 99.77	1.54484
800	1,091.3	204	− 323.72	386.05	− 107.32	1.61926
825	1,070.7	196	− 356.55	420.89	− 115.13	1.69517
850	1,051.6	189	− 391.64	458.00	− 123.18	1.77250
875	1,033.7	183	− 429.08	497.45	− 131.49	1.85121
900	1,016.9	177	− 468.96	539.34	− 140.02	1.93126
925	1,001.1	171	− 511.35	583.74	− 148.79	2.01261
950	986.2	166	− 556.34	630.74	− 157.78	2.09523
975	972.1	162	− 604.00	680.42	− 166.98	2.17908
1,000	958.8	157	− 654.42	732.85	− 176.40	2.26413

Range	Path	Drift
0	− 2.00	+ 0.00
100	2.78	− 1.15
200	0.00	− 4.89
300	− 12.16	− 11.62
400	− 36.06	− 21.85
500	− 74.76	− 36.15

Environmental Conditions

Actual barometric pressure at firing site .... 29.53 inches
Actual speed of sound at firing site .... 1,121 fps
Effective ballistic coefficient at firing site 0.372

Animal Lead Calculations

Average lead for a 3-mph walking target at 100 meters is .... 7 inches
Average lead for a running deer at 100 meters is .... 3 feet
Average lead for a running elk at 100 meters is .... 5 feet
Average lead for a running antelope at 100 meters is .... 8 feet

SIERRA BALLISTICS III						
Data for: 300 Win Mag **Bullet:** 190 grains Match King **BCs:** .533(H), .525(M), .515(L) **Company:** Sierra **Temperature:** 59 **Pressure:** 29.53 **Humidity:** 78% **Zero:** 200 meters **Crosswind:** 10.00 mph **Tail Wind:** + 0.00 mph **Elevation Angle:** 0 degrees **Altitude:** 0 feet **Sight Height:** 1.5 inches						
Range (meters)	Velocity (fps)	Energy (ft-lb)	Bullet Path (inches)	Drop (inches)	Drift (inches)	Time of Flight (sec)
0	2,900.0	3,547	−1.50	0.00	0.00	0.0000
25	2,855.7	3,440	−0.32	0.13	−0.03	0.0261
50	2,811.9	3,335	0.59	0.52	−0.14	0.0525
75	2,768.5	3,233	1.23	1.18	−0.32	0.0794
100	2,725.5	3,133	1.59	2.13	−0.57	0.1067
125	2,683.0	3,036	1.66	3.37	−0.90	0.1344
150	2,640.9	2,942	1.42	4.92	−1.30	0.1626
175	2,599.2	2,850	0.87	6.77	−1.79	0.1912
200	2,558.0	2,760	0.00	8.95	−2.35	0.2203
225	2,517.2	2,673	−1.20	11.45	−3.00	0.2498
250	2,476.6	2,587	−2.75	14.30	−3.74	0.2798
275	2,436.3	2,504	−4.64	17.51	−4.56	0.3104
300	2,396.4	2,422	−6.91	21.07	−5.46	0.3414
325	2,356.9	2,343	−9.55	25.02	−6.47	0.3729
350	2,317.8	2,266	−12.58	29.36	−7.56	0.4050
375	2,279.1	2,191	−16.02	34.10	−8.75	0.4376
400	2,240.8	2,118	−19.87	39.26	−10.04	0.4708
425	2,202.8	2,047	−24.16	44.85	−11.42	0.5046
450	2,165.3	1,978	−28.89	50.89	−12.91	0.5389
475	2,128.1	1,910	−34.09	57.40	−14.51	0.5738
500	2,091.3	1,845	−39.77	64.38	−16.21	0.6094
525	2,054.3	1,780	−45.94	71.85	−18.03	0.6455
550	2,017.8	1,717	−52.62	79.85	−19.96	0.6824
575	1,981.7	1,656	−59.84	88.37	−22.00	0.7199
600	1,945.9	1,597	−67.62	97.45	−24.17	0.7580
625	1,910.6	1,540	−75.96	107.10	−26.46	0.7969
650	1,875.7	1,484	−84.90	117.35	−28.88	0.8365
675	1,841.1	1,430	−94.46	128.21	−31.43	0.8769

Data for: 300 Win Mag (Continued)						
Range (meters)	Velocity (fps)	Energy (ft-lb)	Bullet Path (inches)	Drop (inches)	Drift (inches)	Time of Flight (sec)
700	1,806.9	1,377	−104.66	139.71	−34.12	0.9180
725	1,773.1	1,326	−115.52	151.88	−36.94	0.9599
750	1,739.7	1,277	−127.07	164.74	−39.90	1.0026
775	1,706.7	1,229	−139.34	178.32	−43.01	1.0461
800	1,674.2	1,182	−152.36	192.64	−46.26	1.0904
825	1,642.4	1,138	−166.15	207.73	−49.67	1.1356
850	1,611.2	1,095	−180.75	223.64	−53.23	1.1817
875	1,580.2	1,053	−196.18	240.37	−56.95	1.2287
900	1,549.6	1,013	−212.48	257.98	−60.83	1.2767
925	1,519.6	974	−229.68	276.49	−64.88	1.3255
950	1,490.3	937	−247.83	295.94	−69.10	1.3753
975	1,461.5	901	−266.95	316.37	−73.49	1.4262
1,000	1,433.3	867	−287.09	337.81	−78.05	1.477976

Range	Path	Drift
0	−1.50	−0.03
100	1.59	−0.57
200	0.00	−2.35
300	−6.91	−5.46
400	−19.87	−10.04
500	−39.77	−16.21

Environmental Conditions

Actual barometric pressure at firing site.................................29.53 inches
Actual speed of sound at firing site ....1,121 fps
Effective ballistic coefficient at firing site................................0.530

Animal Lead Calculations

Average lead for a 3-mph walking target at 100 meters....5 1/2 inches

Average lead for a running deer at 100 meters is....................3 feet
Average lead for a running elk at 100 meters is ....5 feet
Average lead for a running antelope at 100 meters is ....8 feet

SIERRA BALLISTICS III

Data for: .338 Lapua **Weight:** 250 grains **BCs:** .675(H), .675(M), .675(L)
Company: Sierra **Temperature:** 59 **Pressure:** 29.53 **Humidity:** 78%
Zero: 200 meters **Crosswind:** 10.00 mph **Tail Wind:** + 0.00 mph
Elevation Angle: 0 degrees **Altitude:** 0 feet **Sight Height:** 2 inches

Range (meters)	Velocity (fps)	Energy (ft-lb)	Bullet Path (inches)	Drop (inches)	Drift (inches)	Time of Flight (sec)
0	2,750.0	4,197	−2.00	0.00	0.00	0.000000
50	2,676.4	3,976	0.76	0.67	−0.14	0.060434
100	2,604.1	3,764	2.06	2.79	−0.57	0.122538
150	2,533.0	3,561	1.84	6.44	−1.31	0.186376
200	2,463.0	3,367	0.00	11.71	−2.36	0.252013
250	2,393.8	3,180	−3.55	18.69	−3.74	0.319534
300	2,325.9	3,002	−8.92	27.49	−5.47	0.389016
350	2,259.1	2,833	−16.20	38.20	−7.56	0.460539
400	2,193.5	2,670	−25.52	50.94	−10.03	0.534189
450	2,129.1	2,516	−36.99	65.85	−12.88	0.610054
500	2,065.8	2,369	−50.76	83.04	−16.14	0.688230
550	2,003.7	2,228	−66.96	102.66	−19.83	0.768814
600	1,942.7	2,095	−85.74	124.88	−23.95	0.851912
650	1,882.9	1,968	−107.27	149.84	−28.54	0.937636
700	1,824.2	1,847	−131.74	177.73	−33.61	1.026101
750	1,766.6	1,732	−159.32	208.74	−39.19	1.117432
800	1,710.2	1,623	−190.23	243.08	−45.29	1.211761
850	1,655.2	1,520	−224.69	280.97	−51.95	1.309221
900	1,602.0	1,424	−262.94	322.64	−59.17	1.409921
950	1,550.6	1,334	−305.24	368.37	−66.98	1.513966
1,000	1,500.9	1,250	−351.85	418.41	−75.40	1.621464
1,050	1,452.9	1,172	−403.08	473.07	−84.45	1.732525
1,100	1,406.4	1,098	−459.23	532.64	−94.15	1.847264
1,150	1,361.9	1,029	−520.63	597.47	−104.51	1.965791

Data for: .338 Lapua (Continued)						
Range (meters)	Velocity (fps)	Energy (ft-lb)	Bullet Path (inches)	Drop (inches)	Drift (inches)	Time of Flight (sec)
1,200	1,320.0	967	– 587.63	667.90	– 115.55	2.088153
1,250	1,280.5	910	– 660.60	744.30	– 127.26	2.214358
1,300	1,243.3	858	– 739.90	827.03	– 139.65	2.344413
1,350	1,208.2	810	– 825.93	916.49	– 152.72	2.478326
1,400	1,175.4	767	– 919.09	1,013.08	– 166.47	2.616090
1,450	1,145.9	729	– 1,019.78	1,117.20	– 180.87	2.757582
1,500	1,118.9	695	– 1,128.39	1,229.24	– 195.90	2.902635
1,550	1,094.3	665	– 1,245.33	1,349.61	– 211.53	3.051101
1,600	1,071.7	637	– 1,370.96	1,478.67	– 227.74	3.202849
1,650	1,050.8	613	– 1,505.67	1,616.82	– 244.51	3.357759
1,700	1,031.4	590	– 1,649.84	1,764.42	– 261.81	3.515722
1,750	1,013.3	570	– 1,803.81	1,921.83	– 279.63	3.676642
1,800	996.5	551	– 1,967.97	2,089.42	– 297.96	3.840426
1,850	980.7	534	– 2,142.66	2,267.55	– 316.78	4.006993
1,900	965.8	518	– 2,328.24	2,456.57	– 336.07	4.176266
1,950	951.9	503	– 2,525.06	2,656.83	– 355.83	4.348176
2,000	938.7	489	– 2,733.46	2,868.67	– 376.04	4.522656

Range	Path	Drift
0	– 2.00	0.00
500	– 50.76	– 16.14
750	– 159.32	– 39.19
1,000	– 351.85	– 75.40
1,500	– 1128.39	– 195.90
2,000	– 2733.46	– 376.04

Average lead for a 3-mph moving target

350 meters	2 feet	1.75 mils
500 meters	3 feet	Heavy 1.75 mils
750 meters	5 feet	2.00 mils
1,000 meters	7 feet	Light 2.25 mils
1,200 meters	9 feet	Heavy 2.25 mils
1,450 meters	12 feet	Heavy 2.50 mils
1,600 meters	14 feet	Light 2.75 mils
1,850 meters	17 feet	Heavy 2.75 mils
2,000 meters	20 feet	3.00 mils

SIERRA BALLISTICS III

Data for: .338 Lapua **Weight:** 300 grains **BCs:** .768(H), .76(M), .75(L)
Company: Sierra **Temperature:** 59 **Pressure:** 29.53 **Humidity:** 78%
Zero: 200 meters **Crosswind:** 10.00 mph **Tail Wind:** + 0.00 mph
Elevation Angle: 0 degrees **Altitude:** 0 feet **Sight Height:** 2 inches

Range (meters)	Velocity (fps)	Energy (ft-lb)	Bullet Path (inches)	Drop (inches)	Drift (inches)	Time of Flight (sec)
0	2,750.0	5,037	−2.00	0.00	0.00	0.000000
50	2,685.2	4,802	0.73	0.67	−0.12	0.060338
100	2,621.5	4,577	2.02	2.78	−0.50	0.122138
150	2,558.7	4,360	1.80	6.40	−1.14	0.185447
200	2,496.9	4,152	0.00	11.61	−2.06	0.250316
250	2,435.6	3,951	−3.47	18.48	−3.26	0.316804
300	2,375.1	3,757	−8.69	27.10	−4.76	0.384974
350	2,315.6	3,571	−15.75	37.56	−6.57	0.454888
400	2,256.6	3,392	−24.75	49.96	−8.69	0.526611
450	2,198.3	3,219	−35.78	64.40	−11.15	0.600225
500	2,141.0	3,053	−48.97	80.98	−13.95	0.675801
550	2,084.6	2,894	−64.42	99.83	−17.11	0.753413
600	2,029.0	2,742	−82.25	121.07	−20.65	0.833137
650	1,974.4	2,596	−102.61	144.83	−24.57	0.915056
700	1,920.7	2,457	−125.64	171.26	−28.89	0.999254
750	1,867.9	2,324	−151.47	200.49	−33.62	1.085820
800	1,816.0	2,196	−180.29	232.71	−38.79	1.174850
850	1,764.5	2,074	−212.25	268.07	−44.42	1.266449
900	1,713.7	1,956	−247.55	306.77	−50.52	1.360749
950	1,664.0	1,844	−286.38	349.01	−57.11	1.457860
1,000	1,615.8	1,739	−328.97	394.99	−64.21	1.557873
1,050	1,569.1	1,640	−375.53	444.96	−71.84	1.660872
1,100	1,523.7	1,546	−426.31	499.14	−80.01	1.766942
1,150	1,479.8	1,458	−481.57	557.80	−88.74	1.876174

| Data for: .338 Lapua (Continued) ||||||||
Range (meters)	Velocity (fps)	Energy (ft-lb)	Bullet Path (inches)	Drop (inches)	Drift (inches)	Time of Flight (sec)
1,200	1,437.1	1,376	– 541.57	621.20	– 98.03	1.988657
1,250	1,395.7	1,297	– 606.60	689.64	– 107.92	2.104487
1,300	1,356.1	1,225	– 676.97	763.40	– 118.41	2.223746
1,350	1,318.6	1,158	– 752.99	842.83	– 129.51	2.346457
1,400	1,283.1	1,096	– 834.99	928.23	– 141.22	2.472628
1,450	1,249.4	1,040	– 923.30	1,019.94	– 153.54	2.602265
1,500	1,217.4	987	– 1,018.28	1,118.33	– 166.46	2.735374
1,550	1,187.2	939	– 1,120.28	1,223.73	– 180.01	2.871960
1,600	1,159.5	895	– 1,229.67	1,336.52	– 194.15	3.011956
1,650	1,134.1	857	– 1,346.80	1,457.06	– 208.86	3.155227
1,700	1,110.7	822	– 1,472.04	1,585.70	– 224.13	3.301647
1,750	1,089.1	790	– 1,605.72	1,722.79	– 239.94	3.451100
1,800	1,069.1	761	– 1,748.21	1,868.68	– 256.26	3.603483
1,850	1,050.5	735	– 1,899.82	2,023.70	– 273.08	3.758703
1,900	1,033.0	711	– 2,060.91	2,188.20	– 290.38	3.916673
1,950	1,016.7	688	– 2,231.79	2,362.49	– 308.16	4.077313
2,000	1,001.4	668	– 2,412.80	2,546.91	– 326.39	4.240551

Range	Path	Drift
0	– 2.00	0.00
500	– 48.97	– 13.95
750	– 151.47	– 33.62
1,000	– 328.97	– 64.21
1,500	– 1,018.28	– 166.46
2,000	– 2,412.80	– 326.39

Average lead for a 3-mph moving target

350 meters	2 feet	1.75 mils
500 meters	3 feet	Heavy 1.75 mils
800 meters	5 feet	Light 2.00 mils
1,000 meters	7 feet	Light 2.25 mils
1,250 meters	9 feet	2.25 mils
1,500 meters	12 feet	Light 2.50 mils
1,750 meters	15 feet	Heavy 2.50 mils
1,900 meters	17 feet	2.75 mils
2,000 meters	18.5 feet	Heavy 2.74 mils

SIERRA BALLISTICS III

Data for: .50 cal MK 211 **Bullet Weight:** 671 grains **BCs:** .701(H), .701(M), .701(L)
Company: Sierra **Temperature:** 59 **Pressure:** 29.53 **Humidity:** 78%
Zero: 500 meters **Crosswind:** −10.00 mph **Tail Wind:** + 0.00 mph
Elevation Angle: 0 degrees **Altitude:** 0 feet **Sight Height:** 3.25 inches

Range (meters)	Velocity (fps)	Energy (ft-lb)	Bullet Path (inches)	Drop (inches)	Drift (inches)	Time of Flight (sec)
0	2,740.0	11,184	−3.25	0.00	0.00	0.00000
50	2,669.4	10,615	4.70	0.68	−0.13	0.06063
100	2,600.0	10,070	11.19	2.81	−0.55	0.12286
150	2,531.7	9,548	16.14	6.48	−1.26	0.18677
200	2,464.4	9,047	19.48	11.76	−2.27	0.25241
250	2,397.8	8,565	21.11	18.76	−3.61	0.31985
300	2,332.4	8,104	20.93	27.57	−5.27	0.38918
350	2,268.1	7,663	18.84	38.28	−7.28	0.46046
400	2,204.9	7,242	14.73	51.01	−9.65	0.53377
450	2,142.7	6,839	8.49	65.87	−12.39	0.60920
500	2,081.6	6,455	−0.01	83.00	−15.51	0.68683
550	2,021.6	6,088	−10.91	102.52	−19.04	0.76675
600	1,962.7	5,738	−24.34	124.58	−22.99	0.84906
650	1,904.8	5,405	−40.48	149.33	−27.38	0.93386
700	1,847.9	5,087	−59.47	176.95	−32.22	1.02124
750	1,792.1	4,784	−81.51	207.61	−37.54	1.11134
800	1,737.4	4,496	−106.77	241.50	−43.36	1.20426
850	1,683.7	4,223	−135.48	278.83	−49.69	1.30012
900	1,631.6	3,966	−167.85	319.82	−56.57	1.39905
950	1,581.3	3,725	−204.12	364.71	−63.99	1.50114
1,000	1,532.5	3,499	−244.54	413.76	−72.00	1.60648
1,050	1,485.3	3,286	−289.39	467.23	−80.59	1.71517
1,100	1,439.6	3,087	−338.94	525.40	−89.79	1.82732
1,150	1,395.4	2,901	−393.50	588.59	−99.62	1.94304
1,200	1,353.2	2,728	−453.40	657.11	−110.10	2.06242
1,250	1,313.4	2,570	−518.97	731.31	−121.22	2.18548

Data for: .50 cal MK 211 (Continued)						
Range (meters)	Velocity (fps)	Energy (ft-lb)	Bullet Path (inches)	Drop (inches)	Drift (inches)	Time of Flight (sec)
1,300	1,275.9	2,425	− 590.57	811.53	− 132.99	2.31222
1,350	1,240.4	2,292	− 668.55	898.14	− 145.41	2.44267
1,400	1,206.8	2,169	− 753.29	991.51	− 158.48	2.57682
1,450	1,175.4	2,058	− 845.18	1,092.02	− 172.20	2.71466
1,500	1,147.0	1,960	− 944.59	1,200.06	− 186.56	2.85608
1,550	1,121.0	1,872	− 1,051.90	1,316.01	− 201.51	3.00093
1,600	1,097.2	1,793	− 1,167.50	1,440.25	− 217.05	3.14906
1,650	1,075.2	1,722	− 1,291.80	1,573.15	− 233.14	3.30036
1,700	1,054.9	1,658	− 1,425.10	1,715.07	− 249.77	3.45471
1,750	1,035.9	1,599	− 1,567.70	1,866.38	− 266.92	3.61201
1,800	1,018.3	1,545	− 1,720.10	2,027.41	− 284.57	3.77217
1,850	1,001.8	1,495	− 1,882.60	2,198.52	− 302.71	3.93511
1,900	986.3	1,449	− 2,055.50	2,380.06	− 321.32	4.10074
1,950	971.8	1,407	− 2,239.20	2,572.37	− 340.40	4.26900
2,000	958.1	1,367	− 2,433.90	2,775.78	− 359.92	4.43981

Range	Path	Drift
0	− 3.25	0.00
500	− 0.01	− 15.51
750	− 81.51	− 37.54
1,000	− 244.54	− 72.00
1,500	− 944.59	− 186.56
2,000	− 2,433.90	− 359.92

Environmental Conditions

Actual barometric pressure at firing site.................................29.53 inches
Actual speed of sound at firing site..1,121 fps
Effective ballistic coefficient at firing site................................0.701

Average lead for a 3-mph moving target

350 meters	2 feet	1.75 mils
500 meters	3 feet	Heavy 1.75 mils
700 meters	4.5 feet	2.00 mils
1,000 meters	7 feet	Heavy 2.00 mils
1,200 meters	9 feet	Heavy 2.25 mils
1,450 meters	12 feet	2.50 mils
1,600 meters	14 feet	Heavy 2.50 mils
1,850 meters	17 feet	Heavy 2.75 mils
2,000 meters	19.5 feet	3.00 mils

SIERRA BALLISTICS III

Data for: .50 cal M8 API **Bullet Weight:** 622.5 grains **BCs:** .701(H), .701(M), .701(L)
Company: Sierra **Temperature:** 60 **Pressure:** 29.53 **Humidity:** 78%
Zero: 500 meters **Crosswind:** −10.00 mph **Tail Wind:** + 0.00 mph
Elevation Angle: 0 degrees **Altitude:** 0 feet **Sight Height:** 3.25 inches

Range (meters)	Velocity (fps)	Energy (ft-lb)	Bullet Path (inches)	Drop (inches)	Drift (inches)	Time of Flight (sec)
0	2,910.0	11,703	−3.25	0.00	0.00	0.0000
50	2,836.5	11,119	3.80	0.60	−0.12	0.0571
100	2,764.3	10,560	9.56	2.49	−0.51	0.1156
150	2,693.3	10,025	13.96	5.74	−1.16	0.1757
200	2,623.4	9,512	16.93	10.41	−2.10	0.2374
250	2,554.8	9,020	18.39	16.60	−3.32	0.3007
300	2,487.3	8,550	18.26	24.38	−4.85	0.3658
350	2,420.4	8,096	16.46	33.84	−6.68	0.4326
400	2,354.6	7,662	12.88	45.06	−8.85	0.5013
450	2,289.9	7,247	7.42	58.16	−11.36	0.5719
500	2,226.3	6,850	−0.010	73.24	−14.21	0.6445
550	2,163.8	6,470	−9.530	90.42	−17.44	0.7192
600	2,102.4	6,108	−21.28	109.81	−21.05	0.7961
650	2,042.0	5,762	−35.37	131.55	−25.05	0.8752
700	1,982.7	5,433	−51.95	155.78	−29.47	0.9567
750	1,924.4	5,118	−71.17	182.65	−34.32	1.0406
800	1,867.2	4,818	−93.20	212.33	−39.62	1.1271
850	1,811.1	4,533	−118.20	244.98	−45.39	1.2163
900	1,755.9	4,261	−146.37	280.80	−51.66	1.3082
950	1,701.9	4,003	−177.91	319.98	−58.43	1.4030
1,000	1,649.2	3,759	−213.03	362.75	−65.73	1.5009
1,050	1,598.3	3,530	−251.97	409.34	−73.59	1.6019
1,100	1,549.0	3,316	−294.98	460.00	−82.01	1.7061
1,150	1,501.3	3,115	−342.31	514.98	−91.01	1.8137
1,200	1,455.1	2,926	−394.25	574.57	−100.62	1.9246

FM 3-05.222

Data for: .50 cal M8 API (Continued)						
Range (meters)	Velocity (fps)	Energy (ft-lb)	Bullet Path (inches)	Drop (inches)	Drift (inches)	Time of Flight (sec)
1,250	1,410.4	2,749	−451.10	639.06	−110.85	2.03914
1,300	1,367.4	2,584	−513.17	708.78	−121.72	2.15726
1,350	1,326.8	2,433	−580.80	784.06	−133.23	2.27907
1,400	1,288.5	2,294	−654.33	865.24	−145.40	2.40456
1,450	1,252.3	2,167	−734.13	952.69	−158.22	2.53375
1,500	1,218.1	2,050	−820.55	1,046.76	−171.68	2.66664
1,550	1,185.8	1,943	−913.98	1,147.85	−185.80	2.80323
1,600	1,156.4	1,848	−1,014.80	1,256.33	−200.56	2.94346
1,650	1,129.7	1,764	−1,123.40	1,372.60	−215.93	3.08716
1,700	1,105.1	1,688	−1,240.20	1,497.03	−231.89	3.23420
1,750	1,082.6	1,620	−1,365.50	1,630.00	−248.41	3.38444
1,800	1,061.7	1,558	−1,499.70	1,771.86	−265.47	3.53777
1,850	1,042.3	1,501	−1,643.20	1,922.99	−283.06	3.69409
1,900	1,024.3	1,450	−1,796.30	2,083.72	−301.16	3.85330
1,950	1,007.4	1,403	−1,959.30	2,254.43	−319.76	4.01531
2,000	991.6	1,359	−2,132.60	2,435.44	−338.83	4.18005

Range	Path	Drift
0	−3.25	0.00
500	−0.01	−15.51
750	−81.51	−37.54
1,000	−244.54	−72.00
1,500	−944.59	−186.56
2,000	−2,433.90	−359.92

Environmental Conditions

Actual barometric pressure at firing site..................................29.53 inches
Actual speed of sound at firing site ....1,121 fps
Effective ballistic coefficient at firing site.................................0.701

Average lead for a 3-mph moving target

350 meters	2 feet	Heavy 1.50 mils
500 meters	3 feet	1.75 mils
700 meters	4 feet	Heavy 1.75 mils
1,000 meters	6.5 feet	2.00 mils
1,300 meters	9.5 feet	2.25 mils
1,500 meters	11.5 feet	Heavy 2.25 mils
1,600 meters	13 feet	2.25 mils
1,800 meters	15.5 feet	Heavy 2.25 mils
950 meters	7.5 feet	.75 mils

Appendix I
Sniper Training Exercises

In all training, trainers stress practical exercises whenever possible. Snipers must achieve certain standards and perform remedial training as required. They must constantly strive to improve their performance to the point that basic skills become instinctive. To maintain this level of proficiency, the snipers periodically conduct the exercises listed in this appendix.

STALKING

I-1. The purpose of stalking exercises is to give the sniper confidence in his ability to approach and occupy a firing position without being observed.

DESCRIPTION

I-2. Having studied a map (and aerial photograph if available), an individual sniper must stalk for a predesignated distance. It could be 1,000 meters or more depending on the area selected. All stalking exercises and tests should be approximately 1,000 meters with a 3-hour time limit. The sniper must stalk to within 200 meters of two trained observers who are scanning the area with binoculars and fire two blanks without being detected.

RECONNAISSANCE BY THE CONDUCTING OFFICER OR NCO

I-3. The area used for a stalking exercise must be chosen with great care. An area in which a sniper must do the low crawl for the complete distance would be unsuitable. The trainer should consider the following:

- As much of the area as possible should be visible to the observer. This level of visibility forces the sniper to use the ground properly, even when far from the observer's location. The stalk lanes should also vary in terrain to give the maximum variations to the sniper, without being a 1,000-meter low crawl.

- Where possible, available cover should decrease as the sniper nears the observer's position. This effect will enable him to take chances early in the stalk and force him to move more carefully as he closes in on his firing position.

- The sniper must start the stalk in an area out-of-sight of the observer.

- The trainer must establish boundaries by means of natural features or the use of markers.

CONDUCT OF THE EXERCISE

I-4. In a location near the jump-off point for the stalk, the sniper receives a brief on the following:

- Aim of the exercise.
- Boundaries.
- Time limit (usually 3 hours).
- Standards to be achieved.

I-5. When the sniper reaches his final firing position, which is closer than 220 meters for the individual stalk and 330 meters for a team stalk, of the observer, he will fire a blank round at an observer. This shot will tell the walker that he is ready to continue the rest of the exercise. The walker will then move to within 3 meters of the sniper. The observer will search a 3-meter radius around the walker for the sniper. If the sniper is undetected, the observer will expose a 6-inch by 6-inch plaque, held directly above or below the observer's binoculars. The sniper will have 30 seconds to correctly identify the letter or number on the plaque. The sniper must remain undetected and the observer will direct the walker to have the sniper fire his second blank round. The observer will look for indicators such as muzzle blast caused by the blank. Use caution so that the muzzle flash caused by the blank round is not confused with the blast of vegetation from a poorly prepared position. If the sniper remains undetected, the walker will then move in and place his hand on the sniper's head. The sniper must then tell the walker his exact range, wind velocity, and windage applied to the scope.

STANDARDS

I-6. If the sniper completes all of these steps correctly, he has passed the stalk exercise. The trainer conducts a critique at the conclusion of the exercise, touching on main problem areas.

CREATING INTEREST

I-7. To create interest and to give the snipers practice in observation and stalking skills, one-half of the class may be positioned to observe the conduct of the stalk. Seeing an error made is an effective way of teaching better stalking skills. When a sniper is caught, he should be sent to the OP to observe the exercise.

RANGE ESTIMATION

I-8. Range estimation exercises are to make the sniper proficient in accurately judging distance.

DESCRIPTION

I-9. The sniper arrives at the OP. The trainer shows him different objects over distances of up to 800 meters. After time for consideration, the sniper writes down the estimated distance to each object. He may use only his binoculars and rifle telescope as aids. He must estimate to within 10 percent of the correct range.

FM 3-05.222

RECONNAISSANCE BY THE CONDUCTING OFFICER OR NCO

I-10. Each exercise must take place in a different area and offer a variety of terrain. The exercise areas should include dead space as well as places where the sniper will be observing uphill or downhill. The trainer should select extra objects in case those originally chosen cannot be seen due to weather conditions or other reasons.

CONDUCT OF THE EXERCISE

I-11. The sniper arrives at the OP, obtains a record card, and receives a review on methods of judging distances and causes of miscalculation. The trainer then briefs him on the following:

- Aim of the exercise.
- Reference points.
- Time limit per object.
- Standard to be achieved.

I-12. The trainer indicates the first object to the sniper. The sniper is allowed 3 minutes to estimate the distance and write it down. He repeats the sequence for a total of eight objects. The trainer collects the card and gives the correct range to each object. He points out in each case why the distance might be underestimated or overestimated. After correction, the card is given back to the sniper. This way, the sniper retains a record of his performance.

STANDARDS

I-13. The sniper fails if he estimates four or more distances incorrectly out of 10 distances.

OBSERVATION

I-14. Observation exercises allow the sniper to practice improving his ability to observe an enemy. They also teach him to accurately record the results of his observations.

DESCRIPTION

I-15. The trainer assigns the sniper an arc of about 1,800 mils to observe identifying the left and right limits.. The first 20 minutes is spent drawing a panoramic sketch. He plots any objects that appear to be out of place. Objects are so positioned as to be invisible to the naked eye, indistinguishable when using binoculars, but recognizable when using the spotting telescope.

RECONNAISSANCE BY THE CONDUCTING OFFICER OR NCO

I-16. When choosing the location for the exercise, the trainer should consider the following points:

- Number of objects in the arc.
- Time limits.

- Equipment that is allowed to be used (binoculars and spotting telescopes).
- Standard to be attained.

I-17. The sniper takes up the prone position on the observation line and spends 20 minutes drawing a panoramic sketch of the area. The staff is available to answer questions about the area if the sniper is unclear. He should focus on one-half of the area for the first 20 minutes and then shift attention to the other half. (This method ensures that he sees all the ground in the arc.) At the end of 40 minutes, the trainer collects the sniper's sheet and shows him the location of each object. This critique is best done by the sniper staying in his position and watching while a member of the staff points out each object. This way, the sniper will see why he failed to find an object, even though it was visible. (A sniper should view first with binoculars and then with spotting telescopes before the trainer picks up the item.)

I-18. The trainer holds a critique session and brings out the main points, noting why the object should have been seen.

SCORING

I-19. The sniper receives half a point for each object correctly plotted and another half a point for naming the object correctly.

STANDARDS

I-20. The sniper fails if he scores fewer than 8 points out of 12 points (12 disguised military objects).

HIDE CONSTRUCTION

I-21. The intent of this exercise is to show the sniper how to build a hide and remain undetected while the area is under observation. The purpose of a hide is to camouflage a sniper or sniper team that is not in movement.

DESCRIPTION

I-22. The trainer gives the sniper 8 hours to build a temporary hide large enough to hold a sniper team with all its necessary equipment.

RECONNAISSANCE BY THE CONDUCTING OFFICER OR NCO

I-23. The hide exercise area should be selected with great care. It can be in any type of terrain, but there should be more than enough prospective spots in which to build a hide. The area should be easily bounded by left and right, far and near limits. If designated properly, the sniper should be able to easily and quickly identify these points. There should be enough tools (for example, axes, picks, shovels, and sandbags) available to accommodate the sniper's entire time. There must also be sufficient rations and water available to the sniper to last the entire exercise, which is about 9 1/2 hours total—8 hours of instruction and 1 1/2 hours of testing.

CONDUCT OF THE EXERCISE

I-24. The sniper receives a shovel, ax, pickax, and approximately 20 sandbags. He is taken to the area and briefed on the purpose of the exercise, time limit for construction, and area limits. The sniper then begins construction of the hide.

NOTE: During the construction, a trainer should be present at all times to act as an advisor.

I-25. At the end of 8 hours, the trainer checks the sniper's hide to ensure it is complete. An infantry officer is brought out to act as an observer. He is placed in an area 300 yards from the hide area, where he starts his observation with binoculars and a 20x M49 spotting scope. The observer, after failing to find a hide, is brought forward 150 yards and again commences observation.

I-26. A trainer in the field (walker with radio) then moves to within 10 yards of a hide and informs the observer. The observer then tells the walker to have the sniper in the hide load and fire his only round (blank). If the sniper's muzzle blast is seen, or if the hide is seen due to improper construction, the sniper fails but remains in the hide. These procedures are repeated for all the sniper teams. The observer is then brought down to within 25 yards of each hide to determine whether the sniper can be seen with the naked eye at that distance. The observer is not shown the hide. He must find it. If the sniper is located at 25 yards, he fails and is allowed to come out and see his discrepancies. If he is not seen, he passes.

OTHER REQUIREMENTS

I-27. The sniper should also fill out a range card and a sniper's logbook and make a field sketch. One way of helping him is to have a trainer show "flash cards" from 150 yards away, beginning when the observer arrives and ending when the observer moves to within 25 yards. The sniper should record everything he sees on the flash cards and anything going on at the OP during the exercise.

STANDARDS

I-28. The sniper must pass all phases to pass the exercise. All range cards, logbooks, and field sketches must be turned in for grading; the trainer makes a final determination of pass or fail.

CAMOUFLAGE AND CONCEALMENT

I-29. Camouflage and concealment exercises help the sniper select final firing positions.

DESCRIPTION

I-30. The sniper conceals himself within 200 yards of an observer. The observer uses binoculars to try to find the sniper. The sniper must be able to fire blank ammunition at the observer without being seen and have the correct elevation and windage on his sight. The sniper must remain unseen throughout the conduct of the exercise.

RECONNAISSANCE BY THE CONDUCTING OFFICER OR NCO

I-31. In choosing the location for the exercise, the trainer ensures that certain conditions are met. They include the following:

- There must be adequate space to ensure snipers are not crowded together in the area. There should be at least twice the number of potential positions as there are snipers. Once the area has been established, the limits should be marked in some manner (for example, flags, trees, and prominent features). Snipers should then be allowed to choose any position within the limits for their final firing position.

- The observer must be able to see the entire problem area.

- As there will be several concealment exercises throughout the sniper course, different types of terrain should be chosen for the sniper to practice concealment in varied conditions. For instance, one exercise could take place in a fairly open area, one along a wood line, one in shrubs, and another in hilly or rough terrain.

CONDUCT OF THE EXERCISE

I-32. The trainer assigns the sniper a specified area with boundaries in which to conceal himself properly. The observer turns his back to the area and allows the sniper 5 minutes to conceal himself. At the end of 5 minutes, the observer turns and commences observation in his search for the concealed sniper. This observation should last approximately one-half hour (more time may be allotted at the discretion of the trainer). At the conclusion of observation, the observer instructs, by radio, one of the two observers (walkers) in the field to move to within 10 meters of the sniper. The sniper is given one blank. If he cannot be seen after the walker moves within the 10 meters, the walker will tell him to load and fire his blank. The observer is looking for muzzle blast, vegetation flying after the shot, and movement by the sniper before and after he fires. If the sniper cannot be seen, the walker then extends his arm in the direction of the sniper, indicating his position. If the sniper remains unseen after indication, the walker goes to the sniper's position and places his hand, palm facing the observer, directly on top of the sniper's head. If the sniper passes all of the above, he must then state his elevation, windage, and what type of movement the observer is making. The sniper must also identify a letter or number plaque held by the observer.

CREATING INTEREST

I-33. To create interest and give the sniper practice in observation, one-half of the class may be positioned with the observer so the other half of the class can profit from the mistakes. When a sniper fails the exercise, he should go to the OP to observe.

Appendix J

Range Estimation Table

Table J-1. Mils for Objects

Target Height (Mils)	6 Feet (1.8 M)	5 Feet, 9 Inches (1.75 M)	5 Feet, 6 Inches (1.7 M)	39 Inches (1 M)	19 Inches (0.5 M)
6.0	300	292	283	167	83
5.9	305	297	288	169	85
5.8	310	302	293	172	86
5.7	316	307	298	175	88
5.6	321	313	304	179	89
5.5	327	318	309	182	91
5.4	333	324	315	185	93
5.3	340	330	321	189	94
5.2	346	337	327	192	96
5.1	353	343	333	196	98
5.0	360	350	340	200	100
4.9	367	357	347	204	102
4.8	375	365	354	208	104
4.7	383	372	362	213	106
4.6	391	380	370	217	109
4.5	400	389	378	222	111
4.4	409	398	386	227	114
4.3	419	407	395	233	116
4.2	429	417	405	238	119
4.1	439	427	415	244	122
4.0	450	438	425	250	125
3.9	462	449	436	256	128
3.8	474	461	447	263	132
3.7	486	473	459	270	135
3.6	500	486	472	278	139
3.5	514	500	486	286	143
3.4	529	515	500	294	147
3.3	545	530	515	303	152
3.2	563	547	531	313	156
3.1	581	565	548	323	161
3.0	600	583	567	333	167

Table J-1. Mils for Objects (Continued)

Target Height (Mils)	6 Feet (1.8 M)	5 Feet, 9 Inches (1.75 M)	5 Feet, 6 Inches (1.7 M)	39 Inches (1 M)	19 Inches (0.5 M)
2.9	621	603	586	345	172
2.8	643	625	607	357	179
2.7	667	648	630	370	185
2.6	692	673	654	385	192
2.5	720	700	680	400	200
2.4	750	729	708	417	208
2.3	783	761	739	435	217
2.2	818	795	773	455	227
2.1	857	833	810	476	238
2.0	900	875	850	500	250
1.9	947	921	895	526	263
1.8	1,000	972	944	556	278
1.7	1,059	1,029	1,000	588	294
1.6	1,125	1,094	1,063	625	313
1.5	1,200	1,167	1,133	667	333
1.4	1,286	1,250	1,214	714	357
1.3	1,385	1,346	1,308	769	385
1.2	1,500	1,458	1,417	833	417
1.1	1,636	1,591	1,545	909	455
1.0	1,800	1,750	1,700	1,000	500
0.9	2,000	1,944	1,889	1,111	556
0.8	2,250	2,188	2,125	1,250	625
0.7	2,571	2,500	2,429	1,429	714
0.6	3,000	2,917	2,833	1,667	833
0.5	3,600	3,500	3,400	2,000	1,000
0.4	4,500	4,375	4,250	2,500	1,250
0.3	6,000	5,833	5,667	3,333	1,667
0.2	9,000	8,750	8,500	5,000	2,500
0.1	18,000	17,500	17,000	10,000	5,000

NOTE: Use of the formula (HT x 1000)/mils will give the range to any known sized object.
Example: a 2-meter door would be: HT = 2(2 x 1,000) = 2,000/mils with a mil reading of 3.5 = 571 meters to the door.

Appendix K

Sniper's Logbook

Nine Steps for a First-Shot Hit

1. Determine the range in meters—set. Take slope into account.
2. Determine the base wind:
    a. In MOA—set. (For iron sights only) or,
    b. Mils for hold-off. 1 mil = 3.5 MOAs, 1/4 mil = .87 MOA, 1/2 mil = 1.75 MOAs, 3/4 mils = 2.62 MOAs.

NOTE: When determining base winds, ensure you know how the mirage looks under the base condition so you can see changes.

3. Determine the spin drift correction:
    a. 600–700 mils—left 1/2 MOA.
    b. 800–900 mils—left 3/4 MOA.
    c. 1,000 mils—left 1 MOA (M118).
4. Determine the temperature change from "0" and set: 100–500 mils +/– 20 degrees = +/– 1 MOA.
    a. 600–900 mils +/– 15 degrees = +/– 1 MOA.
    b. 1,000 mils +/– 10 degrees = +/– 1 MOA.
5. Determine the pressure change versus "0" pressure and set.
6. Determine the altitude change from "0" and set.
7. Determine lead (if a moving target).
8. Assume a good position:
    a. Bone support.
    b. Muscular relaxation.
    c. Natural POA on the aiming point.
9. Fire the shot:
    a. Natural respiratory pause.
    b. Focus on the front sight/reticle.
    c. Follow-through.

NOTE: Ammunition should remain covered so it will stay at a constant temperature. As a weapon heats up it will string rounds high. This is caused by the internal residual heat increasing the chamber temperature that causes increased chamber pressure. This increase results in increased bullet velocity. Log all shots and subsequent changes.

Wind Data

1. Determine direction of—
    a. Average wind.
    b. Gusts
    c. Lulls.

2. Determine velocity of—
    a. Average wind.
    b. Gusts.
    c. Lulls.

3. Determine mil hold for wind call.

4. Observer must be prepared to change his wind call based on gust or lulls. **NOTE:** Lulls are more dangerous than gusts.

5. Refer to your target dimensions in MOAs:
    a. Center = center of target.
    b. Favor = 1/2 between midline and edge of target.
    c. Hold = the edge of the target.

NOTE: You cannot use mirage as a velocity indicator until you know what it looks like for the average wind. This number will change throughout the day.

8. Shoot the condition. Do not chase spotters.

9. Ignore minor fluctuations. Wait for the condition to fully change. Mirage will change before conditions arrive. Boiling mirage indicates change.

10. Grass will give magnitude of the wind but requires practice for direction or velocity.

11. Observer computes correction in **minutes** and gives it to the shooter in **mil** hold-off.

NOTE: This matrix (Figure K-1) is designed to compile data on the individual sniper weapon system's zero at these ranges and temperatures. Figure K-2, pages K-2 through K-36, provides sample components of the SOTIC Shooter's Log as used in the course.

CONSOLIDATED ZERO DATA										
Meters	100	200	300	400	500	600	700	800	900	1,000
Temperature										
50										
55										
60										
65										
70										
75										
80										
85										
90										
95										

Figure K-1. Individual SWS's Data for Zeroing

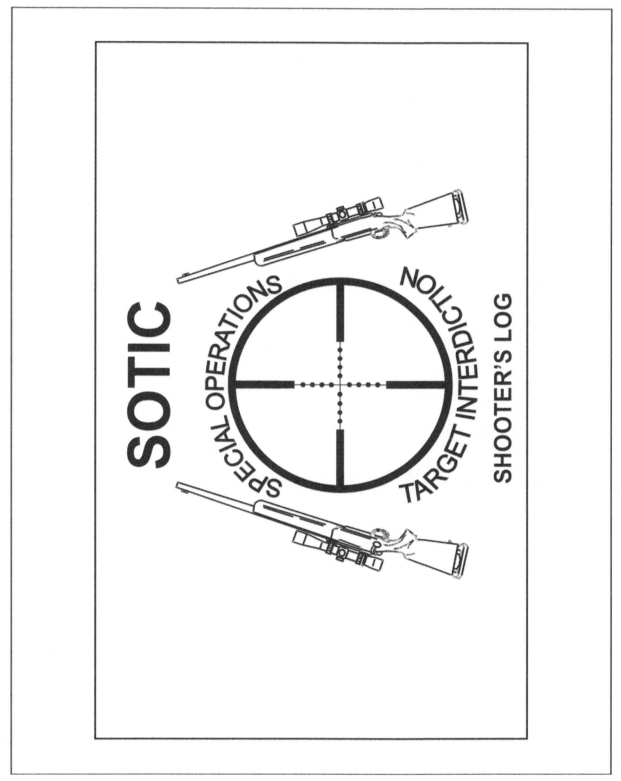

Figure K-2. SOTIC Shooter's Log

Range Y/M	.25 MOA come ups	1 MOA come ups	Norm. come ups	Max Ord ft	Time of flight	MOVERS Base Leads From Leading Edge			
200 Y	-2	-0.5	0	0.18	0.24	Range	1.5mph	2.5mph	5mph
200 M	0	0	0	0.23	0.26	200	0.5	1	2.75
300 Y	9	2.25	2.75	0.5	0.37	300	0.5	1.25	3
300 M	12	3	3	0.64	0.41	400	0.75	1.5	3.25
400 Y	21	5.25	3.25	1	0.52				
400 M	27	6.75	3.75	1.3	0.58	Point Blank Zero 0 - 345			
500 Y	34	8.75	3.5	1.8	0.68	300 on GUN = +/6" from Center			
500 M	44	11	4.25	2.25	0.75				
600 Y	52	13	4	2.8	0.85				
600 M	62	15.5	4.5	3.6	0.95	34 MB = 1" HG			
700 Y	71	17.75	4.5	4.26	1.03	100- 500 1 HG = .25 MOA			
700 M	84	21	5.5	5.4	1.16	600- 800 1 HG = .5 MOA			
800 Y	91	22.75	5	6.11	1.23	900- 1000 1 HG = 1.0 MOA			
800 M	108	27	6	7.85	1.39				
900 Y	115	28.75	5.75	8.5	1.45				
900 M	136	34	7	11.03	1.65	1 MIL = 3.44 MOA(round to 3.5)			
1000 Y	140	35	6.5	11.5	1.69	1 MOA = 1.047(1.0) in/100 yards			
1000M	168	42	8	15.14	1.92	1 MOA = 3 cm/100 meters			
	7.62	NATO	M118LR		175	Grain	2650fps		

MIL	Range			MILs				Hold Off	
MILs	1 M.	MOA =	MIL =	5 mph	10 mph	15 mph	100	4.5	Under
5	200.0	2.2	7.5	0.25	0.5	0.75	200	3.5	Under
4.8	208.3	2.3	7.8	0.25	0.5	0.75	300	2.5	Under
4.5	222.2	2.4	8.4	0.33	0.66	1	400	1.25	Under
4.3	232.6	2.5	8.7	0.33	0.66	1	500	Zero	
4	250.0	2.7	9.4	0.33	0.66	1	600	1.5	Over
3.8	263.2	2.9	9.9	0.4	0.75	1.25	700	3.25	Over
3.5	285.7	3.1	10.7	0.4	0.75	1.25		WIND CHART	
3.3	303.0	3.3	11.4	0.4	0.75	1.25		Deg.	Value
3	333.3	3.6	12.5	0.5	1	1.5		90	FULL
2.8	357.1	3.9	13.4	0.5	1	1.5		75	0.96
2.5	400.0	4.4	15.0	0.5	1	1.5	SPIN	60	0.86
2.3	434.8	4.8	16.4	0.6	1.25	1.8	DRIFT	45	0.7
2	500.0	5.5	18.8	0.75	1.5	2.25	.25moa	30	0.5
1.8	555.6	6.1	20.9	0.8	1.6	2.5		15	0.25
1.65	606.1	6.6	22.8	0.9	1.75	2.7			
1.5	666.7	7.3	25.1	1	2	3		TEMP	=1MOA
1.4	714.3	7.8	26.9	1	2	3	.5moa	1-500	+/- 20
1.3	769.2	8.4	28.9	1.1	2.2	3.3		4-600	+/- 15
1.25	800.0	8.7	30.1	1.25	2.5	3.75	.75moa	1000	+/- 10
1.1	909.1	10	34.2	1.25	2.5	3.75	1moa		

Range Y/M	.25 MOA come ups	1 MOA come ups	TOF sec.	Max Ord ft	Range Max Ord .	MOVERS Base Leads From Leading Edge			
200 Y	-2	-0.5	0.2		115	Range	Slow	Med.	Fast
200 M	0	0	0.2			200	1	1.25	1.75
300 Y	10	2.5	0.3	0.5	165	300	1	1.5	1.8
300 M	14	3.5	0.4			400	1.25	1.75	2
400 Y	24	6	0.5	1	216				
400 M	29	7.25	0.6	1.5		Point Blank Zero - 0 - 375			
500 Y	40	10	0.7	2	270	300 on GUN = +/ - 6" from Center			
500 M	48	12	0.8	2.5	299				
600 Y	58	14.5	0.9	3	328				
600 M	69	17.25	1	4	358	34 MB = 1" HG			
700 Y	78	19.5	1.1	5	388	100 - 500 1 HG = .25 MOA			
700 M	93	23.25	1.2	6	419	600 - 800 1 HG = .75 MOA			
800 Y	102	25.5	1.3	7	443	900 - 1000 1 HG = 1.5 MOA			
800 M	121	30.25	1.4	9					
900 Y	128	32	1.5	10	507				
900 M	154	38.5	1.6	12		1 MIL = 3.375 MOA			
1000 Y	158	39.5	1.7	13	573	1 MOA = 1.047 in/100 yards			
						1 MOA = 3 cm/100 meters			
	7.62	NATO	M118		173	Grain			

MIL	Range			MILs				Hold Off	
MILs	1 M.	MOA =	MIL =	5 mph	10 mph	15 mph	100	5	Under
5	200.0	2.2	7.5	0.25	0.5	0.75	200	2.5	Under
4.8	208.3	2.3	7.8	0.25	0.5	0.75	300	1.5	Under
4.5	222.2	2.4	8.4	0.33	0.66	1	400	1.25	Under
4.3	232.6	2.5	8.7	0.33	0.66	1	500	Zero	
4	250.0	2.7	9.4	0.33	0.66	1	600	1	Over
3.8	263.2	2.9	9.9	0.4	0.75	1.25	700	2.7	Over
3.5	285.7	3.1	10.7	0.4	0.75	1.25		WIND CHART	
3.3	303.0	3.3	11.4	0.4	0.75	1.25		Deg.	Value
3	333.3	3.6	12.5	0.5	1	1.5		90	FULL
2.8	357.1	3.9	13.4	0.5	1	1.5		75	0.96
2.5	400.0	4.4	15.0	0.5	1	1.5	SPIN	60	0.86
2.3	434.8	4.8	16.4	0.6	1.25	1.8	DRIFT	45	0.7
2	500.0	5.5	18.8	0.75	1.5	2.25	.25moa	30	0.5
1.8	555.6	6.1	20.9	0.8	1.6	2.5		15	0.25
1.65	606.1	6.6	22.8	0.9	1.75	2.7			
1.5	666.7	7.3	25.1	1	2	3		TEMP	=1MOA
1.4	714.3	7.8	26.9	1	2	3	.5moa	1-500	+/- 20
1.3	769.2	8.4	28.9	1.1	2.2	3.3		4-600	+/- 15
1.25	800.0	8.7	30.1	1.25	2.5	3.75	.75moa	1000	+/- 10
1.1	909.1	10	34.2	1.25	2.5	3.75	1moa		

Figure K-2. SOTIC Shooter's Log (Continued)

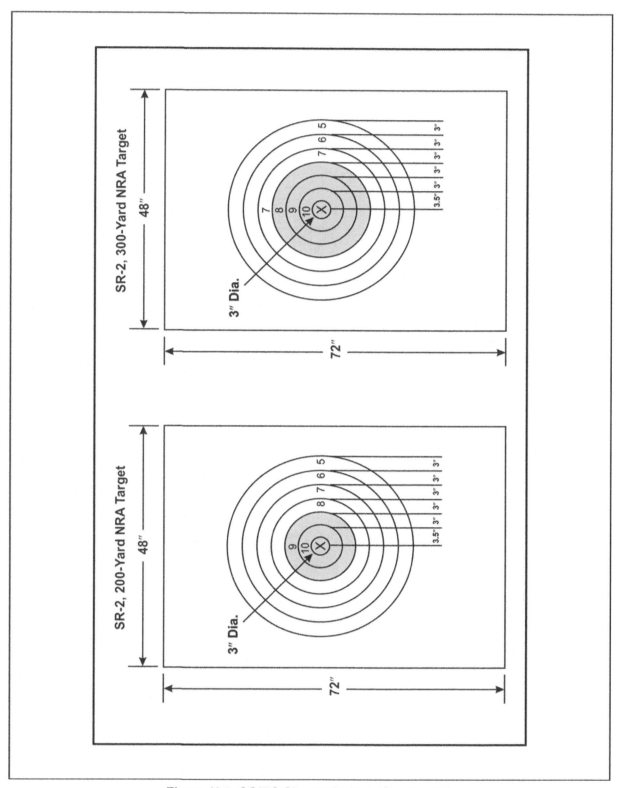

Figure K-2. SOTIC Shooter's Log (Continued)

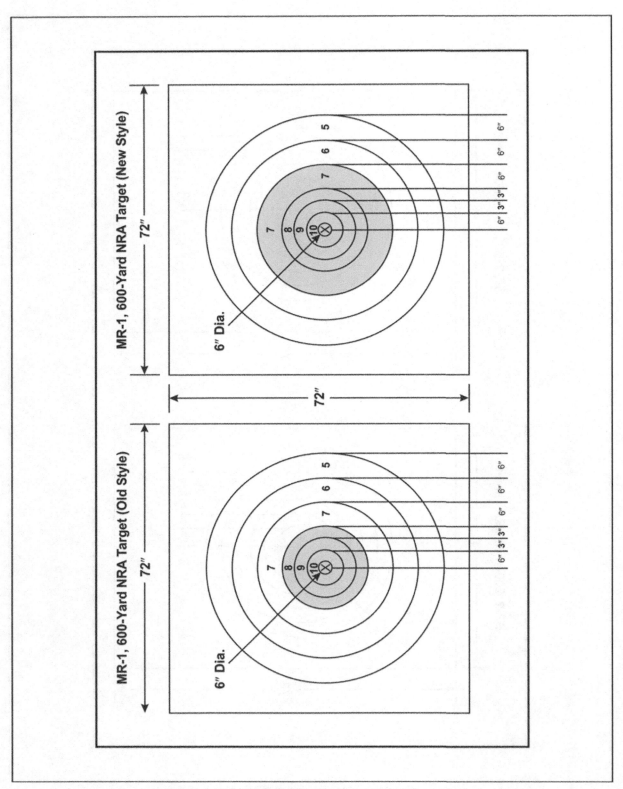

Figure K-2. SOTIC Shooter's Log (Continued)

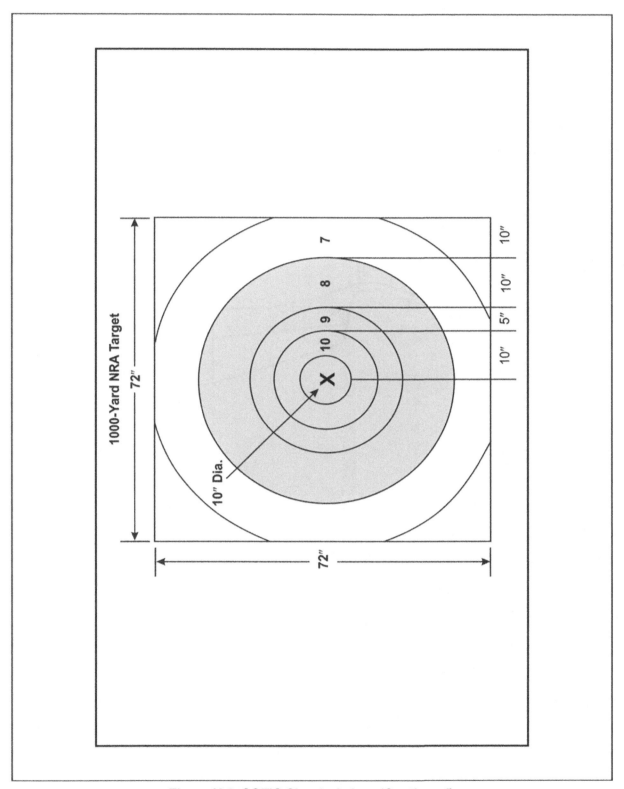

Figure K-2. SOTIC Shooter's Log (Continued)

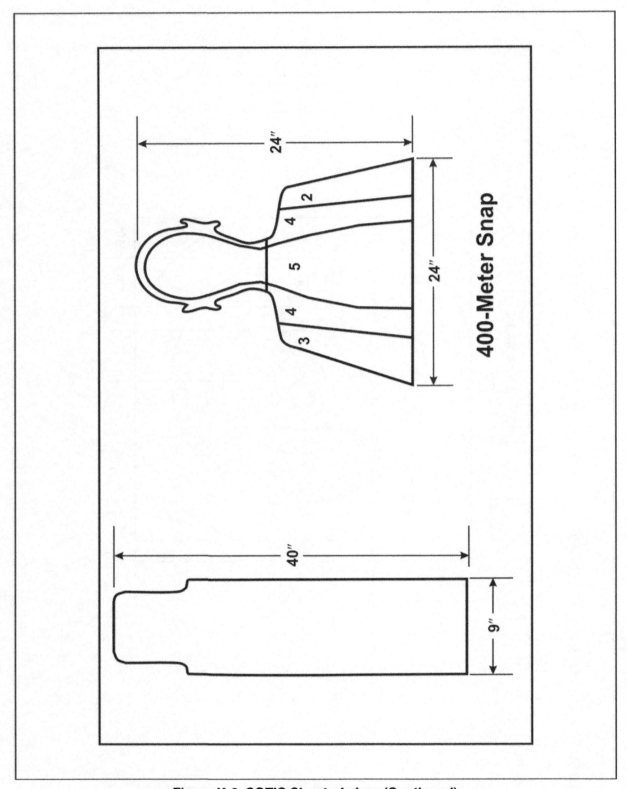

Figure K-2. SOTIC Shooter's Log (Continued)

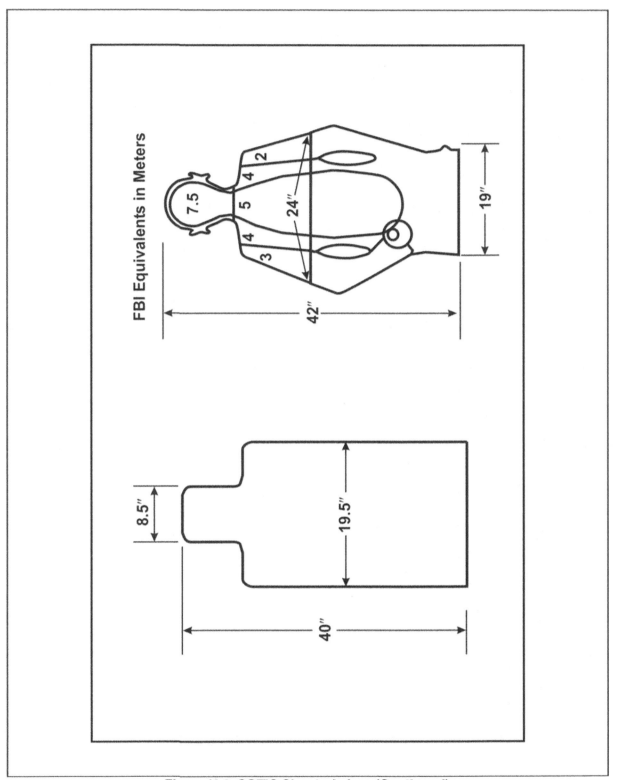

Figure K-2. SOTIC Shooter's Log (Continued)

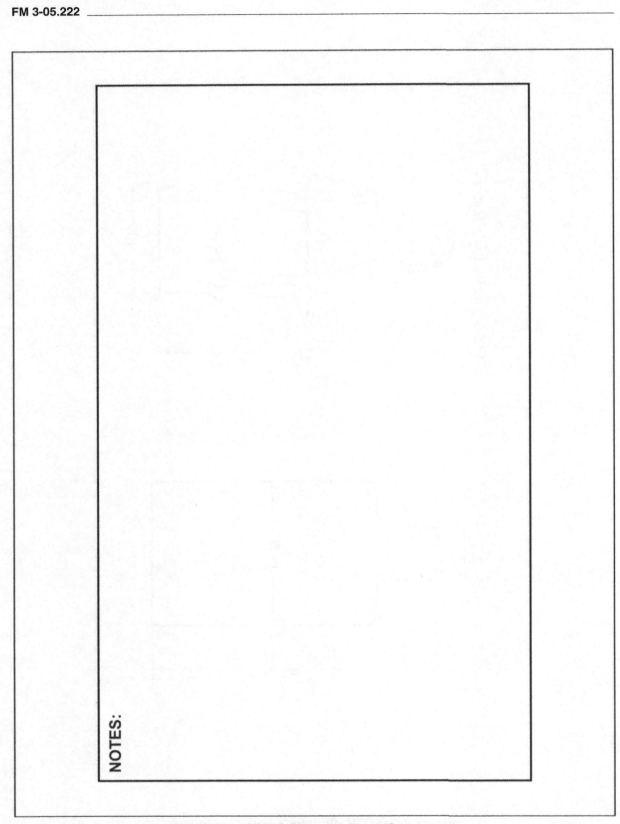

Figure K-2. SOTIC Shooter's Log (Continued)

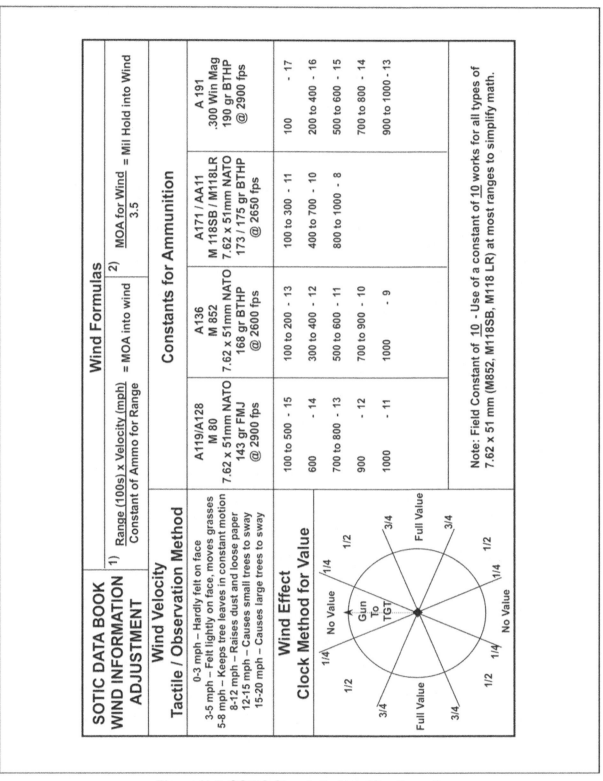

Figure K-2. SOTIC Shooter's Log (Continued)

Wind Chart in MOA for 175 gr M118LR

Meters	2 mph	4 mph	6 mph	8 mph	10 mph	12 mph	14 mph	16 mph	18 mph	20 mph
50	0.00	0.25	0.25	0.50	0.50	0.50	0.75	0.75	1.00	1.00
100	0.25	0.25	0.50	0.50	0.75	1.00	1.00	1.25	1.25	1.50
150	0.25	0.50	0.50	0.75	1.00	1.25	1.50	1.50	1.75	2.00
200	0.25	0.50	1.00	1.25	1.50	1.75	2.00	2.50	2.75	3.00
250	0.50	0.75	1.25	1.50	2.00	2.50	2.75	3.25	3.50	4.00
300	0.50	1.00	1.50	2.00	2.50	3.00	3.50	4.00	4.50	5.00
350	0.50	1.00	1.50	2.25	2.75	3.25	3.75	4.25	5.00	5.50
400	0.50	1.25	2.00	2.50	3.25	3.75	4.50	5.25	5.75	6.50
450	0.75	1.50	2.25	3.00	3.75	4.50	5.25	6.00	6.75	7.50
500	0.75	1.75	2.50	3.50	4.25	5.00	6.00	6.75	7.75	8.50
550	1.00	2.00	2.75	3.75	4.75	5.75	6.75	7.50	8.50	9.50
600	1.00	2.25	3.25	4.50	5.50	6.50	7.75	8.75	10.00	11.00
650	1.25	2.50	3.50	4.75	6.00	7.25	8.50	9.50	10.75	12.00
700	1.25	2.50	4.00	5.25	6.50	7.75	9.00	10.50	11.75	13.00
750	1.50	3.00	4.25	5.75	7.25	8.75	10.25	11.50	13.00	14.50
800	1.50	3.00	4.50	6.25	7.75	9.25	10.75	12.25	14.00	15.50
850	1.75	3.50	5.00	6.75	8.50	10.25	12.00	13.50	15.25	17.00
900	2.00	3.75	5.50	7.50	9.25	11.25	13.00	14.75	16.75	18.50

Figure K-2. SOTIC Shooter's Log (Continued)

Wind Chart in Mils for 175 gr M118LR

Meters	2 mph	4 mph	6 mph	8 mph	10 mph	12 mph	14 mph	16 mph	18 mph	20 mph
50	0.00	0.00	0.00	0.00	0.25	0.25	0.25	0.25	0.25	0.25
100	0.00	0.00	0.00	0.25	0.25	0.25	0.25	0.25	0.25	0.50
150	0.00	0.00	0.25	0.25	0.25	0.50	0.50	0.50	0.50	0.50
200	0.00	0.25	0.25	0.25	0.50	0.75	0.75	0.75	0.75	1.00
250	0.00	0.25	0.25	0.50	0.50	1.00	1.00	1.00	1.00	1.25
300	0.25	0.25	0.50	0.50	0.75	1.00	1.00	1.25	1.25	1.50
350	0.25	0.25	0.50	0.50	0.75	1.00	1.00	1.25	1.50	1.50
400	0.25	0.25	0.50	0.75	1.00	1.25	1.25	1.50	1.75	2.00
450	0.25	0.50	0.50	1.00	1.00	1.50	1.50	1.75	2.00	2.25
500	0.25	0.50	0.75	1.00	1.25	1.50	1.75	2.00	2.25	2.50
550	0.25	0.50	0.75	1.00	1.50	1.75	2.00	2.25	2.50	2.75
600	0.25	0.50	1.00	1.25	1.50	2.00	2.25	2.50	3.00	3.25
650	0.50	0.75	1.00	1.50	1.75	2.00	2.50	2.75	3.25	3.50
700	0.50	0.75	1.25	1.50	2.00	2.25	2.75	3.00	3.50	3.75
750	0.50	1.00	1.25	1.75	2.00	2.50	3.00	3.25	3.75	4.25
800	0.50	1.00	1.25	1.75	2.25	2.50	3.25	3.50	4.00	4.50
850	0.50	1.00	1.50	2.00	2.50	3.00	3.50	4.00	4.50	5.00
900	0.50	1.00	1.50	2.00	2.50	3.25	3.75	4.25	5.00	5.50

Figure K-2. SOTIC Shooter's Log (Continued)

FM 3-05.222

Zero Data

Place	Date	Time	Rifle/Scope #	Ammunition	Temp	Distance

Altitude	Humidity	Baro Press	Mirage	Light	Wind
				Bright ☐ Hazy ☐ Overcast ☐ (·) Direction	Light (3 mph) ☐ Medium (7 mph) ☐ Heavy (15 mph) ☐ mph (·) Direction

Shot #	1	2	3	4	5	6	7	8	9	10
Elev										
Wind										
Call	☐	☐	☐	☐	☐	☐	☐	☐	☐	☐

Shot #	11	12	13	14	15	16	17	18	19	20
Elev										
Wind										
Call	☐	☐	☐	☐	☐	☐	☐	☐	☐	☐

Remarks

Figure K-2. SOTIC Shooter's Log (Continued)

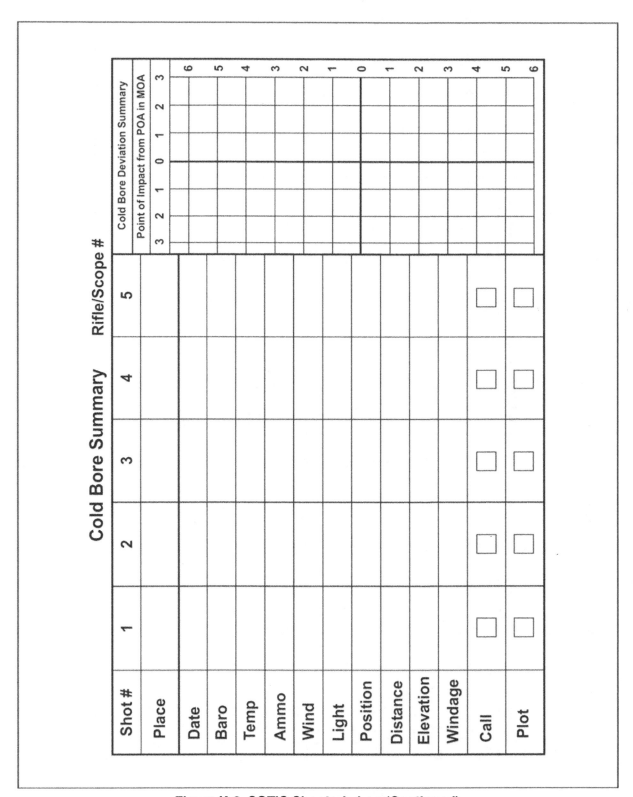

Figure K-2. SOTIC Shooter's Log (Continued)

Zero Summary Chart

Rifle #: _____ Ammunition Type: _____

Meter/Yard	20°/-6°	30°/-1°	40°/4°	50°/10°	60°/15°	65°/18°	70°/21°	75°/24°	80°/26°	85°/29°	90°/32°	95°/35°	100°/37°	105°/40°
50/55														
100/109														
150/164														
200/219														
250/273														
300/328														
350/383														
400/437														
450/492														
500/546														
550/602														
600/656														
650/711														
700/766														
750/820														
800/875														
850/929														
900/984														
950/1039														
1000/1094														

Figure K-2. SOTIC Shooter's Log (Continued)

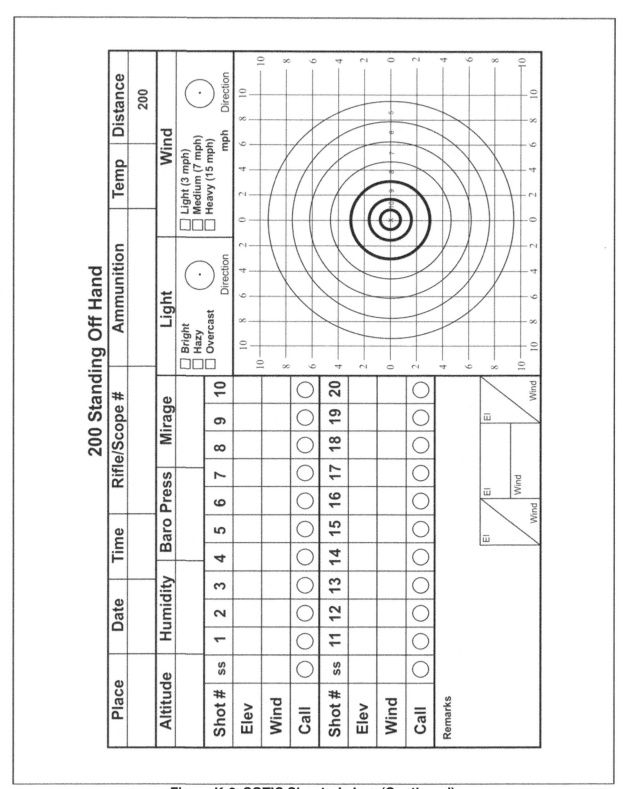

Figure K-2. SOTIC Shooter's Log (Continued)

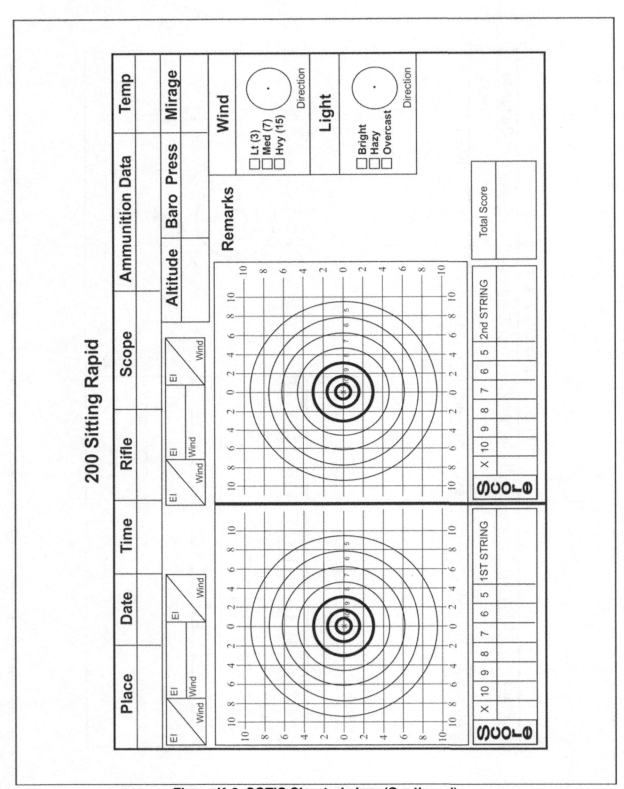

Figure K-2. SOTIC Shooter's Log (Continued)

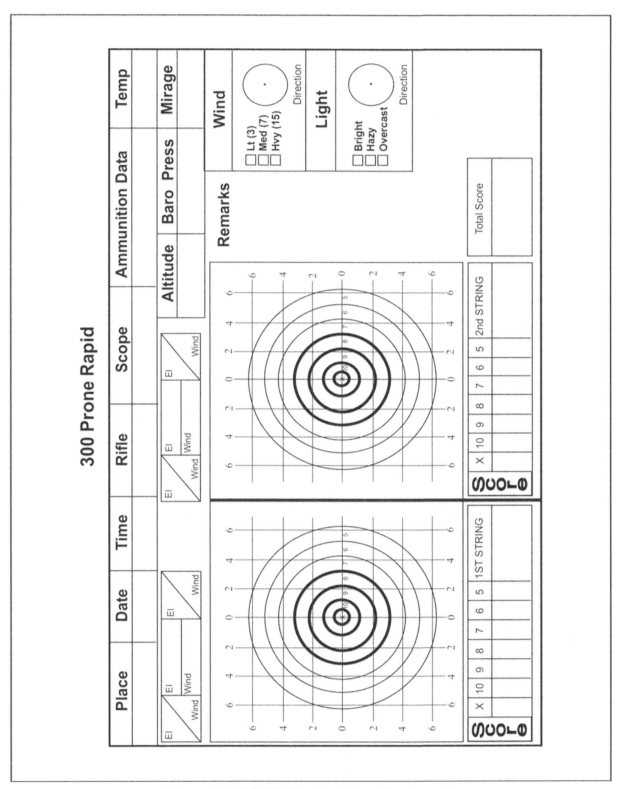

Figure K-2. SOTIC Shooter's Log (Continued)

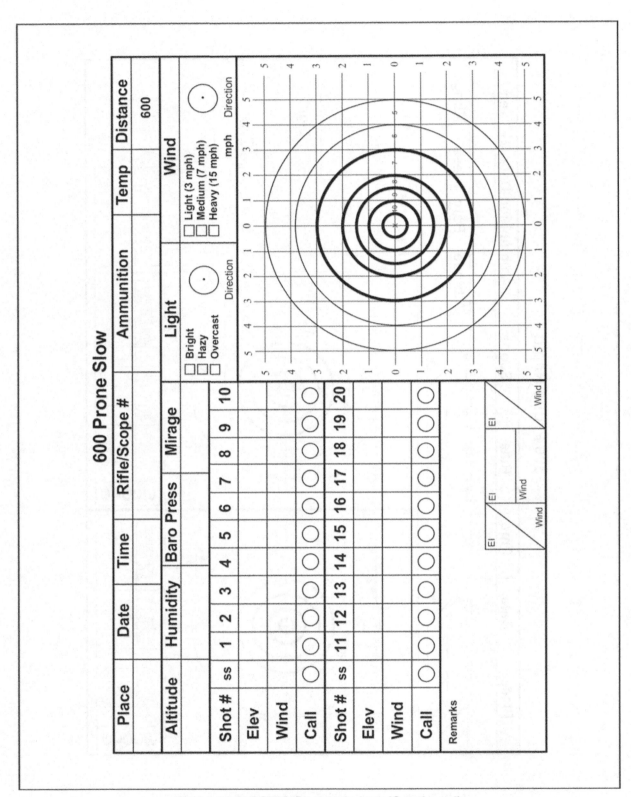

Figure K-2. SOTIC Shooter's Log (Continued)

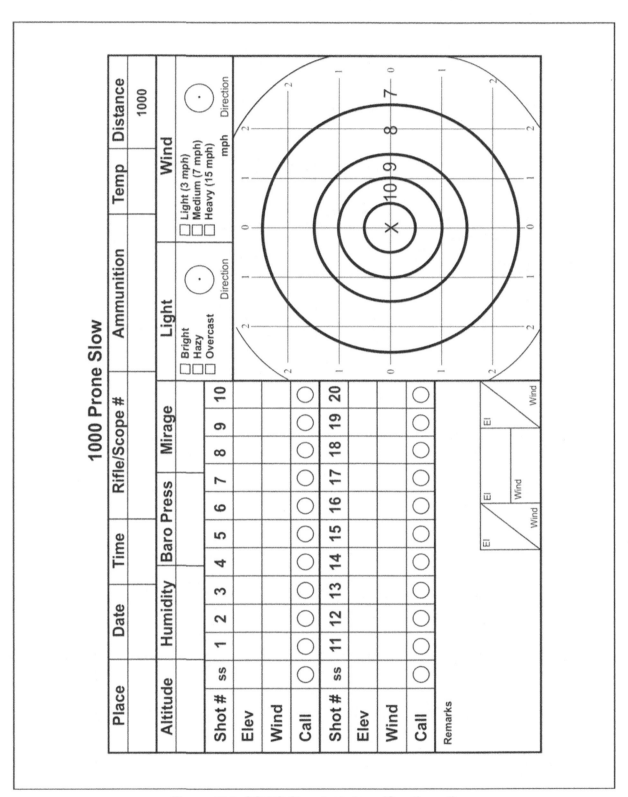

Figure K-2. SOTIC Shooter's Log (Continued)

FM 3-05.222

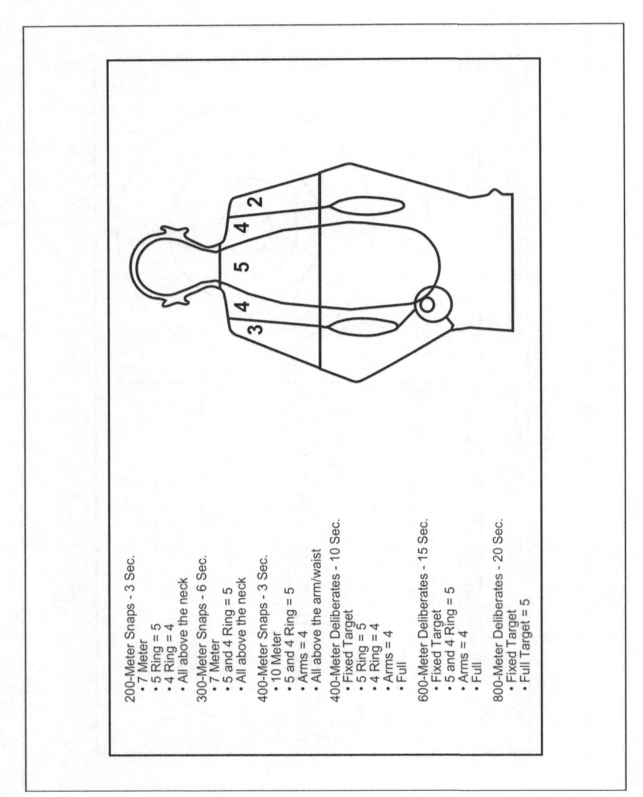

Figure K-2. SOTIC Shooter's Log (Continued)

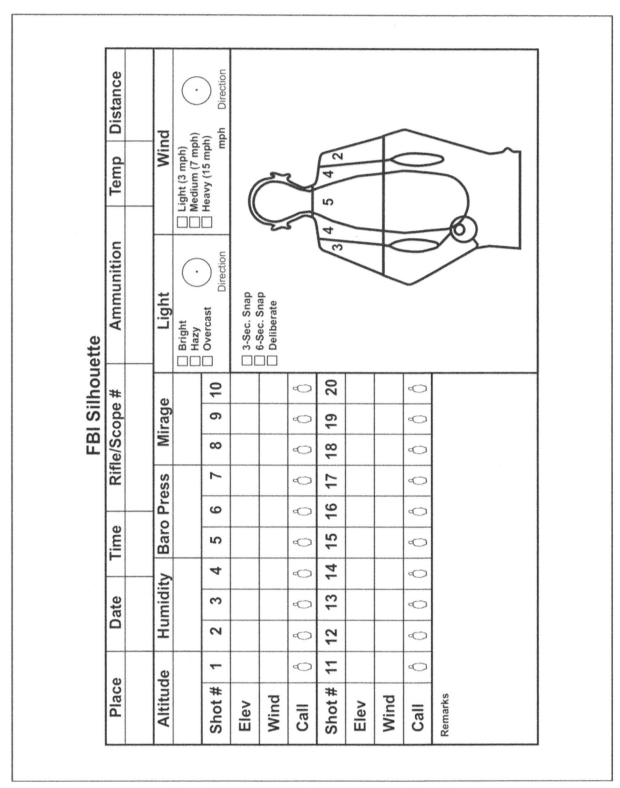

Figure K-2. SOTIC Shooter's Log (Continued)

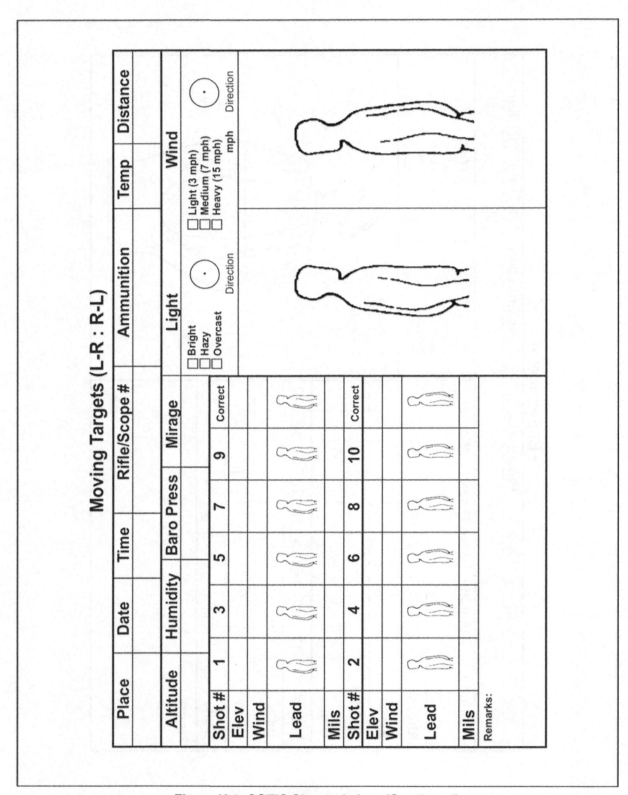

Figure K-2. SOTIC Shooter's Log (Continued)

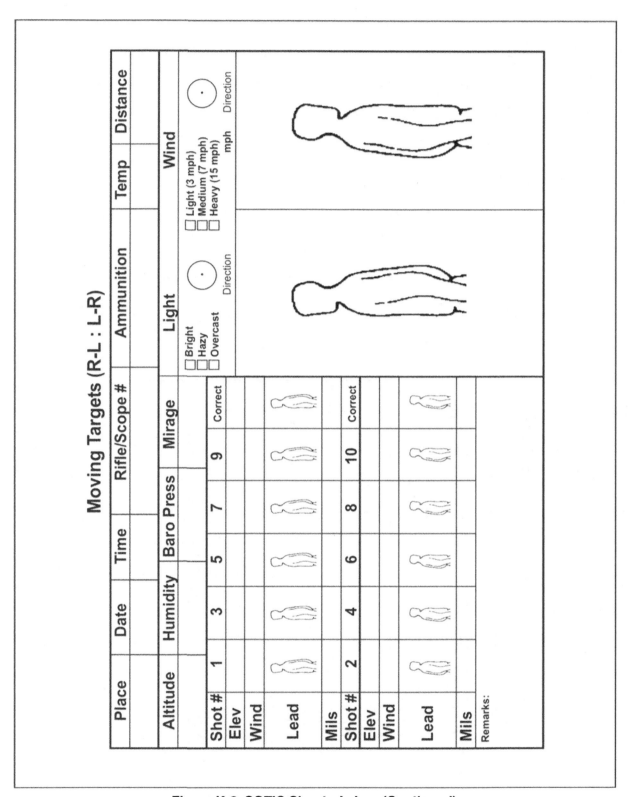

Figure K-2. SOTIC Shooter's Log (Continued)

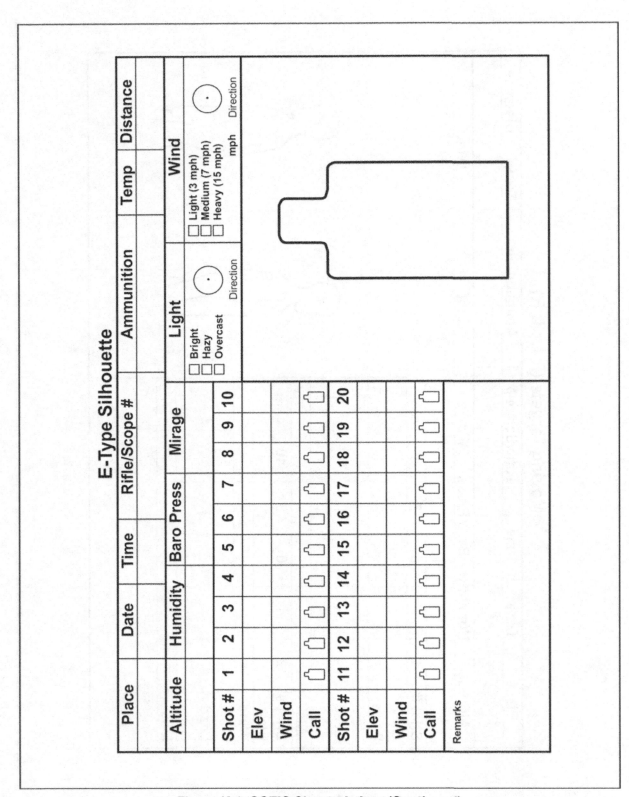

Figure K-2. SOTIC Shooter's Log (Continued)

Place		Date		Time		Rifle/Scope #		Ammunition		Temp	Distance
Altitude		Humidity		Baro Press		Mirage		Light		Wind	

Shot #	1	2	3	4	5	6	7	8	9	10
Elev										
Wind										
Call	○	○	○	○	○	○	○	○	○	○
Shot #	11	12	13	14	15	16	17	18	19	20
Elev										
Wind										
Call	○	○	○	○	○	○	○	○	○	○

Light: ☐ Bright ☐ Hazy ☐ Overcast — Direction

Wind: ☐ Light (3 mph) ☐ Medium (7 mph) ☐ Heavy (15 mph) — mph Direction

Size of Target =

Remarks

Figure K-2. SOTIC Shooter's Log (Continued)

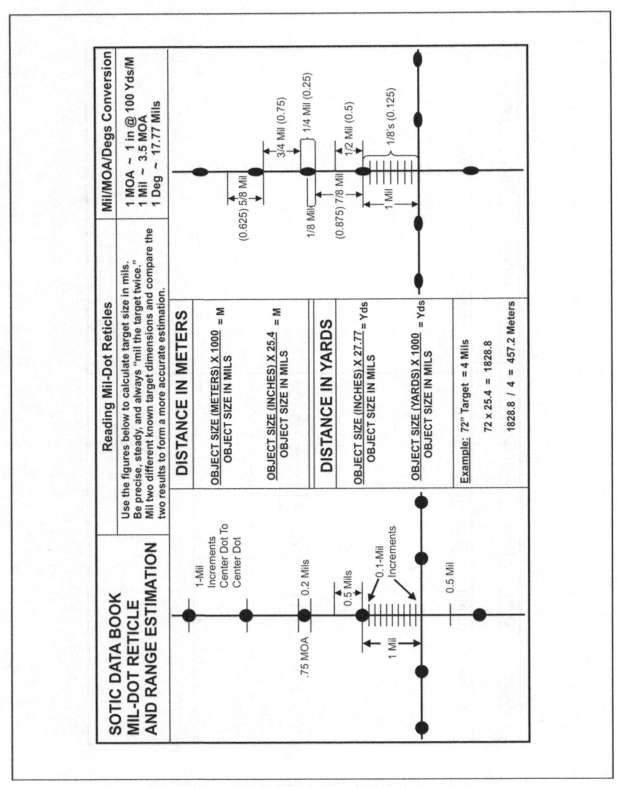

Figure K-2. SOTIC Shooter's Log (Continued)

Range to Common Objects in Meters (Urban)

Target Mils/Inches	Head 9	Shoulders 20	Stop Sign 30	Door Width 36	Door Height 84	Head to Groin 40	55 Gal Drum Ht 34.5	55 Gal Drum Wd 23	Target Mils/Inches	Head 9	Shoulders 20	Stop Sign 30	Door Width 36	Door Height 84	Head to Groin 40	55 Gal Drum Ht 34.5	55 Gal Drum Wd 23
0.7	327	726	1089	1306	3048	1451	1252	835	4.4	52	115	173	208	485	231	199	133
0.8	286	635	953	1143	2667	1270	1095	730	4.5	51	113	169	203	474	226	195	130
0.9	254	564	847	1016	2371	1129	974	649	4.6	50	110	166	199	464	221	191	127
1.0	229	508	762	914	2134	1016	876	584	4.7	49	108	162	195	454	216	186	124
1.1	208	462	693	831	1940	924	797	531	4.8	48	106	159	191	445	212	183	122
1.2	191	423	635	762	1778	847	730	487	4.9	47	104	156	187	435	207	179	119
1.3	176	391	586	703	1641	782	674	449	5.0	46	102	152	183	427	203	175	117
1.4	163	363	544	653	1524	726	626	417	5.1	45	100	149	179	418	199	172	115
1.5	152	339	508	610	1422	677	584	389	5.2	44	98	147	176	410	195	169	112
1.6	143	318	476	572	1334	635	548	365	5.3	43	96	144	173	403	192	165	110
1.7	134	299	448	538	1255	598	515	344	5.4	42	94	141	169	395	188	162	108
1.8	127	282	423	508	1185	564	487	325	5.5	42	92	139	166	388	185	159	106
1.9	120	267	401	481	1123	535	461	307	5.6	41	91	136	163	381	181	156	104
2.0	114	254	381	457	1067	508	438	292	5.7	40	89	134	160	374	178	154	102
2.1	109	242	363	435	1016	484	417	278	5.8	39	88	131	158	368	175	151	101
2.2	104	231	346	416	970	462	398	266	5.9	39	86	129	155	362	172	149	99
2.3	99	221	331	398	928	442	381	254	6.0	38	85	127	152	356	169	146	97
2.4	95	212	318	381	889	423	365	243	6.1	37	83	125	150	350	167	144	96
2.5	91	203	305	366	853	406	351	234	6.2	37	82	123	147	344	164	141	94
2.6	88	195	293	352	821	391	337	225	6.3	36	81	121	145	339	161	139	93
2.7	85	188	282	339	790	376	325	216	6.4	36	79	119	143	333	159	137	91
2.8	82	181	272	327	762	363	313	209	6.5	35	78	117	141	328	156	135	90
2.9	79	175	263	315	736	350	302	201	6.6	35	77	115	139	323	154	133	89
3.0	76	169	254	305	711	339	292	195	6.7	34	76	114	136	318	152	131	87
3.1	74	164	246	295	688	328	283	188	6.8	34	75	112	134	314	149	129	86
3.2	71	159	238	286	667	318	274	183	6.9	33	74	110	133	309	147	127	85
3.3	69	154	231	277	647	308	266	177	7.0	33	73	109	131	305	145	125	83
3.4	67	149	224	269	628	299	258	172	7.1	32	72	107	129	301	143	123	82
3.5	65	145	218	261	610	290	250	167	7.2	32	71	106	127	296	141	122	81
3.6	64	141	212	254	593	282	243	162	7.3	31	70	104	125	292	139	120	80
3.7	62	137	206	247	577	275	237	158	7.4	31	69	103	124	288	137	118	79
3.8	60	134	201	241	561	267	231	154	7.5	30	68	102	122	284	135	117	78
3.9	59	130	195	234	547	261	225	150	7.6	30	67	100	120	281	134	115	77
4.0	57	127	191	229	533	254	219	146	7.7	30	66	99	119	277	132	114	76
4.1	56	124	186	223	520	248	214	142	7.8	29	65	98	117	274	130	112	75
4.2	54	121	181	218	508	242	209	139	7.9	29	64	96	116	270	129	111	74
4.3	53	118	177	213	496	236	204	136	8.0	29	64	95	114	267	127	110	73

Figure K-2. SOTIC Shooter's Log (Continued)

Range to Common Objects in Meters (Vehicles)

Target	HMMV Width	HMMV Top of	HMMV Whl Base	Mid PU Width	Mid PU Top of	Mid PU Whl Base	Whl Base Long	Target	HMMV Width	HMMV Top of	HMMV Whl Base	Mid PU Width	Mid PU Top of	Mid PU Whl Base	Whl Base Long
Mils/Inches	84	72	130	66	64	103	122	Mils/Inches	84	72	130	66	64	103	122
0.7	3048	2613	4717	2395	2322	3737	4427	4.4	485	416	750	381	369	595	704
0.8	2667	2286	4128	2096	2032	3270	3874	4.5	474	406	734	373	361	581	689
0.9	2371	2032	3669	1863	1806	2907	3443	4.6	464	398	718	364	353	569	674
1.0	2134	1829	3302	1676	1626	2616	3099	4.7	454	389	703	357	346	557	659
1.1	1940	1663	3002	1524	1478	2378	2817	4.8	445	381	688	349	339	545	646
1.2	1778	1524	2752	1397	1355	2180	2582	4.9	435	373	674	342	332	534	632
1.3	1641	1407	2540	1290	1250	2012	2384	5.0	427	366	660	335	325	523	620
1.4	1524	1306	2359	1197	1161	1869	2213	5.1	418	359	647	329	319	513	608
1.5	1422	1219	2201	1118	1084	1744	2066	5.2	410	352	635	322	313	503	596
1.6	1334	1143	2064	1048	1016	1635	1937	5.3	403	345	623	316	307	494	585
1.7	1255	1076	1942	986	956	1539	1823	5.4	395	339	611	310	301	484	574
1.8	1185	1016	1834	931	903	1453	1722	5.5	388	333	600	305	296	476	563
1.9	1123	963	1738	882	856	1377	1631	5.6	381	327	590	299	290	467	553
2.0	1067	914	1651	838	813	1308	1549	5.7	374	321	579	294	285	459	544
2.1	1016	871	1572	798	774	1246	1476	5.8	368	315	569	289	280	451	534
2.2	970	831	1501	762	739	1189	1409	5.9	362	310	560	284	276	443	525
2.3	928	795	1436	729	707	1137	1347	6.0	356	305	550	279	271	436	516
2.4	889	762	1376	699	677	1090	1291	6.1	350	300	541	275	266	429	508
2.5	853	732	1321	671	650	1046	1240	6.2	344	295	533	270	262	422	500
2.6	821	703	1270	645	625	1006	1192	6.3	339	290	524	266	258	415	492
2.7	790	677	1223	621	602	969	1148	6.4	333	286	516	262	254	409	484
2.8	762	653	1179	599	581	934	1107	6.5	328	281	508	258	250	402	477
2.9	736	631	1139	578	561	902	1069	6.6	323	277	500	254	246	396	470
3.0	711	610	1101	559	542	872	1033	6.7	318	273	493	250	243	390	463
3.1	688	590	1065	541	524	844	1000	6.8	314	269	486	247	239	385	456
3.2	667	572	1032	524	508	818	968	6.9	309	265	479	243	236	379	449
3.3	647	554	1001	508	493	793	939	7.0	305	261	472	239	232	374	443
3.4	628	538	971	493	478	769	911	7.1	301	258	465	236	229	368	436
3.5	610	523	943	479	464	747	885	7.2	296	254	459	233	226	363	430
3.6	593	508	917	466	452	727	861	7.3	292	251	452	230	223	358	424
3.7	577	494	892	453	439	707	838	7.4	288	247	446	227	220	354	419
3.8	561	481	869	441	428	688	815	7.5	284	244	440	224	217	349	413
3.9	547	469	847	430	417	671	795	7.6	281	241	434	221	214	344	408
4.0	533	457	826	419	406	654	775	7.7	277	238	429	218	211	340	402
4.1	520	446	805	409	396	638	756	7.8	274	234	423	215	208	335	397
4.2	508	435	786	399	387	623	738	7.9	270	231	418	212	206	331	392
4.3	496	425	768	390	378	608	721	8.0	267	229	413	210	203	327	387

Figure K-2. SOTIC Shooter's Log (Continued)

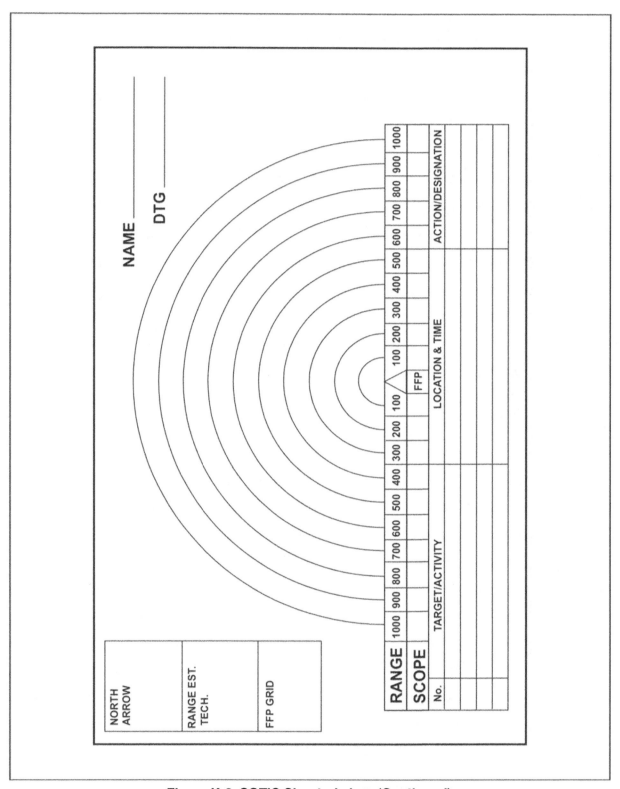

Figure K-2. SOTIC Shooter's Log (Continued)

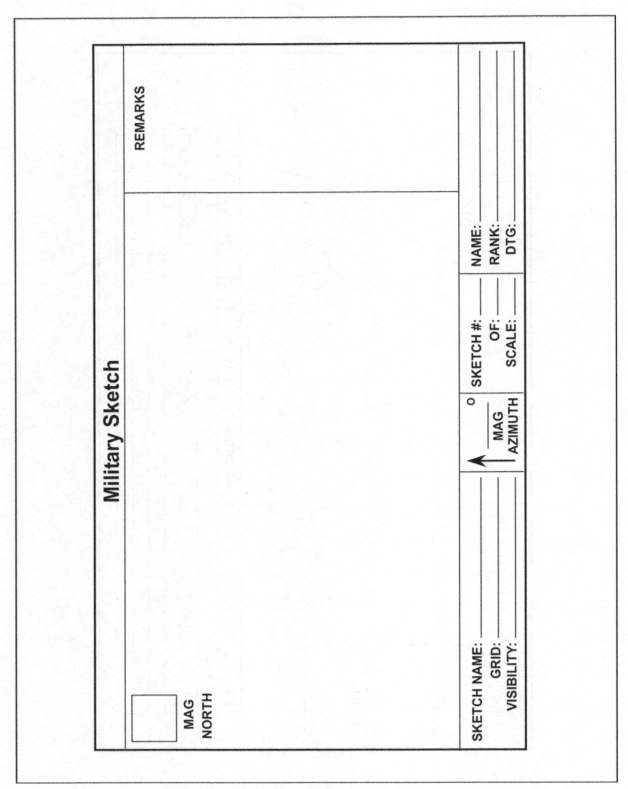

Figure K-2. SOTIC Shooter's Log (Continued)

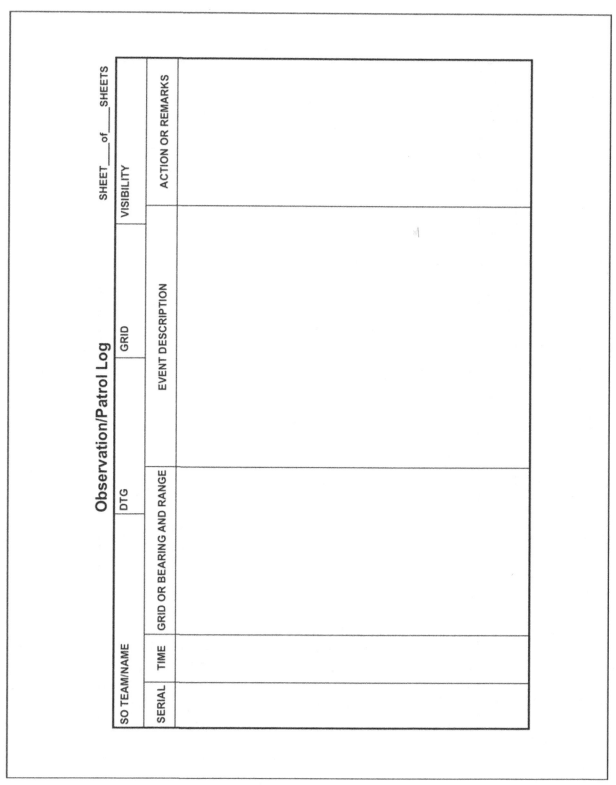

Figure K-2. SOTIC Shooter's Log (Continued)

Mils	Range	Data	Data	Data	Mils	Range	Data	Data	Data
1.25	800				2.15	465			
1.30	769				2.20	455			
1.35	741				2.25	444			
1.40	714				2.30	435			
1.45	690				2.35	426			
1.50	667				2.40	417			
1.55	645				2.45	408			
1.60	625				2.60	400			
1.65	606				2.55	392			
1.70	588				2.60	385			
1.75	571				2.65	377			
1.80	556				2.70	370			
1.85	541				2.75	364			
1.90	526				2.80	357			
1.95	513				2.85	351			
2.00	500				2.90	345			
2.05	488				3.00	333			
2.10	476				3.10	323			

Figure K-2. SOTIC Shooter's Log (Continued)

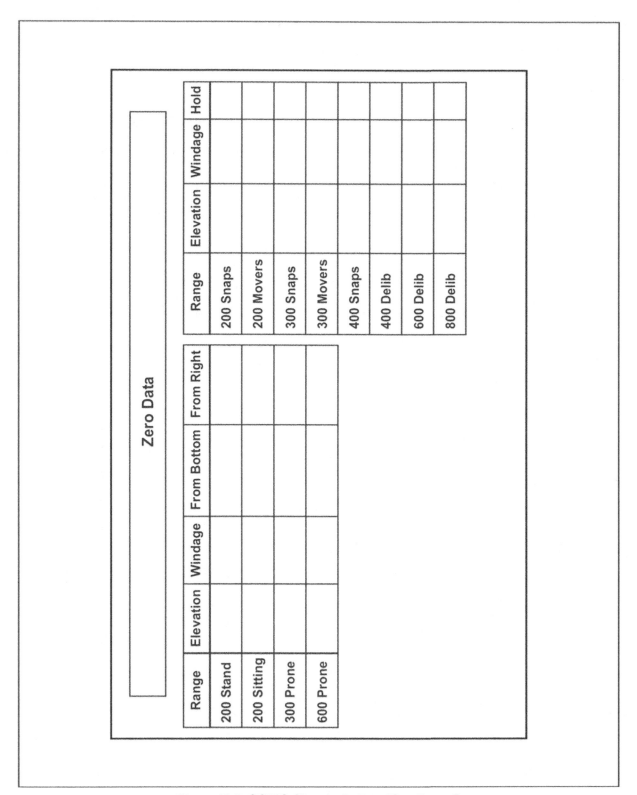

Figure K-2. SOTIC Shooter's Log (Continued)

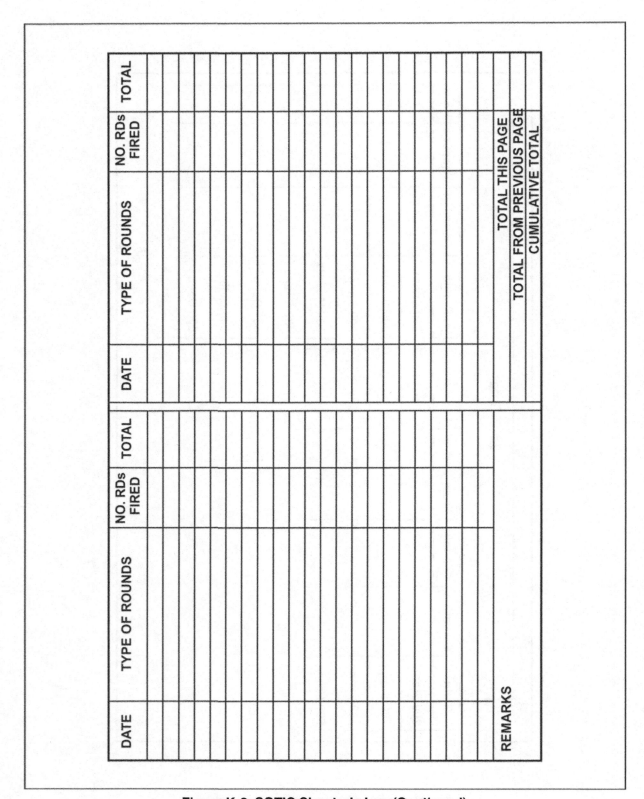

Figure K-2. SOTIC Shooter's Log (Continued)

Figure K-2. SOTIC Shooter's Log (Continued)

Appendix L

Tricks of the Trade

The art of sniping requires learning, repetitious practicing, and the mastering of specific skills. The sniper requires special abilities, training, and equipment. His primary goal is to deliver highly accurate rifle fire against enemy targets.

GOAL

L-1. To achieve this goal and ensure mission success, the sniper should always remember the following:

- For best results, try to use a bulky, lightweight hooded smock in urban areas. Select colors for the smock that will blend with the colors and types of building materials in the area.
- Avoid movement during daylight; if movement is necessary, keep movements slow and deliberate.
- During movements through or occupation of building rooms, be alert to the principles of camouflage and concealment. Do not allow "being inside" to lull you into a reduced awareness of the surroundings.
- Stay in shadows, match clothing to blend with the room or area, hang black sheets to eliminate backlighting against openings or light sources.
- Don't be the only open window in an air-conditioned building. Use existing curtains and leave windows intact. To make a "shooting hole," remove one pane or small corner of the glass.
- Move into the area with help from the host nation.
- Blend into the activities of the area; for example, maintenance crew, civilian clothing, and civilian luggage (guitar cases will always look out of place).
- Try to carry in more equipment and work with multiple teams to cover the entire area.
- Choose a position (if possible) that is naturally in a shadow; if doing so is not possible, make your own shadows by building a "cave" with dark cloth.
- Wear dark clothing to match the background.
- Stay back from the window.
- Don't flag your weapon in your loophole.
- If time allows, make crawl holes from room to room.
- Avoid background light, such as doors opening behind you.

- Be careful of neutral personnel; handle with care.
- Check firing positions (rooftops) and ensure you—
  - Stay below peak line as much as possible.
  - Don't overhang barrel.
  - Put up some type of shade if you are going to be in position a long time.
  - Try to find a position that has a background of some type.
- Have the following equipment and resources for urban operations:
  - Camera.
  - Communications equipment (snipers and command).
  - Food and water.
  - Spotting scope with stand.
  - Binoculars.
  - Dark cloth.
  - Roofing hammer with nails.
  - Tape.
  - Glass cutter.
  - Complete cleaning kit.
  - Multipurpose knife.
  - Silenced pistol.
  - Notebook, pencils, and tape recorder.
  - Sleeping or shooting pad.

L-2. In some cases, the team leader appoints a high-value target to several snipers and they cannot agree on the range (with associated scope adjustments). If this occurs, then each sniper places the range data on his telescope according to his own best estimate. The snipers would then fire simultaneously, thereby increasing the chance of obtaining a hit.

L-3. When operating in a denied area, it is sometimes appropriate to use an indigenous weapon and ammunition. The evidence left (casings or recovered bullets) would disguise the true identity of the sniper and the sponsor. This planning consideration should not be interpreted as "battlefield recovery."

L-4. The sniper must be aware of the ground beneath the muzzle of his rifle. This point is critical when the ground is sandy, dusty, or loose soil. The sniper should either wet the area (urinating will save valuable drinking water) or cover with a suitably sized cloth. In damp conditions (early morning), the sniper should be aware of the possibility of the exhaust smoke indicating the position. An area with broken ground or foliage will help conceal the smoke signature. Also, the sniper should be aware of the muzzle blast moving tall grass and small plants, and therefore, choose his position carefully.

L-5. When in a static position, it is wise to build the sniper hide to provide a direction of fire at an angle to the front of the enemy. This method provides cover and concealment, and the enemy hit by the sniper's fire will look to his front for the sniper's location.

L-6. If the sniper suspects that his system has lost its zero, and the situation allows sighting shots, then he should use "self-marking" targets that do not betray his direction of fire. The sniper may use pools of water, cement walls, or layers of brick. He should not use cans, boxes, or other targets that can be used to sight back on azimuth to his location.

L-7. The sniper's hide should be in a location away from any obvious target reference points. If it looks like an obvious position, it is.

L-8. When firing long ranges where the arc of the bullet will be high, the sniper should always try to visualize the bullet's arc before firing. This practice ensures that there will be no obstacles in the path of the bullet. The sniper should consider this the "mask and overhead clearance" of the sniper rifle.

L-9. Selection of the final firing point is critical to mission success. If the target is expected to be moving, the sniper should select a position that allows a shot at the target as it moves toward or away from him. Relative to the sniper's position, the target will be a stationary one and, therefore, require a no-lead hold.

L-10. If it is necessary to engage a unit of enemy personnel, the sniper should engage the targets that are the greatest threat to him and his team's survival. If this is not a factor, he should engage the targets farthest away from him and not in the front of the enemy formation. If he hits the front-most targets first, the remainder of the unit will deploy and conduct fire and movement to pin the sniper down and engage him. By eliminating the rear-most targets first, the sniper buys himself more time as their numbers will be decreased, possibly without their knowledge. This practice also ensures the sniper the best possible (least suspecting) targets.

L-11. The sniper and his weapon can be of great help in the counterambush immediate-action drill. He should look for target indicators (muzzle flash, disturbed vegetation, ejecting brass) and use a "searching fire" technique. This approach enables him to fire rounds approximately nine inches from the ground, every 6 inches into the suspected enemy location.

L-12. When a sniper and another team's sniper are dealing with multiple targets, such as two hostage-takers, they **must** coordinate to fire simultaneously. Taking them out one at a time may allow the second suspect time to harm the hostages. One technique (if snipers are within earshot or in radio contact with each other on a clear frequency) is for each of them to keep saying aloud in a steady, low voice, "wait...wait...wait..." as long as they do not have a clear shot. When they do, they should stay silent and listen for the moment they are both silent. They should allow a 1-second pause, then open fire together. Another technique is to establish an audible countdown and fire on that number. In some cases, two snipers are assigned to engage a single suspect, particularly if he is behind heavy glass and there is fear that shots may be deflected. One option here is for one sniper to aim for his head and the other for his chest and fire simultaneously.

L-13. In a CBT situation, hostage-takers have been known to switch clothes with the hostages. This trick requires the sniper to distinguish facial features and place top value on higher-powered spotting and rifle scopes. It can also cause him to risk compromise if he decides to move closer to the target.

L-14. The position behind a loophole should be darkened with a drape so that the sniper is not silhouetted and no light comes through the loophole. The sniper should shut his loopholes when anyone enters or exits the hide.

L-15. The observer can tell if the target is hit. The target's response is similar to that of big game. An animal that is fired at and missed always stands tense for a fraction of a second before it bounds away. When an animal is struck by the bullet there is no pause. It bounds away at once on the impact or falls. Thus, a stag shot through the heart commences his death rush at once, to fall dead within 50 yards, whereas a stag missed gives that telltale sudden start. If a human is hit, he falls forward or appears to crumple like a rag doll. Continued activity or falling to the side indicates a superficial hit.

L-16. Speed is important. The sniper should practice for an aimed shot in 2 seconds or less.

L-17. The sniper should use armor-piercing rounds for antimaterial missions to take out the weapon, not the crew. The crew is easier to replace.

L-18. Short of optical or laser range finders and in an offensive role, the mil-relation formula (mortar-crew mil system) will help the sniper determine accurate range. In the defensive role, the surest method of determining precise range is by triangulation.

Appendix M

Sniper Team Debriefing Format

After the mission, the sniper employment officer or S-3 representative directs the sniper team to an area where it prepares for a debriefing. The team remains in the area until called to the operations center.

SNIPER TEAM FUNCTIONS

M-1. The sniper team will—

- Lay out and account for all team and individual equipment.
- Consolidate all captured material and equipment.
- Review and discuss the events listed in the mission logbook from insertion to return, including details of each enemy sighting.
- Prepare an overlay of the team's route, AO, insertion point, extraction point, and significant sighting locations.

S-3 FUNCTIONS

M-2. An S-3 representative controls the debriefing. He directs the team leader to—

- Discuss any enemy sightings since the last communications with the radio base station.
- Give a step-by-step account of each event listed in the mission logbook from insertion until reentry of the FLOT, including the details of all enemy sightings.
- Complete a mission report (Figure M-1, pages M-2 and M-3) and draw an overlay as discussed. The team leader either completes the report or has the observer complete different sections. The team leader then returns the report and overlay to the S-3 representative, while the observer performs postmission maintenance tasks.

M-3. When the debriefing is complete, the S-3 representative releases the sniper team back to its parent unit.

FM 3-05.222 _____

Team Number _____   Date-Time Group (DTG): _____
To _____
Maps Used: 1:25,000: _____
1:50,000: _____
1:250,000: _____
Special: _____

A. Size and Composition of Team: _____
   Team Leader: _____
   Observer: _____

B. Mission: _____
   _____
   _____

C. Priority Intelligence Requirements (PIR) (Use attached sheet):

D. Continuing Intelligence Requirements (CIRs) (Use attached sheet):

E. Time of Departure (DTG):
   Method of Insertion: _____
   Point of Departure (Six-digit grid coordinates): _____
   _____

F. Enemy Spotting En Route (Use attached sheet, if needed):
   1. Ground Activity: _____
   2. Air Activity: _____
   3. Miscellaneous Activity: _____

G. Routes (Out) (Provide overlay): Dismounted—
   By Foot: _____
   By Vehicle (State type): _____
   By Aircraft (State type): _____

H. Terrain (Use attached sheet in the following format):
   Key Terrain   Terrain Compartment
   Significant Terrain   Terrain Corridor
   Decisive Terrain   Map Corrections
   Avenues of Approach (State size)

I. Enemy Forces and Installations (Use attached sheet):

Figure M-1. Sample Mission Report for Debriefing

J. Miscellaneous Information (Use attached sheet, if necessary):
   1. Lack of Animals or Strange Animal Behavior: _____
   2. Mutilated Plants: _____
   3. Uncommon Insects: _____
   4. Abandoned Military Equipment (Check for and include number and type):
      a. Out of Fuel: _____
      b. Unserviceable (Estimate why): _____
      c. Destroyed or Damaged on Purpose by Enemy Forces: _____
      d. Operational Equipment Left Intact: _____
   5. Abandoned Towns/Villages: _____
K. Results of Encounters with Enemy Force and Local Populace: _____
L. Condition of Team, Including Disposition of Dead and Wounded: _____
M. All Maps Returned or Any Other Identifiable Material Returned with Team:
   Yes; No; What Is Missing?; State Item and Where Approximately Lost: _____
N. Conclusions and Recommendations: _____
O. Captured Enemy Equipment and Material: _____
P. Time of Extraction (DTG): _____
   Method of Extraction: _____
   Extraction Point (Six-digit grid coordinates): _____
Q. Routes (Back) (Provide overlay): _____
   1. Dismounted by Ground (E&E): _____
   2. Flight Route Back: _____
R. Enemy Spotting En Route to Base (Use attached sheet, if needed):
   1. Ground Activity: _____
   2. Air Activity: _____
   3. Miscellaneous Activity: _____
S. Time of Return (DTG): _____
   Point of Return (Six-digit grid coordinates): _____
   Team Leader: _____
   (Print Name)   (Grade)

   (Unit)   (Signature)
Additional Remarks by Interrogator/Debriefer: _____

Figure M-1. Sample Mission Report for Debriefing (Continued)

Appendix N

Sniper Range Complex

Sniper training requires closely located ranges designed for conducting initial or sustainment training programs. Individual ranges should allow the sniper to train and test in field fire, observation, range estimation, and stalking exercises. Live-fire ranges should be grouped together to reduce construction costs and land use by combining surface danger areas. Setting targets, scoring, and critiquing students requires moving up and down range while adjacent ranges are being used. Areas for training fieldcraft and other exercises should be close enough to maintain training tempo but not interfere with ongoing live-fire exercises. Ranges should also be self-sustaining, to include integrated administrative, classroom, and storage structures. Figures N-1 through N-3, pages N-2 and N-3, show a recommended sniper range development plan.

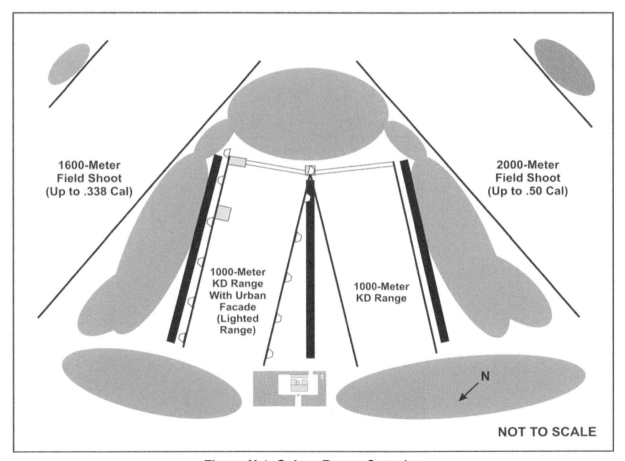

Figure N-1. Sniper Range Complex

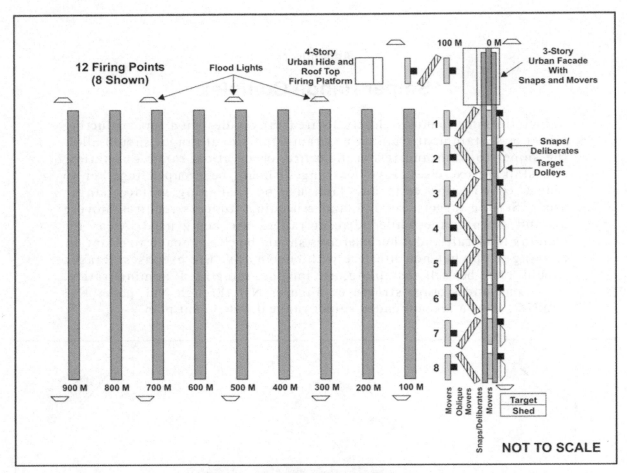

Figure N-2. Proposed KD Ranges

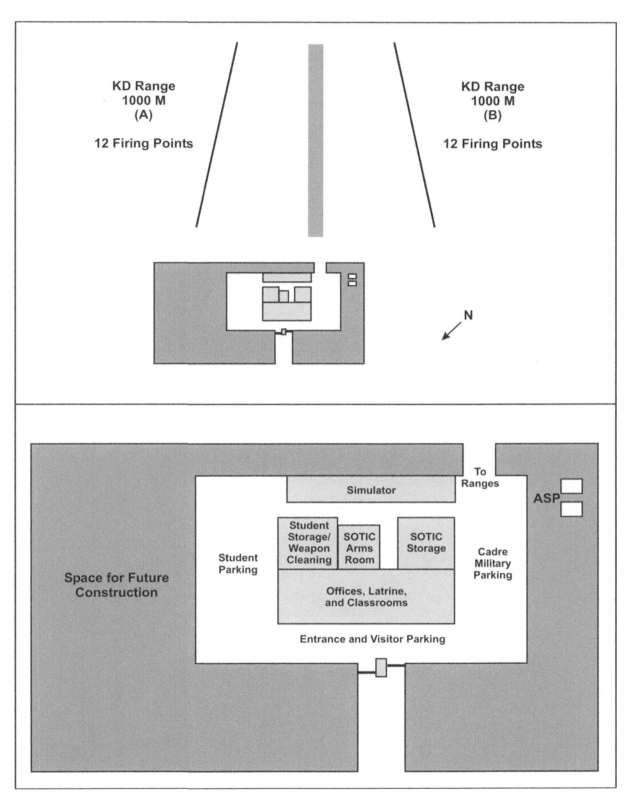

Figure N-3. SOTIC Compound

Appendix O

Aerial Platforms

Sniper teams on today's modern battlefield may occasionally find that staying on the ground to conduct their mission is not feasible or tactically sound. Mission analysis may determine that the aerial platform is the most tactically advantageous method of employment. SOF elements, in conjunction with aviation assets, are called upon to conduct airborne sniper duties in support of airmobile quick reaction force (QRF) operations. These elements provide the ground commander with accurate lethal airborne fire support. To accomplish their mission, snipers must be highly skilled in the art of aerial mission planning. This appendix provides information that will allow units to train and maintain SOF proficiency in airborne sniper duties. This training will enable snipers to provide immediate, safe, accurate, and lethal fires in support of the ground commander during QRF contingency operations.

Safety considerations are paramount and apply to all friendly forces and aircraft (A/C). Critical to this operational capability is the ability to minimize collateral damage to civilian noncombatants and property. The main effort is to maximize the sniper's ability to selectively engage and neutralize high-value point targets from various angles (elevations) at ranges of up to 250 meters, and to suppress area targets at ranges of up to 400 meters. The end state will be SOF snipers and aviation assets trained and proficient in the planning, preparation, and conduct of safe airborne sniper operations.

MISSION PLANNING REQUIREMENTS

O-1. During mission analysis, snipers must determine if conditions require the use of an aerial platform. Situations requiring aerial sniper support usually arise during urban or maritime (for example, visit, board, search, and seizure [VBSS]) operations. Planners must conduct a thorough mission analysis so as not to rule out the possibility of using helicopters as a firing platform during routine SF training and operations. Snipers must also be able to articulate the additional requirements for firing from various rotary-wing A/C, communications (COMM) with the pilot, COMM with other members of the team, and COMM with ground forces.

O-2. The first planning requirement for using aerial platforms is to determine the commander's intent and whether or not there will be helicopters available for use by the snipers. If the commander's intent is to conduct a deliberate assault, then helicopters may not be the platform of choice due to the possibility of loss of surprise, which is paramount during a deliberate assault option. The next step is to look at the capabilities of the given A/C and what is to be accomplished during the mission.

O-3. Each mission objective has specific requirements that must be met to succeed. Therefore, planners must analyze unit capabilities and determine which COAs will meet these requirements. Factors to be considered include—

- Distance from FOB to crisis area.
- Time on station (fuel requirements).
- Number of A/C required to support snipers and assaulters (can A/C do double duty?).
- Can the assault force endure lapses in coverage, or do they require continuous coverage?
- Weather.
- Night-flying capability of supporting A/C.
- A/C crew familiarity with support of SOF-type missions.
- A/C crew familiarity with support of aerial-platform missions.
- Make-up of enemy forces expected in crisis area.
- Weapons capabilities of expected enemy forces.
- CSAR capabilities in case of downed A/C in crisis area.
- A/C type (will it support the weapons and weapons support to be used?).

O-4. If a mission requires an aerial platform with the capability to provide precision fire be present during all phases of the operation, then the number of A/C versus the number of available snipers must be considered. Loiter time for any given A/C will have a major influence on this aspect. Ultimately, the restraints of the A/C will dictate how a mission is accomplished.

O-5. The aerial mission planning requirements can easily be defined using the following five Ws:

- *Who*—is flying the mission and who is the sniper team?
- *What*—type of mission is to be executed (assault, SFAUC) and what type of A/C is to be used?
- *When*—is the mission to be flown and when are rehearsals?
- *Where*—is the mission to be flown (urban or rural)?
- *Why*—is the unit conducting the mission and what are the desired results?

O-6. The unit leader then puts this information into a five-paragraph operations order format and briefs it as an annex during the sniper briefback.

SUPPORT OF GROUND SNIPERS AND GROUND ASSETS

O-7. When supporting ground-based snipers, the aerial platform can act as an observation post relating positions of both friendly and enemy forces, a blocking position to slow either the advance or retreat of enemy personnel, a precision firing platform to reduce point- or crew-served targets encountered by the assault force, and ultimately as an evacuation vehicle for assault force members wounded during the operation and the items recovered. In an urban environment, the aerial platform can act as a crowd deterrent against noncombatants who may be massing for demonstrations against current or ongoing operations, such as rotor downwash from a low-hovering A/C. If extended operations are a possibility or heavy fighting is encountered, the aerial platform can act as an on-call resupply asset if supplies are pre-positioned inside before takeoff.

O-8. During a VBSS, the aerial platform becomes the primary shooting platform and evacuation vehicle for the assault force due to the distance from land that the operation may be taking place. Snipers can orbit the ship that is boarded and provide covering fire for the assault force while they are on the weather deck and engage targets of opportunity that may appear from within the ship while the assault force is clearing the interior of the vessel. The aerial platform may also act as a COMM link to the FOB during the operation due to the reduced COMM capability caused by the metal hull of the ship.

O-9. It is important to remember that helicopters have their own gun systems. These include M-60 and M-240 machineguns and M-134 miniguns. Although normally used in the defensive role, they are equally effective when used offensively. Snipers and air safety officers (ASOs) must be prepared to designate targets for the aircrew. This method allows the sniper team to use an area engagement weapon with a high volume of fire to suppress targets beyond the capabilities of the sniper's precision weapons fire. Snipers must also familiarize themselves with the functioning of the A/C's weapons systems and be prepared to take over their operation if required. Variations in the configurations of the weapons when modified for A/C use means operator training is required. This training is especially important when using the M-134.

URBAN OPERATIONS

O-10. A deliberate urban assault or missions requiring the use of SFAUC techniques are two situations that may require the use of heliborne snipers. In either of the following scenarios, aerial platforms are a highly effective way to provide SOF snipers and their supported assault force with an advantage.

Urban Assault

O-11. Conditions surrounding the use of aerial platforms in urban operations must be examined carefully. Although helicopters are common in many cities, flight routes into the objective area may not coincide with normal aerial traffic patterns. This action may alert the adversary to an impending assault. Establishing flight corridors and conducting flybys with nonthreatening civilian helicopters during the preassault phase may cause the enemy to become used to the idea of helicopters in the area. However, unusual aerial traffic patterns can also arouse suspicion and may lead to a premature compromise of the operation. Using helicopters as a diversion during the assault may also be a possibility.

O-12. Anytime a helicopter is being considered for use as a firing platform, careful attention must be placed on the type of fire it will be required to provide. Wherever possible, the mission to provide precision fire should not be relegated solely to an aerial platform. The less stable position provided by an aerial platform as compared to a ground-based position makes it more difficult to accomplish the sniper mission. Careful consideration must also be given to enemy standoff (for example, RPGs) and ADA capabilities.

SF Advanced Urban Combat

O-13. In this scenario, the aerial platform will most likely be used in support of the SOF mission. Unlike the deliberate urban assault, where the location of the crisis site and possibly even the crisis point is known, this operation finds the assault force moving through an urban environment either looking for the items that they must recover or, after having recovered the items, encountering an enemy force that is hindering their evacuation of the crisis area.

AERIAL PLATFORM TRAINING

O-14. Airborne sniper training will be conducted in five phases. A detailed discussion of each phase follows.

Phase I—Planning

O-15. Planning consists of coordination and deconfliction between SOF, range control (for training operations), and aviation assets. A major point for deconfliction, especially for training, is obtaining the necessary waivers to remove seats and seatbelts from the A/C. When using a UH-60 helicopter, it is possible to leave the four rear seats and one forward-facing center seat without losing any capabilities except troop transport. To remove all seats requires greater planning time to obtain a waiver to remove seats. Depending on the aviation unit and A/C type, a waiver may be required to remove selected seats to optimize rigging configuration. The amount of time necessary to effect a seat-removal waiver depends on the permission authority level.

O-16. Normally, several coordination meetings are necessary and a good projected planning time is 2 to 3 weeks in advance. All resources must be considered, to include available aviation assets, sniper ammunition, door gunner ammunition, transportation, range scheduling, targets, ballistic blankets, rigging materials, harnesses, and ongoing operations. Once trained, planning for a mission can take place in a matter of hours. Phase I ends after final coordination is complete.

Phase II—Preparation

O-17. Premission preparation consists of obtaining targets and ammunition, range and target preparation, A/C rigging, aircrew brief, range safety brief, range operation brief, and rehearsals (Figure O-1, page O-5 through O-7). Sniper personnel will be organized into lifts. The range safety officer (RSO) will designate one individual as the ASO for each lift. The designated ASO will be ASO for the lift following his sniper lift, thus staying on the A/C after firing and being immediately knowledgeable of requirements. The range NCOIC will brief ASOs, snipers, and range personnel on the ground of their duties; the NCOIC is also in charge of conduct of the range. Assistant range NCOIC will issue guidance to ground personnel and prepare targets and ammunition. The A/C commander will receive a copy of the flight manifest from the range NCOIC. Phase II ends after final rehearsals and upon mutual agreement between A/C commander, range OIC, and RSO that all requirements have been met to ensure safe training.

Roll Call/Introductions
- Senior sniper.
- Sniper team leaders and observers.
- Pilot (each A/C).
- Co-pilot (each A/C).
- Crew chiefs.

Questions to Aircrew
- Type of A/C available.
- Number of A/C available.
- Time and location of crew brief.
- Number of packs that can be carried (sniper team/equipment plus ground force/equipment).
- Time to target area.
- Time over target area.

General Information
- Type of operation.
- Safety instructions.
- Primary radio frequency (A/C to ground force C2, A/C to FOB).
- Alternate radio frequencies.
- A/C call signs.
- Ground force C2 call sign.
- Target area description (in general from the air).
- Marking SOP for FLOT of ground force.
- Known obstacles in area to A/C.

Coordination Data
- Time and location of rehearsals.
- Time A/C available for rigging (sniper teams must also tell aircrew approximate time to rig A/C).
- Approximate takeoff time.
- Initial ingress speed.
- Direction of ingress (given in relation to target area, such as from west over building #1).

Figure O-1. Sample Pilot and Sniper Team Air Brief or Pilot's Brief

- Initial altitude on ingress.
- Initial distance from primary target (initial target to be engaged by sniper team).
- Sectors of responsibility (if more than one A/C is used).
- Initial direction of racetrack (clockwise, counterclockwise).
- Preferred location in A/C for primary shooter.

Scheme of Maneuver: A Brief Overview of Operation
- Clearance to engage.
- Infiltration of ground force (if required).
- Race track (time to complete).
- Location of targets.
- Position of A/C to primary target.
- Close on target.
- Engagement of secondary targets (or targets of opportunity).
- Turnover time (if refueling is needed or exfiltration of wounded before mission accomplishment).
- Maximum time on target.
- Exfiltration of ground force or PC.

Operational Information
- Number of passengers on each A/C (to include aircrew and sniper teams).
- Number of snipers per A/C.
- Location of snipers.
- Type of weapons to be employed.
- Location of ammunition in A/C.
- Location of senior sniper (by A/C call sign; or if he is on the ground, a senior sniper will need to be designated for control of aerial platform snipers).

Type and Number of Equipment Needed: Equipment Requirements to Be Provided by Aircrew
- ICS.
- Gunners' belts.
- Head sets.
- Floatation devices (if over water).

Sequence of Events
- Rig A/C.
- Inspect rig (primary sniper rigs, observer inspects).
- Rehearsal positions.

Figure O-1. Sample Pilot and Sniper Team Air Brief or Pilot's Brief (Continued)

- COMM check (radios and intercom systems).
- Senior sniper demonstrates hand-and-arm signals for the following:
    - Time to target (10 minutes, 5 minutes, 2 minutes, and 30 seconds).
    - Clear to load and make ready.
    - Cleared hot.
    - Cease fire.
    - Observe and look.
    - Make safe.
    - No COMM.
    - Rest easy.
    - Slide (commands given to helicopter: forward, aft, right, and left).
    - Nose (right, left).
    - Tail (right, left).
    - Hover.
    - Altitude (increase, decrease).
    - Open distance to target.
    - Close distance to target.
    - Abort engagement.

Emergency Procedures
- A/C problem (snipers will take all commands from aircrew).
- Loss of COMM.
- Weapons malfunction.
- Aid to injured persons while in flight.
- Emergency exit procedures.

Figure O-1. Sample Pilot and Sniper Team Air Brief or Pilot's Brief (Continued)

Phase III—Range Fire

O-18. RSO and A/C crew chief ensure A/C is rigged properly and all personnel are both seated with seat belts fastened (required for takeoff and landing) and secured in the A/C using RSO-approved harnesses and locking carabiners. Sniper equipment and gear will be secured in the A/C using secure quick-release equipment (such as carabiners or fastex buckles as appropriate). After receiving approval from RSO, A/C lifts and training commences. Snipers remove seat belts and assume their shooting positions on command from the ASO. Upon receipt of "CLEAR TO FIRE" command from RSO, the A/C commander gives the "WEAPONS FREE" command to the ASO. The ASO then gives the sniper approval to fire at designated targets, and ensures the sniper adheres to all safety considerations. If COMM link between sniper and pilot is not available, commands relating to A/C attitude and altitude will be relayed via hand-and-arm signals through the ASO (see

COMM section below). Anyone, at any time, is authorized to call "CEASE FIRE" if an unsafe act is observed. On order from the ASO, sniper changeout will occur while airborne between passes. He notifies the A/C commander and ensures personnel remain secured in the A/C with a safety harness at all times. When changing out lifts, the RSO will call "CEASE FIRE" and notify the A/C commander when "CLEAR TO LAND." Lifts will load the A/C upon notification of the crew chief, and the procedure is repeated. To maximize valuable limited airborne training time, changeouts must be safely and swiftly executed. Phase III concludes at the final cease-fire. If possible, the A/C will land and personnel will participate in the after-action review (AAR) immediately following training to capture aircrew perspective, ideas, and techniques while fresh.

Phase IV—After-Action Review

O-19. An immediate AAR is conducted with all participants to continue to refine TTPs, address relevant points, and capture lessons learned. The range OIC administers the AAR with suggestions from senior snipers. The AAR spans Phase I through Phase III, and Phase IV is complete when the AAR is finished.

Phase V—Recovery

O-20. Recovery consists of accountability, range police, clearing the range, and weapons cleaning. Recovery is complete as per detachment SOP, range NCOIC guidance, and per approval from each element commander.

COMMUNICATIONS

O-21. Communications will be one of the most complex aspects of an aerial mission. Radio nets during standard SOF missions can become extremely crowded. Adding one or more A/C to the equation must be carefully thought out. Communications requirements will need to be met by the sniper TOC in coordination with the company COMM section. Compatibility of equipment and communications security (COMSEC) items must be considered, then radio nets and priorities of use must be coordinated. Either C-E or the company COMM section will provide any additional radios that may be required. The company COMM section will ensure the aircrew is fully briefed as to the needs of the sniper team. The sniper commander will receive the COMM plan during the sniper brief so that any conflicts can be corrected before the brief to the ground force commander and the start of rehearsals. Primary responsibility for providing cryptographics needed by the A/C will fall to the sniper TOC radio operator in conjunction with the company COMM section. The aerial sniper team observer (before the first rehearsal) will verify the COMM plan with a complete COMM check.

Special Communications Requirements

O-22. Any COMM plan must include provisions for primary, alternate, contingency, and emergency COMM methods. It must provide COMM systems for the following requirements:

- Ground to air and air to ground.
- Ground to ground.

- Requesting DUSTOFF and aerial MEDEVAC.
- Sniper to ground commander and ground commander to sniper.

O-23. The following radio nets, either sole-use or shared, must be provided for:

- A/C to A/C (if multiple A/C are used).
- A/C to ground force C2.
- A/C sniper team to pilot (can be accomplished using ICS).
- A/C sniper team to ground force C2.
- A/C sniper team to ground-based sniper teams.
- A/C to FOB.
- A/C sniper team to FOB (can be accomplished by using A/C to FOB net via ICS relay).

O-24. Commands and verbiage from sniper or ASO to pilot can originate from either depending on COMM capabilities; for example, COMM helmets. The primary means of COMM between the sniper team and the aircrew should be direct COMM. The aircrew COMM helmet with two COMM feeds will normally satisfy this requirement. These helmets are still new and their issue to U.S. Army SF units is currently limited. A substitute that will meet this need can be fabricated using repair parts obtained from an electronic maintenance shop. The push-to-talk button is ergonomically located using a hook-and-loop fastener to allow the sniper to communicate directly with either the pilot or the ground commander. The basis for this fabrication can be a U.S. Navy crewman helmet or a U.S. Army combat vehicle crewman helmet. These helmets are recommended because their shape does not interfere with the sniper's cheek-to-stock weld.

O-25. The alternate COMM plan is more indirect and, unfortunately, less responsive. The ASO receives targeting information from the ground commander via headset and organic A/C radios. The ASO then designates targets for the sniper directly using a target designator (for example, laser illuminator) or indirectly using reference points or the clock-ray method. The sniper gives the ASO hand-and-arm signals to adjust the attitude of the A/C so that he can engage the target. The ASO relays these commands to the pilot.

Hand-and-Arm Signals From the Sniper to the Air Safety Officer

NOTE: Most pilots prefer to receive the commands in a countdown given in feet. An example is "UP 10, 9, 8, 7, 6, 5..."

O-26. The following are fairly standard commands and are mostly self-explanatory. The commands most confused are "TAIL LEFT" and "TAIL RIGHT," and are used to pivot the A/C. Pilots normally think about the attitude of the A/C in terms of moving the tail because that is how their controls operate. A "TAIL LEFT" command moves the nose of the A/C to the right, and vice versa.

- "UP." Sniper uses thumb pointing up with remaining fingers closed (standard thumbs-up signal). Sniper will normally give this command once, and ASO will make a determination on height and relay this to the pilot. Sniper can halt the A/C by giving the command "HOVER," or give the command "UP" to gain more altitude.

- *"DOWN."* Sniper uses thumb pointing down with remaining fingers closed (standard thumbs-down signal). Sniper will normally give this command once, and ASO will make a determination on height and relay this to the pilot. Sniper can halt the A/C by giving the command "HOVER," or the command "DOWN" to lose more altitude.
- *"HOVER."* Sniper clenches his fist in the standard "FREEZE" command. This command can be used at any time to halt the movement of the A/C. ASO relays to the pilot.
- *"SLIDE LEFT"* or *"SLIDE RIGHT."* Technique 1—Sniper opens palm, fingers and thumb extended and joined, and makes a pushing motion to the left or to the right. Technique 2—Sniper extends index finger from a clenched fist and points left or right. Either technique may be used as long as the sniper and ASO have coordinated beforehand. The ASO relays command to the pilot.
- *"FLY FORWARD"* or *"FLY BACKWARD."* Sniper extends index finger from a clenched fist and points forward or backward. The ASO relays this to the pilot.
- *"PIVOT LEFT"* or *"PIVOT RIGHT."* Sniper, with fingers and thumb extended and joined, palm slightly down in the most natural and comfortable position, points either left or right in the direction he wishes the A/C to turn. Sniper normally uses the "hover" command to stop the A/C's turn. ASO relays this to the pilot.

AERIAL SNIPER EMPLOYMENT

O-27. Inherent to effective sniper marksmanship from airborne platforms is to accurately engage targets at rapidly changing ranges and angles, while moving, and still maintain muzzle and situational awareness. However, numerous methods can be used since the ranges and angles change so rapidly and normal sniper ranging procedures are ineffective. When shooting at an angle, the round strikes higher than normal; the greater the angle, the higher it will strike. An effective technique at 200 meters (m) or below (not the only technique) is to zero the rifle or set the comeups at either 175 m or 200 m and aim (or hold off) for the actual range to target. At ranges greater than 200 m, sniper experience comes into play and hold-offs become critical. There is no secret recipe for doing this, and because ranges and angles change rapidly, changing scope settings while shooting is not recommended. Conducting unknown distance angle shooting (for example, in mountainous terrain) using instinctive ranging techniques is good training to offset these factors. Also knowing the various hold-offs for set zeros (for example, 300, 400, and 500) will allow for a single elevation change to accommodate further ranges. As an example, a shift to the 400-meter setting may permit engagement out 500 meters using stated mil-holds for that zero.

O-28. Target identification or selection can be conducted in two basic manners, depending on the type of COMM available to the sniper. If the sniper is equipped with a helmet that has integrated COMM links, the ground commander can talk directly to the sniper and identify targets. If helmet COMM is not an option, then the ASO can act as the spotter. If the ASO acts as the spotter, it is recommended that he use low magnification, wide-angle binoculars during the day for the widest field-of-view possible. At night, the choice of optics is IAW personal preference and availability. The ASO will receive target identification and report it to the sniper, essentially just pointing the target out. This method requires practice between the sniper and the ASO to establish the "nonverbal sniper dialogue" necessary due to the high background noise present when conducting airborne operations. It can be practiced on the ground to maximize training conducted in the air.

O-29. Airborne snipers essentially have three different positions to fire from, regardless of weapon or A/C type. The prone position can be used, but helicopter vibrations are transmitted directly to the gun via either the bipod or the elbow and arm if using a sling-supported position. The floor-seated position, in conjunction with the rifle sling support, is probably the best position. It allows for both elbows to be supported on the inside of the knees, which helps the legs to act as shock absorbers eliminating the majority of the vibration. This position also affords very good fields of fire without placing the sniper in an awkward position that would sacrifice accuracy. A variation of the seated position that works well for taller snipers onboard UH-60 A/C is to remain in one of the forward facing rear seats adjacent to the door and use the rifle sling-supported sitting position.

O-30. Variations to the seated position include using shooting mats folded and taped into position on the helicopter seat. Use of the crew door in the H-60-series helicopters aids greatly in acquiring targets and accuracy of the sniper. Other position aids are rucksacks, with a partially filled air mattress stuffed inside and covered with shooting mats or sleeping mats, used as a support for the sitting position when a sniper is seated on the floor or ramp of the A/C. This setup must be strapped to the floor to prevent shifting. Snipers can also use a utility strap across the door or ramp for support or place padding on the floor to help eliminate vibrations. Using the strap is the sniper's preference and is also useful for heavy sniper systems. However, unless a heavy sniper system such as the Barret .50 caliber or M500 .50 caliber is used, the use of straps across the door is counterproductive to accuracy. Attempts in the past to use bungee cords or stretchable elastic-type ropes have only been partially successful and generally compromise accuracy for weapon support. Ballistic blankets rigged on the floor and used for sniper protection do provide some vibration dampening characteristics in the floor-seated position, but the use of additional padding is recommended, if possible. Snipers must ensure the ballistic blankets and any additional padding are rigged securely in the A/C to avoid dangerous padding shifts or sliding.

COORDINATING INSTRUCTIONS

O-31. A critical component of using airborne sniping platforms is developing adequate and appropriate rules of engagement (ROE). ROE are normally derived from guidance and promulgated through the OPORD and order-of-ground commander. The following are ROE considerations or methods of reaction that should be reviewed by any unit tasked to perform the aerial sniper mission:

- If A/C receives fire at any time, snipers will return fire IAW ROE in a manner designed to immediately kill or neutralize any enemy personnel.
- If friendly ground elements are engaged at any time, snipers will deliver accurate fire on hostiles IAW the ROE.
- Snipers will selectively engage targets designated by ground commander as meeting ROE.
- If hostile elements should run and are identified as having weapons, snipers will use as much effective fire as needed to deter or suppress hostiles IAW the ROE to prevent friendly casualties.
- If hostile elements run but are not identified as having weapons, warning shots will be fired and A/C snipers will continue to cover the QRF.

AERIAL SWS SELECTION CONSIDERATIONS

O-32. Airborne sniper elements are expected to provide organic direct fire to the ground commander using designated sniper systems. Due to muzzle blast, it is recommended that 7.62-mm sniper systems and smaller be used in lighter, smaller helicopters configured with side doors such as UH-60 and UH-1. The blast from .50-caliber sniper weapons can cause pilot distraction. Sniper systems of up to .50 caliber may be used from A/C that are equipped with a tail ramp, such as the CH-47 and CH-53 series of helicopters.

O-33. The M4 carbine (or the special purpose rifle) works well in an airborne sniper role due to its light weight, high rate of fire, and extensive selection of low-magnification optics. At night, it has a wide variety of night vision accessories and nonvisible lasers available. The primary drawback is its caliber, 5.56 mm, which has poor stopping power and terminal ballistics in comparison with 7.62 mm and larger. It is strongly recommended that the 77-grain, 5.56-mm round be used in lieu of the standard ball round, due to superior ballistics. Its effective range is not considered a liability due to the short ranges inherent in airborne sniping. However, to stabilize the weapon, slings should be modified similar to the leather slings supplied with the M24 SWS. When using the sling for support with the M4 carbine, the SOFMOD kit rail interface system (RIS) should be used to minimize barrel strain, which changes POI considerably. If the use of aftermarket parts is an option, the sniper can use an inexpensive one-piece tubular hand guard that free-floats the barrel and allows the sling swivel to be solid mounted on the hand guard rather than the barrel. Because this is a semipermanent adaptation that prevents M4 RIS accessories (lights, laser sights) from being used, it has limited applications.

O-34. The 7.62-mm, gas-operated sniper rifles (M21, Armalite AR-10, or the Knight SR25) equipped with low or variable magnification or red dot scopes are excellent choices for airborne sniping. They provide a high rate of fire, good external and terminal ballistics, accuracy, medium weight, sling-supported capability, and task-optimized optics. High magnification (10x and above) can be used, but drawbacks similar to those of the 10x-equipped M24

SWS will be encountered (see above). Snipers should conduct extensive seated dry and range fire drills to achieve accurate results.

O-35. Although the M24 SWS (or any manually operated bolt gun with high-magnification optics) works in an airborne sniper role, it has several drawbacks. The system is relatively heavy, which can cause sniper fatigue and difficulty in rapid target acquisition. It has a slow rate of fire that negatively impacts engagement capabilities during multiple target scenarios. It is difficult to keep the 10x (or greater) magnification scope steady on the target due to A/C vibrations and wind gusts. However, the M24 and other guns equipped with high magnification optics can still be made to work very well with highly trained snipers or good instinctive shooters. Using a sling and conducting extensive seated dry and range fire drills during normal ground-range fire can optimize the use of high magnification optics. One key is that the shooter must be able to fire with both eyes open so as to superimpose the reticle over the target as seen by the left eye.

O-36. Effective optics include Aimpoint scopes, Aimpoint scopes with 2x magnifiers, and Advanced Combat Optical Gunsight (ACOG) or Day Optical Scope (4x). If available, durable variable-powered tactical scopes such as the Leupold 3.5–10x or 4.5–14x series are very viable when used at low magnification. With a variable-powered scope, the crosshairs on Leupold 1-inch and 30-mm diameter scopes (and most other U.S.-manufactured scopes) are on the second focal plane, as opposed to the first focal plane. For practical purposes, this means that the crosshairs remain the same size regardless of the magnification. Therefore, the mil dots are normally calibrated to be accurate at only one magnification, normally the scope's highest power. Many European scopes—such as Kahles, Schmidt & Bender, Swarovski, and Zeiss—are first focal plane scopes, so if a European-manufactured, variable-powered tactical scope is used, the reticle and mil dots may change power with magnification. The newer LR series scopes by Leupold, with new reticle, and the newer Gen II reticles by Premier in the Leupold are positioned in the first focal plane, and the mils stay true mils throughout the power range.

SUPPORT AND EQUIPMENT REQUIREMENTS

O-37. Using an aerial platform as a firing position places further requirements on the sniper team due to time and equipment required to set up for and conduct such a mission. Basic equipment will be as per the SFODA SOP. Each individual sniper team will determine their special equipment requirements. Any additional equipment required to support the platform is determined by the type of A/C.

O-38. All sniper personnel wear body armor and a protective helmet. The day pack, which contains ammunition, munitions, and signaling devices, should be secured in the A/C by a snap link or fastex buckle and located so that it is available for quick weapon reloads. The day pack or additional rucksack should also contain the following additional items:

- Night vision optics.
- Maps.
- Emergency COMM equipment.

- Emergency medical kit.
- Food and water.

The sniper should not wear his load-bearing equipment to facilitate accuracy. However, LBEs should be proximally located and secured (attached to day pack or snap linked/secured with fastex buckle in the A/C).

AVIATION

O-39. For training purposes on a small land-based range, one A/C should be used to minimize distractions and maximize safety. The UH-60, UH-1, CH-47, and CH-53 are all viable sniper platforms; the training depends on the mission and assets available. The use of more than one A/C on a small land-based range is feasible, but critical to safe conduct of the range is the use of a person experienced in A/C control.

O-40. For training purposes on a larger land-based range or water-based range, more than one A/C may safely be fired from. If possible, a tactical air control party, combat control team, or SO terminal attack controller should be used to control multi-A/C training missions. If USAF control personnel are not available to assist, a competent Special Forces qualification course (SFQC) graduate familiar with close air support may be the A/C controller IAW the risk assessment, range officer in charge (OIC), and applicable local policies and regulations.

O-41. The sniper planning cell will need to provide a solid timeline to the sniper command element so they can coordinate for the A/C to be available for rigging and rehearsals without interferring with ongoing ground force rehearsals. It is crucial for the snipers to rehearse with the A/C crew to develop the coordination required to function as a team.

AIRCRAFT EMERGENCY PROCEDURES

O-42. All snipers must receive a detailed briefing and conduct rehearsals on emergency shutdown procedures for the A/C (to include fuel supply) and any A/C gun systems. Snipers should know and rehearse the location of all exits, medical equipment, fire extinguishers, A/C COMM systems, ammunition, and survival items. They must also be instructed on A/C destruction, with priority and means of destruction designated for all systems. Special emphasis should be placed on cryptographic items.

AIRCRAFT RIGGING OF A UH-60 (Modify as Appropriate for Other A/C Types)

O-43. **Seats.** If permitted, remove all seats to allow more room for both snipers and any personnel being evacuated. FAA regulations require that a seat be present for each person in the A/C, and that personnel are seated and secured during takeoff and landing. If seat removal is not waiverable, then remove the center (forward) section of seats, leaving one seat in the center between or behind the pilot's seat facing forward, and leave the four seats in the back, which face forward in the A/C. This configures the A/C to accept four snipers in the rear seats and one ASO in the forward-facing (navigator's) seat.

O-44. **Ballistic Blankets.** The number of ballistic blankets able to be secured to the hard points on the floor is determined by the size of the blankets and location of mounting straps. Ideally, the entire floor of the A/C will be covered by a minimum of one thickness of ballistic blanket, with overlaps on adjoining blankets. If available, two thicknesses are preferred. Personnel should ensure the blankets are secured tightly to the floor and use 100-mph tape liberally at the doors to prevent wind gusts from lifting the blankets during flight.

O-45. **Additional Padding.** The ballistic blankets provide some vibration-damping capabilities for floor-seated or prone snipers. This capability can be improved by using additional padding, such as foam rubber or sleeping mats. Depending on the material used, the padding may be placed over or under the ballistic blankets. Personnel should ensure that padding is secured regardless of type or placement; padding shifts during operations can be unsafe.

O-46. **Safety Hookup.** A safety ring constructed of climbing rope or 1-inch tubular nylon connected to no less than four hard points is recommended. The safety ring should consist of two separated ropes or one rope with a bight in it that forms two separated ropes. This provides redundancy in the event of failure. At four locations on the double-looped rope, corresponding to the location of the floor hard points, personnel should isolate the rope further by constructing four bights isolated with knots. Figure-eight loops are very good for this task. Personnel can use locking carabiners to attach each loop to a floor hard point. They must modify the ballistic blanket covering the center of the floor between the doors to allow access to the floor hard points.

O-47. There are various sniper and ASO safety harnesses that are checked and approved by the RSO. The ASO ensures a safe hookup within the A/C before takeoff. The safety harness variations are as follows:

- Aircrew safety harness (monkey harness).
- Commercial climbing harness.
- Seat, hip rappel with safety line (Swiss seat).
- Safety line.

O-48. The safety line connecting the sniper's harness to the safety ring should be constructed of climbing rope or 1-inch tubular nylon, and connected with a locking snap link. It should be long enough to allow freedom of movement within the A/C, but short enough to prevent personnel from exiting the A/C.

PERSONNEL REQUIREMENTS (FOR TRAINING)

O-49. Ground personnel include the following:

- OIC.
- Noncommissioned officer in charge (NCOIC).
- Assistant NCOIC.
- Range safety officer.
- A/C ground controller.

- Range safeties.
- Medic.
- Range personnel.

O-50. Air personnel include the following:
- Airborne OIC or A/C commander (normally the mission air commander or A/C commander).
- Air safety officer (provided by QRF personnel).
- Snipers.
- Aircrew.

O-51. Attachments and detachments include the following:
- Tactical air control party, combat control team, or SO terminal attack controller.
- Snipers not organic to the detachment.
- Additional range personnel.

Tasks to Maneuver Elements

O-52. Duties and responsibilities include the following:
- OIC oversees conduct of the range.
- NCOIC provides organization of the lifts and range operation.
- RSO—
  - Ensures a safe overall range setup and operation.
  - Provides safety brief to all.
  - Briefs the ASO on duties and responsibilities.
  - Checks A/C rigging.
  - Maintains close contact with A/C controller on the ground.
  - Advises and assists the OIC and NCOIC.
- ASO—
  - Ensures the overall safety of snipers and A/C.
  - Provides safety-line hookups of snipers.
  - Tells snipers when to load, go hot, and cease fire. He also visually inspects to ensure weapon clear.
  - Ensures snipers fire in safe direction at designated targets.
  - Depending on available COMM, controls A/C using headset for snipers to effectively engage targets.
- Range assistant NCOIC—
  - Receives direction from range NCOIC.
  - Provides the targets.
- A/C ground controller—
  - Informs A/C when range is hot.
  - Controls A/C as necessary.

- Provides conduit between ground personnel and A/C.
- Range safeties—
  - Ensure ground personnel remain in safe area.
  - Ensure range remains clear.
- Range personnel—
  - Consist of all excess personnel and snipers not currently firing.
  - Assist with targets and range operations as directed.
- Air commander—
  - Oversees entire A/C and aircrew.
  - Ensures detailed A/C safety and crew briefs are issued.

Target Materials and Recommendations

O-53. To achieve maximum benefit from limited airborne training time, it is recommended that targets be used which give immediate feedback to the snipers. A consideration when choosing targets is the effect of rotor wash on target materials and the amount of target maintenance required during training. Targets that require minimum maintenance during range fire will eliminate wasted time due to target changeout or servicing. The following are suggested target materials in order of preference from best to worst:

- *Steel/Iron Maidens.* Excellent choice for targets. Provide immediate sniper feedback, require minimal attention once set up, and good for incorporating "instinctive" distance judging during training. Primary drawbacks are heavy weight and difficulty in initial fabrication. Personnel should spray paint targets to assist with feedback. They must also ensure that A/C remains a safe distance from steel targets.
- *Clay Pigeons and Clay Roof Tiles.* Good choice for targets; however, they must be replaced as they are destroyed. Good feedback to the sniper, inexpensive cost, and availability make these a good choice. Primary drawbacks are the small size and time spent replacing targets during training.
- *Balloons, Inflated Surgical Gloves, and Inflated Condoms.* Fair choice for targets, but several drawbacks make them less preferred. Balloons are normally small in size, which is not representative of the center mass of a man-sized target. A lot of time is also required to inflate the balloons. Additionally, the rotor wash or very hot weather tends to kill many balloons, which can give improper feedback to the snipers and definitely cause lost training time due to target maintenance.
- *Paper Targets.* Poor choice; least preferred. Provides no feedback to sniper and requires time-consuming target maintenance during training.

Glossary

AAR	after-action review
A/C	aircraft
accuracy	In sniping, the ability of the sniper and his weapon to deliver precision fire on a desired target. Accuracy can easily be measured as the ability to group all shots close to a desired impact point. The deviation from the desired impact point or the size of the group is a function of range. Accuracy is the product of *uniformity*.
action	The mechanism of a sniper rifle or other firearm that normally performs loading, feeding, locking, firing, unlocking, extracting, and ejection. Also known as the receiver or frame.
ADA	air defense artillery
adjustable objective	Fine-focusing ring on the objective lens of a telescope that helps to eliminate parallax.
adjusted aiming point	An aiming point that allows for gravity, wind, target movement, zero changes, or MOPP firing. Also known as a "hold."
ammunition lot	A quantity of cartridges made by one manufacturer under uniform conditions from the same materials. Ammunition within a lot is expected to perform in a uniform manner.
ammunition lot number	Code number that identifies a particular quantity of ammunition from one manufacturer. It is usually printed on the ammunition case and the individual boxes in which the ammunition comes.
AO	area of operations
APFT	Army Physical Fitness Test
API	armor-piercing incendiary
ART	automatic ranging telescope
ASO	air safety officer
ATB	appears to be
ball ammunition	General-purpose standard service ammunition with a solid core (usually of lead) bullet.
ballistic coefficient	A number used to measure how easily a bullet slips through the air (aerodynamic efficiency). Most bullets have ballistic coefficients between .100 and .700. Higher ballistic coefficients are required for long-range shooting.
ballistics	A science that deals with the motion and flight characteristics of projectiles in the weapon, internal; out of the weapon, external; and effect on the target, terminal.

BC	ballistic coefficient
BDC	bullet drop compensator
BDU	battle dress uniform
beat	The sniper's operational area where established control measures (boundaries, limits) define his territory.
berdan primer	Form of primer that does not have an integral anvil. Still found in Europe, it is reloaded with difficulty.
BMCT	beginning morning civil twilight
BMNT	beginning morning nautical twilight
boat tail bullet	A bullet with a tapered base to reduce aerodynamic drag. Drag partly comes from the effects of cavitation (turbulence) and the progressive reduction of the diameter toward the rear of the bullet allows the air to fill in the void.
boxer primer	Standard primer with an integral anvil.
BRASS	breathe, relax, aim, slack, squeeze
brass	Empty cartridge case.
breech	The chamber end of the barrel.
bullet drop	The amount that a bullet falls horizontally due to the effect of gravity.
bullet drop compensator	Any device that is integral to the rifle telescope that is designed to compensate for the bullet's trajectory.
C	centigrade
C2	command and control
CA	Civil Affairs
cal	**caliber**—The measurement taken within the barrel from groove to groove or from the outside diameter of the bullet.
CARVER	criticality, accessibility, recuperability, vulnerability, effect, recognizability
CAS	close air support
CAT	category
CBT	combatting terrorism
CD	counterdrug
chamber	Part of the bore, at the breech, formed to accept and support the cartridge.
chronograph	An instrument used to measure the velocity of a projectile.
CIR	continuing intelligence requirement

clandestine operation	An activity to accomplish intelligence gathering, counterintelligence, or other similar activities sponsored or conducted by governmental departments or agencies in such a way as to assure secrecy or concealment of the operation. It differs from covert operations in that the emphasis is placed on the concealment of the operation, rather than on the concealment of the sponsor's identity.
CLP	cleaner, lubricant, preservative
CM	countermine
CO	combat operations
cold-bore shot	The first shot from a clean, unfired weapon.
collimator	Boresighting device.
COMM	communications
COMSEC	communications security
concealment	Protection from view. This is not necessarily the same as cover. Cover provides concealment, but concealment does not always provide cover.
COOR	coordination
cover	Protection from hostile gunfire. Cover is a relative term. Cover that is thick enough to stop pistol bullets may not be adequate protection against rifle bullets. This is a crucial fact to keep in mind when selecting cover.
covert operation	An operation that is planned and executed as to conceal the identity of, or permit plausible denial by, the sponsor(s). This differs from a clandestine operation in that emphasis is placed on the concealment of the sponsor's identity, rather than on the concealment of the operation.
CP	command post
crimp	The bending inward of the mouth of the case in order to grip the bullet or around the primer to seal it.
cross dominance	A soldier with a dominant hand and a dominant eye that are not on the same side; for example, a right-handed firer with a dominant left eye.
crown	The technique used to finish the barrel's muzzle. The rifling at the end of the barrel can be slightly relieved, or recessed. The purpose is to protect the forward edge of the rifling from damage, which can ruin accuracy.
CS	Stands for 0-chlorobenzalmalnonnitrile—is actually a white solid powder usually mixed with a dispersal agent, like methylene chloride, which carries the particles through the air. CS is more stable, more potent, and less toxic than the more commonly used CN agent.
CSAR	combat search and rescue

CT	counterterrorism
CUP	copper units of pressure
DA	Department of the Army; direct action
deflection	The change in the path of the bullet due to wind or passing through a medium.
detailed search	A systematic observation of a target area in detail, using overlapping observation in a 180-degree arc, 50 meters in depth, starting in and working away from the observer.
DOD	Department of Defense
drag	The aerodynamic resistance to a bullet's flight.
drift	The horizontal deviation of the projectile from its line of departure due to its rotational spin or the effects of the wind.
drop	The distance that a projectile falls due to gravity measured from the line of departure.
dry firing	Aiming and firing the weapon without live ammunition. This is an excellent technique to improve marksmanship skills and does not cause any damage to a center-fire firearm. It is best done with an expended case in the chamber to cushion the firing pin's fall.
DTG	date-time group
E-4	specialist
E&E	evasion and escape
E&R	evasion and recovery
EECT	ending evening civil twilight
EENT	ending evening nautical twilight
effective wind	The average of all of the varying winds encountered.
EFL	effective focal length
EST	estimate
exit pupil	The small circle of light seen coming from the ocular lens of an optical device when held at arm's length. The exit pupil can be determined mathematically by dividing the objective lens diameter (in millimeters) by the magnification. The result will be the diameter of the exit pupil in millimeters. (Example: for a 6- x 42-mm telescope: 42 mm divided by 6 = 7 mm.) The size of the exit pupil will help in determining the effectiveness of the optical device in low-light conditions. The human pupil dilates to approximately 7 mm under low-light conditions, and a telescope with a 7-mm exit pupil will provide the maximum light possible to the sniper's eye.
exterior or external ballistics	What happens to the bullet between the time it exits the barrel and the time it arrives at the target.

eye relief	The distance that the eye is positioned behind the ocular lens of the telescopic sight. A two- to three-inch distance is average. The sniper adjusts the eye relief to ensure a full field of view. This distance is also necessary to prevent the telescope from striking the sniper's face during recoil.
F	fahrenheit
FDC	fire direction center
FEBA	forward edge of the battle area
FFP	forward firing position
FID	foreign internal defense
FLIR	forward-looking infrared
FLOT	forward line of troops
FM	field manual
FN	Fabrique Nationale
FO	forward observer
FOB	forward operational base
follow-through	The continued mental and physical application of marksmanship fundamentals after each round has been fired.
fouling	Buildup of copper and powder residue in the bore. These two types of fouling require different cleaning solvents for complete removal.
fps	feet per second
free-floating barrel	A barrel that is completely free of contact with the stock. This is critical to accuracy because of barrel harmonics. As the bullet is traveling down the barrel, the barrel is vibrating like a tuning fork. Any contact with the barrel will dampen or modify these vibrations with (usually) a negative impact on shot-group size or point of impact.
ft	feet
gn	grain
GPS	global positioning system
grain	A unit of measure; 7,000 grains are equal to 1 pound. Used to describe bullet weight (for example, 173 grains) or powder charge.
grooves	The low point of rifling within a barrel.
group	Formed from numerous shots fired at a target using the same point of aim, for checking accuracy. For standardization, it is best to fire five-shot groups with the same aiming point. It is a statistical fact that group size will increase with the number of shots fired.
GT	general technical

GV	given variable
GW	guerrilla warfare
HA	humanitarian assistance
hand load	Also called reload. Nonfactory-manufactured ammunition.
hand stop	A device attached to the weapon's fore-end (modified with a metal rail) designed to prevent the supporting hand from sliding forward.
hasty search	A very rapid check for enemy activity; primarily used as a security measure to determine immediate threats or danger to the sniper when occupying positions.
HD	humanitarian demining
headspace	In rifles, the distance from the shoulder of the cartridge case to the head of the case. For bottle-necked cases, the measuring point is centered on the shoulder and is known as the datum line. For belted magnum cases, the headspace is measured from the front of the belt to the head of the case. This dimension is critical for the safety of the shooter, as well as the accuracy of the weapon system.
HELR	high-efficiency, low-reflection
hide	The term used to describe sniper positions, normally concealed from the enemy.
hold-off	A shooting technique used to compensate for bullet trajectory by using a modified point of aim above or below the desired point of impact. Also used to describe the modified point of aim used to compensate for wind or target movement.
hold-over	The modified point of aim used above the target to compensate for bullet trajectory.
hold-under	The modified point of aim used below the target to compensate for a projectile on its upward axis of its trajectory. This is also used when shooting at angles (slopes).
hollow-point	Describes a bullet with a hollow cavity in the tip designed to enhance terminal ballistics. Although the Sierra MatchKing bullets are hollow points, their design is intended to improve accuracy, not terminal ballistics. This bullet type has been approved by the Judge Advocate General for combat use.
HUMINT	human intelligence
IAD	immediate action drills
IAW	in accordance with
ID	identification
indexing targets	The method that a sniper team employs to identify targets within its effective field of fire.

interior ballistics	What happens to the bullet before it leaves the muzzle of the rifle. Calculations are used to measure pressure forces inside the cartridge and barrel during firing.
IO	information operations
IR	infrared, information requirement
jacket	The copper covering over the lead core of a bullet.
KD	known distance
keyhole	When the bullet hits the target other than point first. Usually indicated by an elliptical bullet hole. Caused by inadequate rotational stabilization of the bullet (usually due to insufficient barrel twist; the twist is "too slow"), deflection of the bullet by objects in the bullet's path, or other factors.
KIM	A memory game named after the main character in Kipling's novel *Kim*.
L&S	Leupold and Stevens
lands	The high points in the rifling of a barrel. This is the part of the barrel that actually engraves the bullet, imparts the spin to the bullet, and ultimately stabilizes the bullet.
LBE	load-bearing equipment
lead	The modified point of aim in front of a moving target needed to ensure a hit. This depends on the range to, and the speed of, the target.
LOC	line of communications
loophole	Firing port. A hole cut to conceal the sniper but allow him to engage targets in his sector.
LOS	line of sight
LP	listening post
LR	long range
LSR	light sniper rifle
mean radius	The average radius of shot dispersion from the center of a shot group.
METL	mission-essential task list
METT-TC	mission, enemy, terrain and weather, troops and support available—time available and civil considerations
mid-range trajectory	The highest point in the bullet's flight. This occurs, technically, slightly beyond the halfway mark of the distance at which the rifle is zeroed. This is the highest vertical distance of the bullet above the line of sight.
Mil	Milradian, 6283 in a circle, 3.44 (3.5 rounded) MOA in a mil. Only a military compass has 6400 in a circle.

mil	An angular unit of measurement equal to 1/6283 of a complete revolution (there are 6283 mils in 360 degrees). The mil is used to estimate distance and size based on the mil-relation formula: 1 mil equals 1 meter at 1,000 meters. There are 3.44 MOA in 1 mil. (However, rounding to 3.5 eases calculations and causes an acceptable error that is within 1 inch at 1000 meters.)
mil dot	Used to describe the reticle in telescopic sights. (For example, the M3A that has 3/4 MOA dots that are 1 mil apart.)
MILES	multiple integrated laser engagement system
minute of angle	A unit of angular measurement equal to 1/60th of a degree. Although usually approximated as 1 inch per 100 yards of range, it is actually equal to 1.0472 inches per 100 yards of range or 3 cm at 100 meters.
mirage	The heat waves or the reflection of light through layers of air of different densities and temperatures. With optical aids, mirage can be seen even on the coldest days. Mirage is used to estimate the effective wind to be applied to the sight of the SWS.
mm	millimeter
MMPI	Minnesota Multi-Phasic Personality Inventory
MOA	minute of angle (1.0472 inches at 100 yards or 2.9 centimeters at 100 meters)
MOPP	mission-oriented protective posture
MOUT	military operations in urban terrain
mph	miles per hour
MRE	meals, ready to eat
MSR	main supply route
MSS	mission support site
MTT	mobile training team
muzzle	The end of the barrel where the bullet leaves the barrel.
muzzle velocity	The speed of a projectile as it leaves the muzzle of the weapon.
NADS	night augmented day scope
NATO	North Atlantic Treaty Organization
natural point of aim	The direction that the body and rifle combination is oriented while in a stable, relaxed firing position.
natural respiratory pause	The temporary cessation of breathing after an exhalation and before an inhalation.
NBC	nuclear, biological, and chemical
NCO	noncommissioned officer
neck	The portion of a cartridge case that holds the bullet.

NM	National Match
No.	number
NVD	night vision device
OAKOC	observation and fields of fire, avenues of approach, key terrain, obstacles, and cover and concealment
objective lens	The lens at the front of the telescope. It is usually larger in diameter than the ocular lens.
ocular lens	The lens at the rear of the telescope, nearest the sniper's eye.
OIC	officer in charge
OP	observation post
OPCON	operational control
OPFOR	opposing forces
ORP	objective rally point
parallax	The apparent movement of the target in relation to the reticle when the sniper moves his eye in relation to the ocular lens's exit pupil. When the target's image is not focused on the same focal plane as the telescope's reticle, parallax is the result. Current issue U.S. Army rifle telescopes have a field parallax adjustment that when properly used makes parallax error an insignificant factor.
PIR	priority intelligence requirements
POA	**point of aim**—The exact location on a target with which the rifle sights are aligned.
POI	**point of impact**—The point that a bullet strikes; usually considered in relation to the point of aim.
powder	The propellant material used in most ammunition.
PRC	populace and resources control
primer	A small explosive cap in the center of the head of the cartridge case that is struck by the firing pin to fire the round. It consists of a small cup filled with a detonating mixture that provides the flame (actually, a shock wave) that converts the propellant powder into a gas.
primer pocket	The recess in the base of the cartridge case that accepts the primer. In military ammunition, it is usually crimped and sealed with a lacquer sealant for waterproofing.
probability of hit	Refers to the chance (denoted as a percentage) that a given round will hit the target at a given range. Probability of hit values range from 0 to 1.0.
PSG	platoon sergeant
PSYOP	psychological operations
QRF	quick reaction force

rail	A metal track installed in the fore-end of weapon to accept a hand stop or sling.
ranging	The technique that a sniper uses to compensate for bullet trajectory by adjusting the ballistic cam of an adjustable or ranging telescope.
RAP	rear area protection
RBC	rifle bore cleaner
recoil lug	The heavy metal protrusion beneath the front of the action designed to stabilize the action in the stock and transfer the recoil to the stock.
reload	Hand-loaded ammunition.
reticle (reticule)	The sighting image, usually crosshairs, in a telescopic sight.
retina	The light-sensitive layer at the back of the eye. It consists of rod (black and white sensitive for night vision) and cone (color sensitive for day vision) cells.
rifle cant	Any leaning of the rifle to the left or right from a vertical position during firing. Rifle cant should be eliminated because of the potential for increasing misses at longer ranges.
rifling	The spiral grooves in the bore of firearms that spin the bullet to provide it with rotational stability. This will ensure that the bullet flies true with a point-first attitude.
rimfire	A cartridge whose priming compound is located in the rim of the cartridge case and generally of caliber .22. This type of ammunition is discharged by a strike of the firing pin to the rim. This ammunition is generally considered nonreloadable.
rimless	The rim of the cartridge is the same diameter as the body of the case.
rimmed	The rim of the cartridge is larger in diameter than the body of the cartridge case.
rings	The metal devices used to support the scope. They are usually 1 inch or 30 mm in diameter.
RIS	rail interface system
ROE	rules of engagement
round	Refers to a complete cartridge.
RPG	rocket-propelled grenade
RSO	range safety officer
S-3	battalion or brigade operations staff officer
SA	security assistant
SALUTE	size, activity, location, unit, time, and equipment
SAR	search and rescue

SB	Special Ball
SC	survey and communications
SCBS	Shooter's Choice Bore Solvent
scout	An individual who is usually ahead of his parent organization to conduct surveillance on the enemy, conduct reconnaissance, and report information to his parent organization.
SEO	sniper employment officer
service rifle	The primary rifle of a military force.
SF	Special Forces
SFAUC	Special Forces advanced urban combat
SFODA	Special Forces operational detachment A
SFQC	Special Forces qualification course
silencer	See suppressor.
SIMRAD	Commonly used euphemism referring to the AN/PVS-9(KN-200), AN/PVS-9A(KN-200F), and KN-250 family of "clip-on" night vision scopes that mount to day optics (M3A) by using a modified front ring mount.
sniper specialist	An individual trained in sniper employment (preferably sniper qualified) who advises the commander or operations officer (S-3) on proper sniper employment.
sniper team	Two snipers of equal training and ability; the foundation of sound sniper employment.
SO	special operations
SOF	special operations forces
SOI	signal operation instructions
SOP	standing operating procedure
SOPMOD	special operations program modifications
SOTIC	Special Operations Target Interdiction Course
speed of sound	1,120.22 fps at standard conditions. Projectiles traveling faster than this pass through the sound barrier twice: once when they exceed the sound barrier (within the barrel) and once when they reenter subsonic speeds. This effect causes a sonic crack caused by the compressed air waves as the round passes through the air.
SR	special reconnaissance
SRC	sniper range complex
SSO	sunrise/sunset overlay
stalking	The sniper's art of moving unseen into a firing position, engaging his target, and then withdrawing undetected.

STANO	surveillance, target acquisition, and night observation
stock weld	The contact of the cheek with the stock of the weapon.
suppressor	A device designed to muffle or minimize the sounds of the discharging of a firearm. It will also reduce the visual muzzle blast of the weapon as well. It is usually fitted onto the muzzle but can also be an integral assembly with the barrel. This usually works with subsonic ammunition to eliminate the bullet's sonic crack as well; however, the shooter will need to get very close to use this ammunition. The sniper can use full-load ammunition as the sonic crack can confuse the enemy as to the true location of the shooter. The subsonic ammunition will give a "whirring" noise that is a direct line to the shooter.
surveillance	The systematic observation of areas, places, persons, or things by visual, aural, electronic, photographic, or other means. The sniper makes extensive use of fixed and roving surveillance to acquire targets or assess target vulnerabilities.
swivel	The attachment point for the sling to the stock.
SWS	sniper weapon system
target indicators	Any sign that can enable an observer to detect the location of the enemy, his installations, or his equipment.
TC	training circular
terminal ballistics	What happens to the bullet when it comes into contact with the target. The study of the effect of a bullet's impact on the target.
terminal velocity	The speed of the bullet upon impact with the target. This will determine the effectiveness of the bullet because of its direct contribution to energy and energy transfer.
TM	technical manual
TOC	tactical operations center
torque	The turning force applied to screws or bolts.
trace	The air turbulence created by the shock wave of a bullet as it passes through the air. This air turbulence can be observed (with an optical aid) in the form of a vapor trail as the bullet travels toward the target.
tracer	Type of ammunition that is visible at night due to its phosphorous compound in the base of the bullet.
tracking	Engaging moving targets where the lead is established and maintained; moving with the target as the trigger is squeezed. Also used to describe the technique of following the enemy by his markings left on the terrain.
trajectory	The flight path the bullet takes from the rifle to the target. The path of a bullet in flight.

trapping	A technique for engaging moving targets. The aiming point is established forward of the target. The rifle is held stationary and fired as the target approaches the aiming point. Also known as the Ambush Technique.
TRP	target reference points
TTP	tactics, techniques, and procedures
twist	The rate of pitch of the rifling in a firearm's bore. Usually measured by the length of barrel in inches required for the bullet to make one complete revolution, and expressed as a twist rate (for example, 1 turn in 11.2 inches: 1/11.2).
U.S.	United States
USA	United States Army
USAF	United States Air Force
USAJFKSWCS	United States Army John F. Kennedy Special Warfare Center and School
USASFC	United States Army Special Forces Command
USC	United States Code
USMC	United States Marine Corps
USN	United States Navy
UW	unconventional warfare
VBSS	visit, board, search, and seizure
velocity	The speed of the projectile.
windage	The distance or amount of horizontal correction that a sniper must use to hit his target, due to the effects of wind or drift. The adjustment on the telescope or iron sights to compensate for horizontal deflection of the bullet.
x	The power of optical magnification (for example, 10x, 3x–9x).
zero	The range at which the point of aim and the point of impact are one and the same.

Bibliography

AR 40-501. *Standards of Medical Fitness.* 28 March 2002.

AR 385-63. *Policies and Procedures for Firing Ammunition for Training, Target Practice, and Combat.* 15 October 1983.

ARTEP 7-8-MTP. *Mission Training Plan for the Infantry Rifle Platoon and Squad.* 29 September 1994.

ARTEP 7-92-MTP. *Mission Training Plan for the Infantry Reconnaissance Platoon and Squad.* 26 May 2002.

ARTEP 31-807-30-MTP. *Mission Training Plan for the Special Forces Operational Detachment A (SFODA).* 1 June 2001.

Avery, Ralph. *Combat Loads for the Sniper Rifle.* Desert Publications, Cornville, AZ 86325. 1981. (ISBN 0-87947-544-7).

Chandler, Norman A. *Death From Afar.* Iron Brigade Publishing, P.O. Box D, St. Mary's City, MD 20686. 1 March 1992.

DA Pam 350-38. *Standards in Weapons Training.* 1 October 2002.

DA Pam 350-39. *Standards in Weapons Training (Special Operations Forces).* 1 October 2002.

Davis, Jr., Wm. C. *Handloading.* National Rifle Association of America, 1600 Rhode Island Ave., N.W., Washington, DC 20036. 1981. (ISBN 0-935998-34-9).

Ezell, Edward C. *Small Arms of the World, 12th Edition.* Stackpole Books, Cameron and Kelker Streets, P.O. Box 1831, Harrisburg, PA 17105. 1983. (ISBN 0-8117-1687-2).

FM 3-05.20. *Special Forces Operations.* 26 June 2001.

FM 7-8. *Infantry Rifle Platoon and Squad.* 22 April 1992.

FM 20-3. *Camouflage, Concealment, and Decoys.* 30 August 1999.

FM 21-75. *Combat Skills of the Soldier.* 3 August 1984.

FM 23-9. *M16A1 and M16A2 Rifle Marksmanship.* 3 July 1989.

FM 23-10. *Sniper Training.* 17 August 1994.

FM 90-10. *Military Operations on Urbanized Terrain (MOUT).* 15 August 1979.

FM 3-06.11 (FM 90-10-1). *Combined Arms Operations in Urban Terrain.* 28 February 2002.

George, Lt. Col. John. *Shots Fired in Anger.* National Rifle Association of America, 1600 Rhode Island Ave., N.W., Washington DC 20036. 1987. (ISBN 0-935998-42-X).

Henderson, Charles. *Marine Sniper: 93 Confirmed Kills.* Stein and Day, Scarborough House, Briarcliff Manor, NY 10510. 1986. (ISBN 0-425181-650).

Hogg, Ian V. *The Cartridge Guide*. Stackpole Books, Cameron and Kelker Streets, P.O. Box 1831, Harrisburg, PA 17105. 1982. (ISBN 0-8117-1048-3).

Hunnicutt, Robert W., ed. *Semi-Auto Rifles: Data and Comment*. National Rifle Association of America, 1600 Rhode Island Ave., N.W., Washington, DC 20036. December 1988. (ISBN 0-935998-54-3).

Idriess, Ion L. *The Australian Guerrilla: Sniping*. Paladin Press, P.O. Box 1307, Boulder, CO 80306. January 1989. (ISBN 0-87364-104-3).

Long, Duncan. *Modern Sniper Rifles*. Paladin Press, P.O. Box 1307, Boulder, CO 80306. June 1988. (ISBN 0-87364-470-0).

Lonsdale, Mark V. *Advanced Weapons Training for Hostage Rescue Teams*. S.T.T.U. Training Division, P.O. Box 491261, Los Angeles, CA 90049. 1988. (ISBN 0-939235-01-3).

Lonsdale, Mark V. *CQB: A Guide to Unarmed Combat and Close Quarter Shooting*. S.T.T.U. Training Division, P.O. Box 491261, Los Angeles, CA 90049. June 1991. (ISBN 0-939235-03-X).

Lonsdale, Mark V. *Sniper Counter Sniper*: A Guide for Special Response Teams. S.T.T.U. Training Division, P.O. Box 491261, Los Angeles, CA 90049. August 1993. (ISBN 0-939235-00-5).

Mace, Boyd. *The Accurate Varmint Rifle*. Precision Shooting, Inc., 37 Burnham Street, East Hartford, CT 06108. 1991.

McBride, Herbert W. *A Rifleman Went to War*. Lancer Militaria, P.O. Box 886, Mt. Ida, AR 71957. 1987. (ISBN 0-935856-01-3).

McBride, Herbert W. *The Emma Gees*. Lancer Militaria, P.O. Box 886, Mt. Ida, AR 71957. 1988. (ISBN 0-935856-03-X).

NRA Highpower Rifle Rules. National Rifle Association of America, 1600 Rhode Island Ave., N.W., Washington, DC 20036. 1992.

Page, Warren. *The Accurate Rifle*. Stoeger Publishing Company, 55 Ruta Court, South Hackensack, NJ 07606. 1973. (ISBN 0-88317-023-X).

Presidential Executive Order 12333, *United States Intelligence Activities*, Part II, paragraph 2-11. 4 December 1981.

Ramage, C. Kenneth, ed. *Lyman Reloading Handbook, 46th Edition*. Lyman Publications, Rt. 147, Middlefield, CT 06455. 1982.

Ross, Ellen, ed. *Highpower Rifle Shooting, Volume III*. National Rifle Association of America, 1600 Rhode Island Ave., N.W., Washington DC 20036. 1985.

Sasser, Charles W., and Craig Roberts. *One Shot - One Kill (Snipers)*. Pocket Books/Simon & Schuster Inc., 1230 Avenue of the Americas, New York, NY 10020. April 1990.

Senich, Peter R. *Limited War Sniping*. Paladin Press, P.O. Box 1307, Boulder, CO 80306. 1977. (ISBN 0-87364-126-4).

Senich, Peter R. *The Complete Book of U.S. Sniping*. Paladin Press, P.O. Box 1307, Boulder, CO 80306. April 1988. (ISBN 0-87364-460-3).

Senich, Peter R. *The German Sniper: 1914–1945*. Paladin Press, P.O. Box 1307, Boulder, CO 80306. September 1982. (ISBN 0-87364-223-6).

Shore, Captain C. *With British Snipers to the Reich*. Paladin Press, P.O. Box 1307, Boulder, CO 80306. 1988. (ISBN 0-87364-475-1).

Skennerton, Ian. *The British Sniper: British & Commonwealth Sniping & Equipments, 1915–1983*. Ian D. Skennerton, P.O. Box 56, Margate Q. 4019, Australia. 1983. (ISBN 0-949749-03-6).

Stevens, R. Blake. *U.S. Rifle M14 - From John Garand to the M21*. Collector Grade Publications Incorporated. P.O. Box 250, Station "E," Toronto, Canada M6H 4E2. January 1991. (ISBN 0-88935-110-4).

Thompson, Leroy. *The Rescuers: The World's Top Anti-Terrorist Units*. Paladin Press, P.O. Box 1307, Boulder, CO 80306. 1986.

TM 9-1005-223-20. *Organizational Maintenance Manual (Including Repair Parts and Special Tools List) for Rifle, 7.62-mm, M14 W/E (NSN: 1005-589-1271), M14A1 W/E (NSN: 1005-072-5011) and Bipod, Rifle, M2 (NSN: 1005-711-6202)*. 2 August 1972.

TM 9-1005-306-10. *Operator's Manual for 7.62-mm M24 Sniper Weapon System (SWS) (NSN: 1005-01-240-2136)*. 23 June 1989.

TM 9-1265-211-10. *Operator's Manual for Multiple Integrated Laser Engagement System (MILES) Simulator System, Firing, Laser: M89 (NSN: 1265-01-236-6725) for M16A1/M16A2 Rifle and Simulator System, Firing, Laser: M90 (NSN: 1265-01-236-6724) for M249, Squad Automatic Weapon (SAW)*. 28 February 1989.

TM 11-5855-203-10. *Operator's Manual for Night Vision Sight, Individual Served Weapon, AN/PVS-2 (NSN: 5855-00-087-2947), AN/PVS-2A (5855-00-179-3708), and AN/PVS-2B (5855-00-760-3869)*. 29 August 1974.

TM 11-5855-213-10. *Operator's Manual for Night Vision Sight, Individual Served Weapon, AN/PVS-4 (NSN: 5855-00-629-5334)*. 1 February 1993.

TM 11-5855-262-10-1. *Operator's Manual for Night Vision Goggle, AN/PVS-7A (NSN: 5855-01-228-0939)*. 15 March 1993.

Truby, J. David. *Silencers, Snipers and Assassins*. Paladin Press, P.O. Box 1307, Boulder, CO 80306. June 1987. (ISBN 0-87364-012-8).

U.S. Army Special Operations Command (USASOC) Regulation 350-1. *Training*. 28 July 1995.

Ward, Joseph T. *Dear Mom: A Sniper's Vietnam*. Ivy Books (Ballantine Books), New York, NY. October 1991.

Index

A
administrative prerequisites, 1-1, 1-2
aerial platforms, O-1
aiming, 3-19
ammunition, 2-10
    alternatives, 2-12, 2-13
    types and characteristics, 2-10 through 2-12
art of sniping, L-1
assessment tests
    objective, 1-4
    subjective, 1-4

B
ballistics, 3-37, 3-38, 3-43, H-1
binoculars, 2-14 through 2-16
bolt assembly, 2-2
breath control, 3-28

C
camouflage, 4-39
    clothing, 4-6 through 4-8
    equipment, for, 4-8, 4-9
    fundamentals, 4-1
    materials, 4-6
    types of, 4-5
CARVER process, 5-12, 5-13
civil affairs, 5-18
civil disturbance assistance, 5-37 through 5-40
cleaning the SWS, 2-25 through 2-28
collateral activities, 5-18
combat search and rescue, 5-30
conventional defensive operations, 5-35 through 5-37
conventional offensive operations, 5-32 through 5-34
counterterrorism operations, 5-29
countertracking, 4-25, 4-38, 4-39
cover and concealment, 4-10, 4-11, 5-39, I-5

D
debriefing format, M-1
deception techniques
    backward walking, 4-40
    big tree, 4-40, 4-41
    cut the corner, 4-41
    slip the stream, 4-42
    Artic circle, 4-39, 4-42
    fishhook, 4-43
direct action, 5-24 through 5-28
dog-tracker teams, 4-34 through 4-38

E
employment
    methods, 5-1
    planning, 5-2
    organization, 5-5
    command and control, 5-7
environmental effects on firing
    humidity, 3-65
    light, 3-65
    temperature, 3-65
    wind, 3-59 through 3-61
eye relief, 3-19, 3-20

F
final focus point, 3-33, 3-34
firing phases, 3-31 through 3-33
firing positions, 3-1 through 3-4, 6-17
    Hawkins, 3-6
    other supported positions, 3-13, 3-14
    prone backward, 3-7
    prone supported, 3-5
    sitting supported, 3-8
    sling-supported kneeling, 3-10, 3-11
    sling-supported prone, 3-6, 3-7
    sling-supported sitting, 3-8, 3-9
    squatting, 3-11
    standing unsupported, 3-12, 3-13
    supported kneeling, 3-9, 3-10
    supported standing, 3-11, 3-12
foreign internal defense, 5-23
foreign/nonstandard sniper weapons systems, F-1

G
ghillie suit, 2-30, 4-7, 4-8

H
hides, 4-67 through 4-69, 6-13
    construction of, 4-77 through 4-79, I-4
    types of, 4-70 through 4-77
hold-off, 3-69, 3-71, 3-72

I
iron sights, 3-19, 3-51 through 3-53

K
keep-in-memory (KIM) games, 4-91, 4-92

Index-1

L

Leupold & Stevens M3A telescope, 2-7
logbook, K-1

M

maintenance, 2-30, 2-33, 2-34, 2-38
marksmanship training, 3-1
methods of estimation
    100-meter unit-of-measure, 4-62, 4-63
    appearance-of-objects, 4-63
    combination, 4-64
military sketch, 4-84 through 4-87, 5-13
mil relation (worm formula), 4-60, 4-61
minute of angle, 3-37 through 3-39
mission-essential tasks list, B-1
mission packing list, D-1
movement, types of
    backward, 4-16
    hands and knees crawl, 4-13
    high crawl, 4-13, 4-14
    low crawl, 4-14, 4-15
    medium crawl, 4-14, 4-15
    night, 4-17
    turning while crawling, 4-15, 4-16
    walking, 4-13
moving targets, 3-72 through 3-76
M24 sniper rifle, 2-1
M3A telescope, 2-7
M49 observation telescope, 2-16, 3-58
M82A1 caliber .50 SWS, E-1

N

night vision goggles
    AN/PVS-5, 2-26
    AN/PVS-7, 2-26, 2-27
night vision sights
    AN/PVS-2, 2-18, 2-19, 3-52
    AN/PVS-4, 2-19, 2-20, 3-56

O

observation, 4-43, 4-45, 4-46, I-3
    devices, 2-13 through 2-28
    techniques, 4-53
observation log, 4-82 through 4-84
obstacles and barriers, 3-77, 3-78
operation order, 5-14 through 5-16
organization, sniper team, 1-5, 1-6

P

parallax, 3-34
personal qualities, 1-1, 1-3
point interdiction, 5-1
point of aim, 3-26, 3-27
psychological operations, 5-18

R

range card, 4-80, 4-81
range complex, N-1
range estimation, 4-57 through 4-59, J-1
range finders, 4-61
    AN/GVS-5, 2-28
    AN/PVS-6, 2-28
reconnaissance and surveillance, 4-17, 5-1

S

safety, 2-2, 4-17, 5-7
shot groups, 3-46 through 3-48
sight alignment, 3-20 through 3-22
sighting, 3-19
sight picture, 3-21, 3-22
slope firing, 3-66 through 3-69
sniper data book (cards), 3-48 through 3-51, K-1
sniper rifle telescope
    characteristics, G-1, G-2
    adjustments, G-2
sniper weapon system (SWS), 2-1
SOTIC *(special operations target interdiction course)* graduates, 1-2, 1-5, 1-7
Soviet telescopes, F-5 through F-8
special operations missions, 5-16
special reconnaissance, 5-26
stalking, 4-17 through 4-20, C-13
stock adjustment, 2-4
surveillance devices, 2-32, 2-33
sustainment, C-1

T

target analysis, 5-10
team firing techniques, 3-17
telescopes, G-1
telescopic sight, 2-6, 3-24, 3-25, G-4
tracking, 3-75, 3-76, 4-25 through 4-33
training exercises, I-1
trigger assembly, 2-2
trigger control, 3-29 through 3-31
tripod, 2-14, 3-15, 3-16
troubleshooting, 2-41

U

urban hides
    hasty, 6-14, 6-15
    prepared, 6-15 through 6-17
urban terrain, 6-1
    movement techniques in, 6-7, 6-8
    building entry techniques in, 6-9
    engagement techniques in, 6-24, 6-25

unconventional warfare, 5-21
United States telescopes, F-4 through F-8

W

weights, measures, and conversion tables, A-1

wind classification, 3-59, 3-60

wind velocity, 3-60 through 3-64

Z

zeroing, 3-51
   with iron sights, 3-52, 3-53
   with telescopic sights, 3-54, 3-55